NanoScience and Technology

NanoScience and Technology

Series Editors:
P. Avouris   B. Bhushan   D. Bimberg   K. von Klitzing   H. Sakaki   R. Wiesendanger

The series NanoScience and Technology is focused on the fascinating nano-world, mesoscopic physics, analysis with atomic resolution, nano and quantum-effect devices, nanomechanics and atomic-scale processes. All the basic aspects and technology-oriented developments in this emerging discipline are covered by comprehensive and timely books. The series constitutes a survey of the relevant special topics, which are presented by leading experts in the field. These books will appeal to researchers, engineers, and advanced students.

Applied Scanning Probe Methods II
Scanning Probe Microscopy
Techniques
Editors: B. Bhushan and H. Fuchs

Applied Scanning Probe Methods III
Characterization
Editors: B. Bhushan and H. Fuchs

Applied Scanning Probe Methods IV
Industrial Application
Editors: B. Bhushan and H. Fuchs

Scanning Probe Microscopy
Atomic Scale Engineering
by Forces and Currents
Editors: A. Foster and W. Hofer

**Single Molecule Chemistry
and Physics**
An Introduction
By C. Wang and C. Bai

**Atomic Force Microscopy, Scanning
Nearfield Optical Microscopy
and Nanoscratching**
Application
to Rough and Natural Surfaces
By G. Kaupp

Applied Scanning Probe Methods V
Scanning Probe Microscopy
Techniques
Editors: B. Bhushan, H. Fuchs,
and S. Kawata

Applied Scanning Probe Methods VI
Characterization
Editors: B. Bhushan and S. Kawata

Applied Scanning Probe Methods VII
Biomimetics
and Industrial Applications
Editors: B. Bhushan and H. Fuchs

**Roadmap
of Scanning Probe Microscopy**
Editors: S. Morita

Nanocatalysis
Editors: U. Heiz and U. Landman

Nanostructures
Fabrication and Analysis
Editor: H. Nejo

**Fundamentals of Friction and Wear
on the Nanoscale**
Editors: E. Gnecco and E. Meyer

**Lateral Alignment
of Epitaxial Quantum Dots**
Editor: O. Schmidt

Nanostructured Soft Matter
Experiments, Theory, Numerical
Simulations and Perspectives
Editor: A.V. Zvelindovsky

Charge Migration in DNA
Perspectives from Physics, Chemistry
and Biology
Editor: T. Chakraborty

Tapash Chakraborty

Charge Migration in DNA

Perspectives from Physics, Chemistry, and Biology

With 130 Figures and 15 Tables

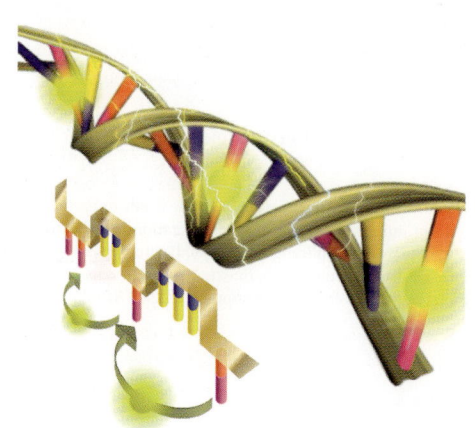

 Springer

Editor:

Prof. Dr. Tapash Chakraborty
Canada Research Chair in Nanoscale Physics
University of Manitoba
Department of Physics and Astronomy
Winnipeg, Manitoba, R3T 2N2, Canada *e-mail:chakrabt@cc.umanitoba.ca*

Series Editors:

Professor Dr. Phaedon Avouris
IBM Research Division
Nanometer Scale Science & Technology
Thomas J. Watson Research Center, P.O. Box 218
Yorktown Heights, NY 10598, USA

Professor Bharat Bhushan
Nanotribology Laboratory for Information
Storage and MEMS/NEMS (NLIM)
W 390 Scott Laboratory, 201 W. 19th Avenue
The Ohio State University, Columbus
Ohio 43210-1142, USA

Professor Dr. Dieter Bimberg
TU Berlin, Fakultät Mathematik,
Naturwissenschaften,
Institut für Festkörperphysik
Hardenbergstr. 36, 10623 Berlin, Germany

Professor Dr., Dres. h.c. Klaus von Klitzing
Max-Planck-Institut für Festkörperforschung
Heisenbergstrasse 1, 70569 Stuttgart, Germany

Professor Hiroyuki Sakaki
University of Tokyo
Institute of Industrial Science,
4-6-1 Komaba, Meguro-ku, Tokyo 153-8505, Japan

Professor Dr. Roland Wiesendanger
Institut für Angewandte Physik
Universität Hamburg
Jungiusstrasse 11, 20355 Hamburg, Germany

Library of Congress Control Number: 2007926315

ISSN 1434-4904

ISBN 978-3-540-72493-3 Springer Berlin Heidelberg New York

Springer is a part of Springer Science+Business Media

springer.com

© Springer-Verlag Berlin Heidelberg 2007

Typesetting and production: LE-TEX Jelonek, Schmidt & Vöckler GbR, Leipzig
Cover: eStudio Calamar S.L., F. Steinen-Bro, Pau/Girona, Spain

SPIN 12038404   57/3180/YL - 5 4 3 2 1 0   Printed on acid-free paper

Preface

This book is based on some of the invited talks presented at the international symposium, Charge migration in DNA: Physics, chemistry and biology perspectives, held at the University of Manitoba, Winnipeg during June 6–9, 2006. Charge migration through DNA has been the focus of considerable interest in recent years. It is now well established that excess charges in DNA, created either by irradiation (UV) or by chemical reaction, migrate along the stacked base pairs of the DNA duplex. Understanding the nature of charge transfer and transport along the double helix is important for fields as diverse as biology, chemistry, and nanotechnology. At a fundamental level, it is also an interesting challenge for physicists to understand the electronic properties of DNA [1], that is crucial for understanding the nature of charge migration. Although there has been a vast amount of work reported in the past decade, the original idea that DNA may act as a molecular wire dates back to 1962 [2] when it was proposed that the π-orbital overlap between the stacked base pairs 0.34 nm apart along the axis of duplex DNA could provide a one-dimensional pathway for migration of electrical charge. Intense experimental and theoretical activities in the past decade have provided us with a wealth of information about the important characteristics of the charge motion in DNA. It is well known that among the four common bases of DNA, guanine (G) has the lowest ionization energy (7.75 eV) [3]. Therefore, in most instances, G is the initial oxidation site and its radical cation (created by the loss of an electron) is commonly involved in the oxidation reactions. Similarly, an electron-less center created somewhere in the chain eventually moves through the DNA π-stack and ends up at a guanine site, usually comprising a pair (GG) or triplet (GGG) of guanine, that has even lower energies (7.28 eV and 7.07 eV respectively). Charge transfer through DNA can result in so-called "chemistry-at-a-distance" [4], where oxidative DNA damage occurs at a site located far from the bound oxidant.

The chemistry-at-a-distance by charge transfer was indeed demonstrated by forming a radical guanine cation at one end of a DNA strand with a GGG unit at the other end separated by the adenine sites [5]. The hole is accepted by the GGG unit which neutralizes the radical G. The charge migration however showed unique sensitivity to A/T bases interspersed between the G sites, which behave as a potential barrier due to their higher ionization energies. A commonly accepted picture of charge hopping is that, for short distances, the holes hop between the G "stepping stones" by coherent tunneling through the intervening A/T bridges. However, when guanines are separated by longer distances, the holes progress via an incoherent, multi-step charge transfer process, where the holes are thermally activated onto the A/T bridges. Once there, the hole supposedly hops along the adenines in an essentially distance-independent manner, until it reaches the GGG trap.

Understanding the intricacies of charge migration in DNA is far from being an academic endeavor, but rather it has important implications in biology, particularly in unraveling the mechanisms of DNA damage that are linked to many diseases. As discussed above, a guanine radical cation (hole) produced by one-electron oxidation of DNA due to carcinogenic agents, ionizing radiation, etc., can migrate to a remote guanine through the DNA π stack. The holes can react with water and/or oxygen to produce guanine-damaged sites in DNA that are known to play an important role in the processes of aging, carcinogenesis, and mutagenesis [6]. Clearly, a better knowledge of the itinerary of a charge through DNA would provide us with valuable information about the perils of the DNA damage.

Finally, with rapid advances in nanofabrication techniques and the resulting rapid pace of miniaturization of electronic devices, molecular electronic devices that employ self-organization of biological molecules could soon become a reality. In this context, DNA may play a crucial role because of its two interesting properties: the complimentarity-based recognition of a nucleobase pair and its ability to self-replicate by complimentarity of its bases. In fact, DNA based molecular electronic devices are expected to operate within the picoseconds range. Understanding charge migration through DNA is essential for development of DNA-based molecular technologies, such as electrochemical sequencing techniques and nanoscale electronic devices. However, as the following chapters clearly indicate, we have a lot to learn yet in order to achieve the goals stated above.

In Chap. 1, Cuniberti et al. have presented a review of theoretical models that are used for simple, tight-binding-based analysis of charge transport in DNA. These simplified models for the DNA strand can offer insights albeit qualitatively, into the intrinsic transport characteristics, statistical properties, sequence dependence and also the effects of solution and the environment.

In Chap. 2, Grozema and Siebbeles explain the experimental data from the literature on the distance and sequence dependence of the rate of charge transfer through DNA with a quantum mechanical model based on a tight-binding description of the charge. Site-energies and charge transfer integrals were calculated for all combinations of adjacent nucleobases using density-functional theory. To reproduce quantitatively the absolute values of experimental rate constants, the effect of the reorganization energy, due to structural rearrangements within the DNA helix and the surrounding water, had to be taken into account. The experimental rates could be reproduced with reorganization energies near 1 eV. The theoretical framework is used to discuss the mobility of charge carriers in DNA.

In Chap. 3, Berlin and Ratner describe charge migration in DNA within a theoretical framework of a variable-range hopping model which has been successfully used to analyze steady-state measurements of the charge transfer efficiency for this molecule. According to the model proposed, the ability of DNA to serve as the medium for very long-range (up to 200 – 300 Å) charge transfer is caused by the energetics of the base pairs stacked in the interior of the double helix. The energy landscape for charge migration along the stack of the nucleobases is shown to exhibit features typical for complex disordered systems. They also show that a charge moving in this landscape can be transferred over large distances via a series of short quantum hops with typical length of 13 – 18 Å alternating with relatively long thermally activated jumps between "resting" sites of the stack. The physical nature of the hopping charges and the issues of dynamic and static disorder are also discussed in the context of the transport properties of DNA systems.

In Chap. 4, Koslowski and Cramer address the phenomenon of charge transport in DNA using a simple, but chemically specific approach intimately related to the Su-Schrieffer-Heeger model. In that model, the Hamiltonian is carefully parameterized using the ab-initio density-functional calculations. In the presence of an excess positive charge, the emerging potential energy surfaces for hole transfer are found to correspond to the formation of small polarons localized mainly on the individual bases. Thermally activated hopping between these states is analyzed using the Marcus theory of charge transfer. Their results are fully compatible with the conjecture of long-range charge transfer in DNA via two competing mechanisms, and the computations provide the corresponding charge-transfer rates both in the short-range superexchange and in the long-range hopping regime as the output of a single atomistic theory. Furthermore, it reproduces the order of magnitude of the current flow in DNA-gold nanojunctions, the overall shape of the current-voltage curves and their dependence upon the DNA sequence.

In Chap. 5, Apalkov, Wang and Chakraborty have explored the geometry effects on charge transfer in a DNA molecule where they view the molecule as two strands of nucleotide bases with interstrand coupling. For a charge to migrate from one end of the molecule to the other, there exists several dominant channels in this two-strand model, as opposed to the standard assumption that only one such channel exists, according to the single-chain model. In this duplex-geometry picture, a weak distance dependence of the charge transfer was found to occur for pure quantum transport through DNA because there are many more available tunneling channels in DNA. The observed crossover between the strong and weak distance dependence may therefore be attributed to a crossover from unichannel to multichannel tunneling transport.

Another aspect of charge transfer through DNA is related to the transverse tunneling through DNA. In this case the transport is determined by the bias voltage applied in the transverse direction of DNA. Since the transverse tunneling occurs through a finite region of DNA the local energetics of the DNA molecule can be extracted from the tunneling current - bias voltage characteristics. One of the interesting aspects of the local structure of DNA is the property of the hole/electron trap. The trap occupies a finite region of the molecule. The energetics within the trap are determined by charge hopping between the sites of the trap, i.e., between the base pairs, and the charge-phonon interactions. A detailed discussion of the transverse tunneling through a DNA trap is presented in this chapter. The transverse transport through the traps of the DNA molecule can also be used to extract information about the two-charge bound state. Formation of such a bipolaronic state is possible for a strong charge-phonon interaction, when the phonon-mediated attraction between the charges becomes stronger than the Coulomb repulsion between them. In the transverse tunneling current the presence of bound states results in pair tunneling of the charges and the specific current-voltage dependence. The condition for formation of the bipolaron bound state and manifestation of such a state in the current-voltage characteristics of the transverse current are also discussed in detail.

In Chap. 6, Asai and Shimazaki discuss the vibronic mechanisms of charge transport and migration in a single DNA molecule. They discuss in detail theoretical studies in both the weak and in the strong coupling limit. Comparative arguments between transport theory and hole transfer reaction theory follow these discussions. While both the elastic and the hopping conduction mechanisms are found in DNA, the former may be very difficult to observe unless the DNA molecule could be short enough, because of the large energy gap between the metallic electrode and DNA.

High energy radiation damage to DNA results in direct ionization of DNA and its immediate surroundings. Holes are generated throughout the DNA and its first hydration layer in accord with the electron density and the electrons produced add randomly to the DNA bases. Within a short time frame the holes move to the most stable site, the guanine base, or react by deprotonation thus localizing the damage. Electrons rapidly transfer to the DNA bases of highest electron affinity, thymine and cytosine. From these initial events the major products of radiation damage to DNA result. In Chap. 7, Becker, Adhikary and Sevilla have reviewed the recent efforts that have elucidated hole and electron transfer processes within DNA and from its hydration layer. In addition recent results are presented and discussed in this chapter. demonstrating that visible light induces hole transfer to other bases, as well as, most significantly, to the sugar phosphate backbone resulting in sugar radicals and ultimately strand breaks, i.e., a significant DNA damage.

Macia in Chap. 8 discusses the thermoelectric performance of short DNA chains connected between metallic contacts at different temperatures on the basis of effective model Hamiltonians. In case of the single-stranded oligonucleotides composed of three nucleobases (codons) the presence of resonance effects leads to a significant enhancement of the thermoelectric power. This result suggests the possible existence of a thermoelectric signature for different codons of biological interest. The thermoelectric performance of PolyG-PolyC and PolyA-PolyT double-stranded chains connected between organic contacts also reveal the existence of important resonance effects, leading to a significant enhancement of the Seebeck coefficient depending on the Fermi level position. High thermoelectric power factors can be obtained close to the resonance energy. The results suggest that significantly high values of the thermoelectric figure of merit may be attained for synthetic DNA samples at room temperature. The possibility of combining p-type and n-type synthetic DNA chains in the design of a nanoscale Peltier cell is considered, taking into account the environmental effects.

In recent years, the proliferation of large-scale DNA sequencing projects for applications in clinical medicine and health care has driven the search for new methods that could reduce the time and cost. The commonly used Sanger sequencing method relies on the chemistry to read the bases in DNA and is far too slow and expensive for reading personal genetic codes. There were earlier attempts to sequence DNA by directly visualizing the nucleotide composition of the DNA molecules by scanning tunneling microscopy (STM). However, sequencing DNA based on directly imaging DNA's atomic structure has not yet been successful. In Chap. 9, Xu, Endres, and Arakawa report a potential physical alternative by detecting unique transverse electronic signatures of DNA bases using ultrahigh vacuum STM. Supported by the principles, calculations and statistical analyses, these authors argue that it would be possible to directly sequence DNA by the STM-based technology without any modification of the DNA.

In Chap. 10, Wang and Fiebig discuss about a new field, DNA photonics that is important to understand the role of DNA as a functional building block in molecular nanoscale devices, and is also expected to shed light on the complex interactions between structural and electronic properties of DNA. The latter is important for biomedical applications such as DNA-targeted drug design. In this chapter, the authors present experimental data from several different classes of functionalized DNA systems and illustrate the relationship between the structural dynamics and charge injection/migration using state-of-the art femtosecond broadband spectroscopy. They also highlight the importance of the initial electronic excitation for modelling electron transfer rates and point out that ultrafast electronic energy migration, dissipation, and (de)localization must be included into the theoretical description of light-induced dynamics in DNA.

Conductance measurements on short DNA wires were found to display various types of behavior that range from insulating to semi-conducting, and even to quasi-metallic, depending on the experimental set up, the environment and the nature of the DNA molecule . The variance of the results as well as the ab-initio calculations suggest that the environment and vibrational modes of DNA play an important role in the transport properties. In Chap. 11, Schmidt et al., report on their study of the electron transport through simple tight-binding models of short double-stranded DNA wires strongly coupled to the vibrational modes (vibrons) of the DNA. The vibrational modes can dissipate energy to the surrounding environment, represented by a bath. By applying equation-of-motion techniques they address the influence of specific DNA vibrational modes on the transport process, with parameters motivated by the ab-initio calculations. For homogeneous DNA sequences such as the polydeoxyguanosine-polydeoxycytidine (poly(dG)-poly(dC)) wires, the vibrons strongly enhance the linear conductance at low temperatures. Beyond the 'semiconducting' gap the finite bias conductance is only qualitatively affected. The transport through such homogeneous DNA can be understood as quasi-ballistic transport through the extended states, which are modified by the coupling to the vibrational modes.

In Chap. 12, Fischler and Simon provide an overview of the current state of the art of DNA-based assembly of metal nanoparticles in one, two and three dimensions. They have summarized different methods of liquid-phase synthesis of metal nanoparticles as well as their functionalization with DNA. The examples selected in this chapter show that the interdisciplinary research at the frontier between biomolecular chemistry, inorganic chemistry, and materials science leads to new materials with unique properties. Based on these properties one may anticipate a broad scope of applications for designing nucleic acid scaffolds to be used for both the assembly of surface-bound nanoparticle architectures as well as three-dimensional aggregates for bioanalytical and advanced materials research. When DNA is used as a template for the assembly of nanoparticles, the examples given in this chapter show that nanowires with metallic conductivity can be obtained. These results have already prompted exciting research on the set-up of functional devices of higher complexity. However, it is still a great challenge to develop these processes further in order to develop devices or even device architectures that are robust enough to be applied in nanoelectronic circuitry.

There is a huge number of papers published in the literature on many of these topics. However, we hope that the articles in this book to some extent reflect the achievements of the present times and future directions of research on the fascinating subject of charge migration in DNA. I would like to express my sincere thanks to all the authors for their help and cooperation that made this book a reality. Many thanks to my secretary Mrs. Cheri Raban for her superb assistance in preparing the chapters in a coherent form from the manuscripts that were originally created in a wide variety of styles. Thanks are also due to all my collaborators, in particular, Dr. Vadim Apalkov, Dr. Xue-Feng Wang, Dr. Hong-Yi Chen and Dr. Julia Berashevich for their help at various stages during the preparation of this book. I also would like to thank the Dean's Office, Faculty of Science of the University of Manitoba, for partial financial support for the symposium. Finally, my sincere thanks to Dr. Claus E. Ascheron, and Dr. Angela Lahee from Springer for their help with the publication of the book.

Winnipeg, Canada   *Tapash Chakraborty*
March 2007

References

1. M. Taniguchi and T. Kawai, Physica E **33**, 1 (2006); E. Braun and K. Keren, Adv. Phys. **53**, 441 (2004); R. Endres, D.L. Cox, and R.R.P. Singh, Rev. Mod. Phys. **76**, 195 (2004).
2. D.D. Eley and D.I. Spivey, Trans. Faraday Soc. **58**, 411 (1962).
3. H. Sugiyama and I. Saito, J. Am. Chem. Soc. **118**, 7063 (1996).
4. P.J. Dandliker, R.E. Holmlin, and J.K. Barton, Science **275**, 1465 (1997); S.M. Gasper and G.B. Schuster, J. Am. Chem. Soc. **119**, 12762 (1997).
5. E. Meggers, M.E. Michel-Beyerle, and B. Giese, J. Am. Chem. Soc. **120**, 12950 (1998).
6. A.P. Grollman and M. Moriya, Trends in Genetics **9**, 246 (1993); K.B. Beckman and B.M. Ames, J. Biol. Chem. **272**, 19633 (1997); S. Loft and H.E. Poulsen, J. Mol. Med. **74**, 297 (1996).
7. C. Wan, T. Fiebig, S.O. Kelley, C.R. Treadway, and J.K. Barton, Proc. Natl. Acad. Sci. USA **96**, 6014 (1999).

Contents

11 Vibrons in DNA: Their Influence on Transport

Benjamin B. Schmidt, Evgeni B. Starikov, Matthias H. Hettler,

12 DNA-Based Assembly of Metal Nanoparticles: Structure and Functionality

List of Contributors

Amitava Adhikary
Department of Chemistry
Oakland University
Rochester
MI 48309
USA

Vadim Apalkov
Department of Physics
and Astronomy
Georgia State University
Atlanta, Georgia 30303
USA

Yasuhiko Arakawa
Nanoelectronics Collaborative
Research Center IIS & RCAST
The University of Tokyo
4-6-1 Komaba, Meguro-ku
Tokyo 153-8505
Japan

Yoshihiro Asai
National Institute of
Advanced Industrial Science
and Technology (AIST)
Umezono 1-1-1
Tsukuba Central 2
Tsukuba, Ibaraki 305-8568
Japan

and
Core Research for Evolutional
Science and Technology (CREST)
Japan Science and Technology
Corporation (JST)
Kawaguchi 332-0012
Japan
yo-asai@aist.go.jp

David Becker
Department of Chemistry
Oakland University
Rochester
MI 48309
USA

Yuri A. Berlin
Department of Chemistry
Center for Nanofabrication
and Molecular Self-Assembly
and Materials Research Center
Northwestern University
2145 Sheridan Road, Evanston
Illinois 60208-3113
USA
berlin@chem.northwestern.edu

Tapash Chakraborty
Department of Physics
and Astronomy
University of Manitoba
Winnipeg Canada R3T 2N2
tapash@physics.umanitoba.ca

Tobias Cramer
Institut für
Physikalische Chemie
Universität Freiburg
Albertstraße 23a
D-79104 Freiburg im Breisgau
Germany

G. Cuniberti
Institute
for Theoretical Physics
University of Regensburg
D-93040 Regensburg
Germany
g.cuniberti@physik.uni-R.de

Robert G. Endres
Department of Molecular
Biology
Princeton University
Princeton, NJ 08544-1014
USA
rendres@Princeton.EDU (RGE)

Torsten Fiebig
Eugene F. Merkert Chemistry
Center
Boston College
Chestnut Hill, MA 02467
USA
Fiebig@bc.edu

Monika Fischler
RWTH Aachen
Institute of Inorganic
Chemistry
Landoltweg 1
52074 Aachen
Germany

Ferdinand C. Grozema
Opto-Electronic Materials
Section DelftChemTech
Delft University of Technology
Julianalaan 136
2628 BL Delft
The Netherlands

Matthias H. Hettler
Forschungszentrum Karlsruhe
Institut für Nanotechnologie
Postfach 3640
76021 Karlsruhe
Germany
matthias.hettler@int.fzk.de

Thorsten Koslowski
Institut für
Physikalische Chemie
Universität Freiburg
Albertstraße 23a
D-79104 Freiburg im Breisgau
Germany
Thorsten.Koslowski
    @physchem.uni-freiburg.de

Enrique Maciá
Departamento de Física
de Materiales
Facultad Ciencias Físicas
Universidad Complutense
de Madrid
E-28040 Madrid
Spain
emaciaba@fis.ucm.es

Mark A. Ratner
Department of Chemistry
Center for Nanofabrication
and Molecular Self-Assembly
and Materials Research Center
Northwestern University
2145 Sheridan Road, Evanston
Illinois 60208-3113
USA

A. Rodríguez
Dpto. Matemática Aplicada
y Estadística, E.U.I.T.
Aeronáutica U.P.M.
Pza Cardenal Cisneros s/n
Madrid 28040
Spain
antonio.rodriguezm@upm.es

R.A. Römer
Department of Physics
and Centre for Scientific
Computing
University of Warwick
Coventry CV4 7AL
United Kingdom
Rudolf.Roemer@warwick.ac.uk

Benjamin B. Schmidt
Institut für Theoretische
Festkörperphysik
and DFG-Center for Functional
Nanostructures (CFN)
Universität Karlsruhe
76128 Karlsruhe
Germany
    and
Forschungszentrum Karlsruhe
Institut für Nanotechnologie
Postfach 3640
76021 Karlsruhe
Germany

Michael D. Sevilla
Department of Chemistry
Oakland University
Rochester
MI 48309
USA
sevilla
    @ouchem.chem.oakland.edu

Tomomi Shimazaki
National Institute of
Advanced Industrial Science
and Technology (AIST)
Umezono 1-1-1
Tsukuba Central 2
Tsukuba, Ibaraki 305-8568
Japan

    and
Core Research for Evolutional
Science and Technology (CREST)
Japan Science and Technology
Corporation (JST)
Kawaguchi 332-0012
Japan
t-shimazaki@aist.go.jp

Laurens D.A. Siebbeles
Opto-Electronic Materials
Section DelftChemTech
Delft University of Technology
Julianalaan 136
2628 BL Delft
The Netherlands
l.d.a.siebbeles@tudelft.nl

Ulrich Simon
RWTH Aachen
Institute of Inorganic Chemistry
Landoltweg 1
52074 Aachen
Germany
ulrich.simon@ac.rwth-aachen.de

Evgeni B. Starikov
Institut für Theoretische
Festkörperphysik
and DFG-Center for Functional
Nanostructures (CFN)
Universität Karlsruhe
76128 Karlsruhe
Germany
    and
Forschungszentrum Karlsruhe
Institut für Nanotechnologie
Postfach 3640
76021 Karlsruhe
Germany

Qiang Wang
Eugene F. Merkert
Chemistry Center
Boston College
Chestnut Hill, MA 02467
USA

Xue-Feng Wang
Department of Physics
and Astronomy
University of Manitoba
Winnipeg
Canada R3T 2N2

Wolfgang Wenzel
Forschungszentrum Karlsruhe
Institut für Nanotechnologie
Postfach 3640
76021 Karlsruhe
Germany

Mingsheng Xu
Nanoelectronics Collaborative
Research Center IIS & RCAST
The University of Tokyo
4-6-1 Komaba, Meguro-ku
Tokyo 153-8505
Japan
msxu@iis.u-tokyo.ac.jp (MSX)

1 Tight-Binding Modeling of Charge Migration in DNA Devices

G. Cuniberti[1], E. Maciá[2], A. Rodríguez[3], and R.A. Römer[4]

[1] Institute for Theoretical Physics, University of Regensburg, D-93040 Regensburg, Germany
g.cuniberti@physik.uni-R.de
[2] Departamento de Fisica de Materiales, Universidad Complutense de Madrid, E-28040 Madrid, Spain
emaciaba@fis.ucm.es
[3] Dpto. Matemática Aplicada y Estadística, E.U.I.T. Aeronáutica U.P.M., Pza Cardenal Cisneros s/n, Madrid 28040, Spain
antonio.rodriguezm@upm.es
[4] Department of Physics and Centre for Scientific Computing, University of Warwick, Coventry CV4 7AL, United Kingdom
Rudolf.Roemer@warwick.ac.uk

1.1 Introduction and Motivation

Within the class of biopolymers, DNA is expected to play an outstanding role in molecular electronics [1]. This is mainly due to its unique self-assembling and self-recognition properties which are essential for its performance as carrier of the genetic code. It is the long-standing hope of many scientists that these properties might be further exploited in the design of electronic circuits [2–6]. In the last decade of the 20th century, *transfer* experiments in natural DNA in solution showed unexpected high charge transfer rates [3, 7–10]. That would imply that DNA might support charge transport. In contradistinction, *electrical transport* experiments carried out on single DNA molecules displayed a variety of possible behaviors: insulating [11, 12], semiconducting [13,14] and ohmic-like [15–18]. This variation might be traced to the high sensitivity of charge propagation in DNA to extrinsic (interaction with hard substrates, metal-molecule contacts, aqueous environment) as well as intrinsic (dynamical structure fluctuations, base-pair sequence) influences. Recently, experiments on single poly(GC) oligomers in aqueous solution [17] as well as on single suspended DNA with a more complex base sequence [14] have shown unexpectedly high currents of the order of 100–200 nA. Again these results, if further confirmed, suggest that DNA molecules may support rather high electrical currents given the right environmental condition.

The theoretical interpretation of these recent experiments and, in a more general context, the elucidation of possible mechanisms for charge transport in DNA has not, however, been unequivocally successful so far. While *ab initio* calculations [19–28] can give at least in principle a detailed account of the electronic and structural properties of DNA, the huge complexity of the molecule and the diversity of interactions present preclude a complete treatment for realistic molecule lengths. When interactions with counter ions and hydration shells or vibrational degrees of freedom are to be considered, the situation easily becomes intractable. On the other hand, model-based Hamiltonian approaches to DNA [29–44] have been already discussed in great detail and can play a complementary role by addressing single factors that influence charge transport in DNA. However, here it is of course clear neither a-priori nor a-posteriori (given the aforementioned experimental situation) which model should be used. Somewhat mirroring the experimental situation, a large variety of models exists and the results are not necessarily consistent across different models.

In this chapter, we review the tight-binding models of DNA which have been proposed in the literature and argued to reproduce experimental [29] as well as *ab-initio* results [45]. We first concentrate on simple one- and two-channel models of DNA in which the main transport mechanism is concentrated in the effects of π-overlap in the base or base-pair sequences. The models are usually constructed to take into account the HOMO-LUMO gap of the single base pairs similar to many of the DFT-based studies. A main feature of the next class of models is the presence of sites which represent the sugar-phosphate backbone of DNA but along which no electron transport is permissible. These models construct a gap due to transversal perturbation of the π-stack, i.e. even when the on-site energies are constant. The aim of this review is thus to explain the present state of affairs in the tight-binding-model-based approach and we will be very brief on many others aspects of the charge migration problem, as these are already well treated in the other chapters of this book.

1.2 The Electronic Structure of DNA

DNA is a macro-molecule consisting of repeated stacks of bases formed by either AT (TA) or GC (CG) pairs coupled via hydrogen bonds and held in the double-helix structure by a sugar-phosphate backbone. In Fig. 1.1, we show a schematic drawing. The electronic energetics of a double-stranded DNA chain should take into account three different contributions coming from (i) the nucleobase system, (ii) the backbone system and (iii) the environment, as sketched in Fig. 1.2.

Attending to the energies involved in the different interactions, the resulting energy network can be hierarchically arranged, starting from the high energy values related to the on-site energies of the bases and sugar-phosphate groups (8–12 eV) [46, 47] passing through intermediate energy values related to the hydrogen bonding between Watson-Crick pairs (~ 0.5 eV) [46] and the coupling between the bases and the sugar moiety (~ 1 eV) [47] and ending with the aromatic base stacking low energies (0.01–0.4 eV) [46, 48]. The energy scale of environmental effects (1–5 eV) is related to the presence of counter ions and water molecules, interacting with the nucleobases and the

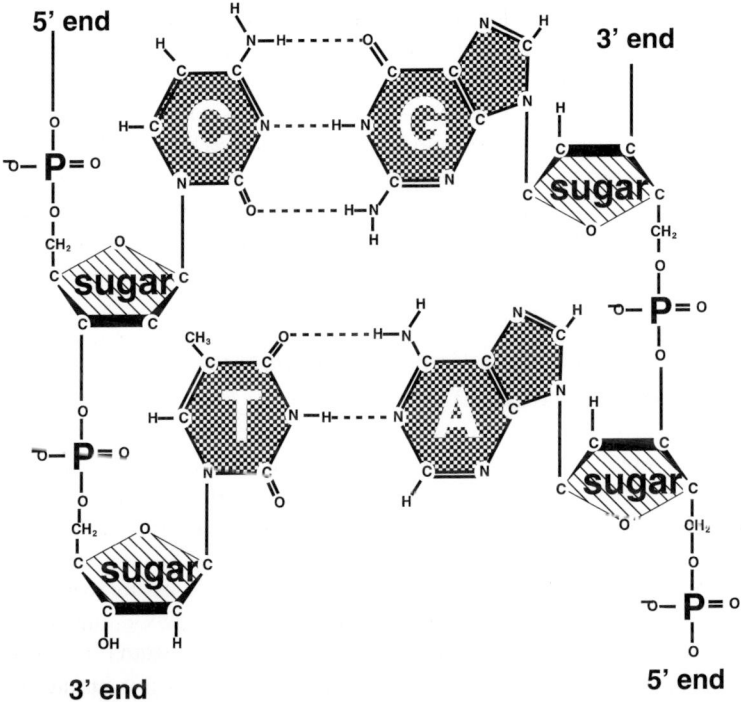

Fig. 1.1. The chemical composition of DNA with the four bases Adenine (A), Thymine (T), Cytosine (C), Guanine (G) and the backbone. The backbone is made of phosphorylated sugars shown as *shaded*, the nucleobases are indicated in *dark grey*

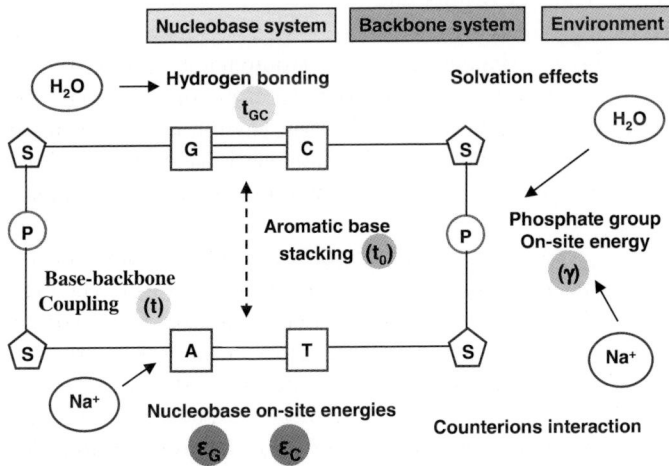

Fig. 1.2. Sketch illustrating the overall energetics of a double-stranded DNA chain

backbone by means of hydration, solvation and charge transfer processes. It is about one order of magnitude larger than the coupling between the complementary bases, and about two orders of magnitude larger than the base stacking energies.

We emphasize that in many of the models to be reviewed later in this chapter, simplified assumptions about these energy scales are employed. Mostly, however, the ionization energies $\epsilon_G = 7.75\,\text{eV}$, $\epsilon_C = 8.87\,\text{eV}$, $\epsilon_A = 8.24\,\text{eV}$ and $\epsilon_T = 9.14\,\text{eV}$, [48–52] are taken as suitable approximations to the on-site energetics at each base.

1.3 Numerical Techniques for Charge Transport in the Quantum Regime

Before we, in the following, turn our attention to the variety of simplified models which have been proposed to capture the essential charge transport features of DNA, let us briefly recall some of the techniques used to investigate these.

There are several approaches suitable for studying the transport properties of quasi-one-dimensional tight-binding models for long DNA and these can be found in the literature on transport in solid state systems, or, perhaps more appropriately, quantum wires [53]. Since the variation in the sequence of base pairs precludes a general solution, one normally uses methods well-known from the theory of disordered systems [54, 55]. The main advantage of these methods is that they work reliably (i) for the relatively short DNA strands ranging from 13 base pairs (as in the DFT studies [56]) up to 30 base pairs length which are being used in the nanoscopic transport measurements [13] as well as (ii) for somewhat longer DNA sequences as modeled in the electron transfer results and (iii) even for complete DNA sequences which contain, e.g., for human chromosomes up to 245 million base pairs [57]. We measure the effectiveness of the electronic transport by various measures such as the *localization length* ξ, participation numbers, etc. These, roughly speaking, parameterize whether an electron is confined to a certain region of the DNA (resulting in the insulating behavior) or can proceed across the full length L of the DNA molecule (the metallic behavior).

1.3.1 Recursive Green Function Technique

The first method one can use is the recursive Green function approach pioneered by MacKinnon [58, 59]. It can be used to calculate the dc and ac conductivity tensors and the density of states (DOS) of a d-dimensional disordered system and has been adapted to calculate all kinetic linear-transport coefficients such as the thermoelectric power, the thermal conductivity, the Peltier coefficient and the Lorenz number [60, 61]. Briefly, the approach utilizes the advanced and retarded Green's functions, $\mathcal{G}^-(E - i0^+)$ and

$\mathcal{G}^+(E+i0^+)$, respectively, and the usual definition $[(E \pm i0^+)\delta_{ij} - H_{ij}] G_{ij}^{\pm} = \delta_{ij}$, where $G_{ij}^{\pm}$ is the matrix element $\langle i|\mathcal{G}^{\pm}|j\rangle$ and H_{ij} is similarly the matrix element of the Hamiltonian [61]. δ_{ij} denotes the Kronecker δ between basis states $\{|i\rangle\}$. If one considers only the nearest-neighbor connections, then these expressions can be written recursively as

$$-H_{ii+1}G_{i+1j}^{\pm} = \delta_{ij} - [(E \pm i0)\delta_{ij} - H_{ij}] G_{ij}^{\pm} + H_{ii-1}G_{i-1j}^{\pm} . \qquad (1.1)$$

Here $H_{ii\pm1}$ are the terms in the Hamiltonian connecting slice i with its neighboring slices $i \pm 1$. If we now reinterpret the left index i as a pseudo time, then we see that the future Green function slice $i + 1$ can be constructed by the present slice at i and the previous slice at $i - 1$. The method is well suited to study coherent transport properties and can be extended to include incoherent processes as well [62].

1.3.2 Transfer and Transmission Matrix Approach

The next method of choice is the iterative transfer-matrix method (TMM) [54, 63–66] which allows us in principle to determine the localization length ξ of the electronic states in systems with varying cross section M and length $L > M$. This localization length describes the decay of the wave function for transport along a quasi one-dimensional system and ξ may be used as a rough guide of the extent of electronic states.

For disordered systems, typically a few million sites $L \gg M$ are needed to achieve a reasonable accuracy for ξ [54]. However, in the present situation we are interested in finding ξ also for DNA strands of typically only a few hundred or a few ten thousand base-pair long sequences. Thus in order to restore the required precision, one modifies the conventional TMM and now performs the TMM on a system of fixed length L_0. This modification has been previously used [67–69] and may be summarized as follows: After the usual forward calculation with a global transfer matrix T_{L_0}, we add a backward calculation with transfer matrix $T_{L_0}^b$. This forward-backward-multiplication procedure is repeated K times. The effective total number of TMM multiplications is $L = 2KL_0$ and the global transfer-matrix is $\tau_L = \left(T_{L_0}^b T_{L_0} \right)^K$. It can be diagonalized as for the standard TMM with $K \to \infty$ to give $\tau_L^{\dagger}\tau_L \to \exp[\mathrm{diag}(4KL_0/\xi_i)]$. The largest ξ_i for all $i = 1, \ldots, M$ then corresponds to the localization length of the electron on the DNA strand and will be measured in units of the DNA base-pair spacing $(0.34\,\mathrm{nm})$. Let us emphasize that the above approach converges even for $L < \xi$. However, in that case, the values of ξ clearly are dominated by finite-size and boundary effects and their significance is no longer quantitative, but qualitatively indicates extended states smeared out over the finite system length L. Last, the transmission coefficient $T_L(E)$, related to the Landauer conductance g via $g = (2e^2/h)T_L(E)/(1 - T_L(E))$ [70–72], is defined in terms of the matrix elements of τ_L.

1.3.3 Attaching Leads

Let us assume that, as a first approximation, we can consider a DNA model in terms of a linear chain with a single orbital per site, where each lattice site represents a base pair. The ends of the chain are connected to leads modeled as semi-infinite one-dimensional chains of atoms with one orbital per site. Broadly speaking, one expects the binding to metallic leads would affect the electronic structure of the molecule. If so, we should consider the states belonging to the coupled molecular-metallic system rather than those of the molecular subsystem alone [73]. Thus we shall consider henceforth that the coupling between the contacts and the molecule is weak enough so that the lead-molecule-lead junction can be properly described in terms of three non-interacting subsystems [74, 75], according to

$$H = H_{\text{DNA}} - t_{\text{Contact}} \left(|0\rangle\langle 1| + |N\rangle\langle N+1| + h.c. \right)$$

$$+ \sum_{k=0}^{-\infty} \varepsilon_{\text{Lead}} |k\rangle\langle k| - t_{\text{Lead}} |k-1\rangle\langle k| + h.c.$$

$$+ \sum_{l=N+1}^{+\infty} \varepsilon_{\text{Lead}} |l\rangle\langle l| - t_{\text{Lead}} |l\rangle\langle l+1| + h.c. . \qquad (1.2)$$

In (1.2), H_{DNA} is the DNA Hamiltonian, the second term describes the DNA-lead contact, and the last two terms describe the contacts at both sides of the DNA chain, where N is the number of base pairs, ε_n are the on-site energies of the base pairs, t_{Contact} is the hopping strength between the leads and the end nucleotides, $\varepsilon_{\text{Lead}}$ is the leads on-site energy and t_{Lead} is their hopping term.

The Green function methods are well-suited to include contact effects since their boundary conditions at the contacts require specification of a suitable Green function in the leads which can be chosen to model the geometry of contacts. The TMM usually starts assuming a particle-like injection of carriers into the transport channels and a proper treatment of the leads is lacking, but the extracted localization lengths at least for long chains are largely independent of the exact choice. Irrespective of the numerical methods used, most earlier tight-binding studies assumed perfect coupling to metallic leads or simply ignored the issue altogether. The role of contact effects within the TMM framework was recently reported for a poly(GACT) tetra-nucleotide in [76] in terms of two contact matrices which explicitly take into account the presence of the t_{Contact} hopping integral. Depending on the value adopted for t_{Contact}, the obtained transmission coefficient does not reach, in general, the full transmission condition $T_{\text{L}}(E) = 1$ due to the symmetry breaking related to the coupling of the G (T) end nucleotides at the left (right) leads, respectively. This extreme sensitivity is due to the interference effects between the DNA energy levels and the electronic structure of the leads at the metal-DNA interface, and indicates that the *optimal* system configuration for efficient charge transfer is determined by the resonance condition $t_{\text{Contact}} = \sqrt{t \cdot t_{\text{Lead}}}$. Quite interestingly, one realizes that, due to the resonance effects a stronger coupling to the leads does not always result in a larger conductance through the system. That is in agreement with the results obtained by Guo and co-workers for the transmission coefficient of poly(G)-poly(C) molecules, who made use of the Green function technique [77]. Subsequent works have exploited the existence of this optimal charge injection condition to study the charge migration efficiency through more realistic duplex chains (see Chap. 5 in this volume).

In general, modeling the geometry and bonding character of the contact at the interface is a very delicate issue, since detailed information on both the metal geometry and DNA chemical bonding at the contacts is poorly known to date. Consequently, in most modeling of the DNA-contact interface, the parameter t_{Contact} deals with the tunneling probability between the frontier orbitals, thus roughly encompassing the bonding effects at the interface. Recent transport experiments have shown that deliberate *chemical bonding* between DNA and electrodes is a prerequisite for achieving reproducible conductivity results [12, 78, 79]. Accordingly, the study of contact effects on the charge migration efficiency is an important issue to be considered in realistic models of DNA transport.

1.4 Tight-Binding Model Approaches

The ab initio methods are clearly very powerful. However, from a physics perspective, the question immediately arises if an even simpler, effective model approach might capture the essentials of charge migration equally well. This strategy is known as the tight-binding approach to DNA – note that in this language the term *tight-binding* is employed somewhat differently from theoretical chemistry. It has been used right from the start of the physics involvement in DNA research. The idea is to capture the main path-ways of charge migration along the DNA molecule stack in a simple model of *site* and *hopping strengths*. Charge transport along this model is then described by simple tight-binding orbitals on the sites and suitably parametrized hopping onto the neighboring sites. The advantage of this approach is clear: once the appropriate on-site energies and hopping strengths are known, much larger system sizes can be studied than with the ab initio methods. The downside of course is that the determination of the effective parameters and in particular the choice of what to leave out completely will be at least to some degree a matter of personal preference and thus open to criticism.

1.4.1 Importance of the DNA Sequence: One-Dimensional Models

The simplest TB model of the DNA stack can be constructed as a one-dimensional model as given in Fig. 1.3. There is a single central conduction channel in which the individual sites represent a base-pair. Every link between

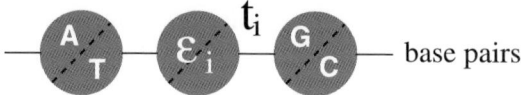

Fig. 1.3. The wire model for electronic transport along DNA corresponding to the Hamiltonian given in (1.3). *Lines* denote hopping amplitudes and *circles* give the central (*grey*) nucleobase pairs

the sites implies the presence of a hopping amplitude. The Hamiltonian for this *wire* model (H_W) is given by

$$H_W = \sum_{i=1}^{L} -t_i|i\rangle\langle i+1| - t_{i-1}|i\rangle\langle i-1| + \varepsilon_i|i\rangle\langle i| \, , \qquad (1.3)$$

where t_i is the hopping between nearest-neighbour sites $i, i+1$ along the central branch and we denote the on-site energy at each site along the central branch by ε_i. L is the number of sites/bases in the sequence. For constant $t_i = t$, $\epsilon_i = 0$ and open boundary conditions, the spectrum of the model is given by

$$E = -2t \cos\left(\frac{\pi k}{L+1}\right) \qquad (1.4)$$

with $k = 1, 2, \ldots, L$. For random choice of on-site energies or hopping strengths, this model is well-known as the Anderson model [80] with diagonal or off-diagonal disorder and its transport properties are governed by one-dimensional Anderson localization [81].

In order to use this Hamiltonian to model DNA, one needs to know the appropriate parameters for on-site energies and hopping strengths [48–52], or, alternatively, one argues that mostly the statistical properties of these quantities determine the transport. For natural DNA sequences, a useful choice for the on-site energies might be the ionization potentials mentioned in Sect. 1.2. But since base-pairs are modelled by a single site, the DNA is effectively described as a sequence of GC (identical to CG) and AT (or TA) pairs with links between like (GC-GC or AT-AT) or unlike (GC-AT, AT-GC) pairs. Thus the model parameters for the pairs should be computed as suitable estimates based on the ionization potentials of individual bases [48–52].

Already such a simple model as (1.3) allows to study various aspects of charge transport in DNA. Electrical transport through individual DNA molecules was studied in [82], using poly(G)-poly(C) DNA. Individual molecules are coupled to external baths [83], thus leading to partial decoherence. Good agreement with the experimental results of [13] was demonstrated. A twist angle in the hopping parts of (1.3) was used in [84] to model the effect of thermal fluctuations on transport in DNA. The participation ratio was used to estimate the extent of the electronic states. Assuming that inelastic effects due to the temperature can be ignored, the authors then computed the temperature dependence of the conductivity. The transmission spectrum for a chain of poly(G)-poly(C) DNA molecules was studied in [77] where also disorder and contact effects were taken into account. The model contains various parameters according to the HOMO/LUMO structure of DNA. Furthermore, charging effects, i.e. the Coulomb blockade were studied within a mean-field approach. For a DNA chain consisting of AT and GC pairs, [85] investigated the structural and the dynamical disorder. Here, in addition to the on-site energies in (1.3), also the hopping elements t_i are chosen according to the specific DNA sequence, which itself, however, consists of random sequences of A,T,G,C nucleotides. It was shown that both types of disorder can significantly influence the transport properties. In [86], both (quasi-)coherent and incoherent transport regimes were studied using the Landauer and the Kubo formalism via a continued fraction approach for poly(G)-poly(C) and also poly(A)-poly(T) DNA chains. Superexchange-like exponential length dependence was found for the coherent and Ohmic-like behavior for the incoherent regimes.

The next group of studies focused on the influence that possible correlations in both artificial and natural DNA sequences might have on the transport. Natural λ-phage DNA was investigated in [33] within a transfer-matrix approach. Transmission spectra were shown to be very different from that in poly(G)-poly(C) DNA. The results were argued to be roughly consistent with those from electron *transfer* studies. The influence of long-range correlations in DNA sequences was studied in [32]. Natural DNA of the first completely sequenced human chromosome 22 (Ch22) was compared to artificial sequences such as random and Fibonacci sequences. Is was found that long-range correlations induce coherent charge transfer over longer length scales, at least for Ch22. In [87] the authors used the same numerical method as in [32] and corroborated the results for Ch22 by comparing to a Rudin-Shapiro sequence. An intriguing relation between the length of a region in coding DNA versus the non-coding DNA and a repeatedly higher transport characteristic in coding DNA was reported in [88, 89].

The influence of temperature and the associated structural fluctuations of DNA and thus the on-site and hopping parameters have been studied in the next group of papers. In [90], Conwell and Rakhmanova investigated a polaronic model in which the hopping elements were influenced by vibrations along the chain. It was shown that for reasonable values of the parameters a polaron can indeed form. In [91] the authors studied a similar situation but also had taken into account the rotation between the base pairs along the DNA stack. That paper was actually aimed at the charge transfer and proposed that thermal fluctuations were the limiting step for site-to-site charge transfer. Polarons, which have a "twist" and can thus model the double-helix structure of DNA better, were investigated in [92]. Non-linear effects were taken into account and it was shown that these lead to different polaronic behavior. In [93], it was argued that polaronic transport can be trapped by the thermal denaturation of poly(G)-poly(C) DNA. Thermal effects were modelled by an anharmonic Morse potential. Semi-emperical quantum-chemical calculations were performed in [94] for poly(G)-poly(C) and also for poly(A)-poly(T) DNA using a polaron model. Localization lengths of charge states larger than 2000 base pairs have been computed and it was shown that significant differences between poly(G)-poly(C) and poly(A)-poly(T) DNA exist.

In addition to the temperature, the solution in which DNA is prepared or measured, its geometry and bend, as well as the properties of the contacts to external leads will influence the measured transport characteristics. The influence of disorder for (1.3) has been investigated in [85]. Contact effects were studied by [76] for poly(GACT) chains. Resonance conditions were identified which showed that a strong coupling to the leads does not always result in larger conductance.

The simple wire model (1.3) has also been used for studies of charge transfer. Briefly, DNA bridges containing only AT base pairs were investigated in [95] and decay lengths comparable with single-step tunneling were found. The presence of Kondo bound states [96] leads to long tunneling lengths above 100 nm. Similarly, time-dependent random hopping strengths were studied in [97] and analyzed in a charge transfer context. Finally, a soliton-based explanation for charge transfer in long segments of DNA was given in [98].

1.4.2 Importance of Base-Pairing: Two-Channel Models

A central simplification of the wire model is the description of a DNA base-pair as a single site. By doing so, one looses the distinction between a pair with G (or A) on the 5′ end of the DNA and a C (or T) on the 3′ side and one where C sits on the 5′ and G on the 3′, i.e. GC is equal to CG. This distinction becomes important when considering hopping between the base-pairs, e.g., the hopping from GC to AT is different from CG to AT because of the different size of the DNA bases and thus the different overlap between G to A and C to A (and similarly for C to T and G to T) [99]. Furthermore, the relevant electronic states of DNA (highest-occupied and lowest-unoccupied molecular orbitals with and without an additional electron) are localized on one of the bases of a pair only [100]. The reduction of the DNA base-pair architecture into a single site per pair, as in the wire model (1.3), is obviously a highly simplified approach.

This deficiency of the wire model may be overcome by modelling each DNA base as an independent site. The hydrogen-bonding between the base-pair is then described as an additional hopping perpendicular to the DNA stack as shown in Fig. 1.4. There are two central branches linked with one another, with interconnected sites where each represents a complete base. This *two-channel* model is a planar projection of the structure of the DNA with its double-helix unwound, and still without regard for the backbone. We note that the results for electron transfer also suggest that the

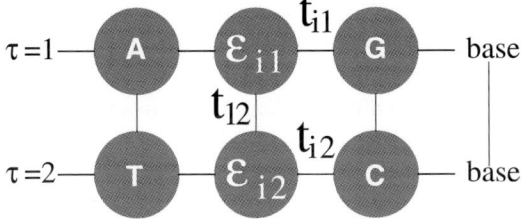

Fig. 1.4. The ladder model for electronic transport along DNA. The model corresponds to the Hamiltonian (1.5). Electronic pathways are shown as *lines*, whereas the nucleobases are given as (*grey*) *circles*

transfer proceeds preferentially down one strand [3]. The Hamiltonian now reads

$$
H_{\mathrm{L}} = \sum_{i=1}^{L} \left[\sum_{\tau=1,2} \left(t_{i,\tau} |i, \tau\rangle\langle i+1, \tau| + \varepsilon_{i,\tau} |i, \tau\rangle\langle i, \tau| \right) + t_{1,2} |i, 1\rangle\langle i, 2| \right] + h.c. ,
$$

$$(1.5)$$

where $t_{i,\tau}$ is the hopping amplitude between sites along each branch $\tau = 1, 2$ and $\varepsilon_{i,\tau}$ is the corresponding on-site potential energy. The new parameter t_{12} represents the hopping between the two central branches, i.e., perpendicular to the direction of conduction. As before, we may now attempt to use ab-initio methods to compute t_{12} or simply model it relative to the strength of the parallel hopping $t_{i,\tau}$. For the ordered system with $t_{i,\tau} = t$ and $\epsilon_{i,\tau} = 0$, the two channel model is just a special case of the two-dimensional rectangular system with spectrum $-2t_x \cos\left(\dfrac{\pi k_x}{L_x + 1}\right) - 2t_y \cos\left(\dfrac{\pi k_y}{L_y + 1}\right)$, $k_x = 1, 2, \ldots, L_x$, $k_y = 1, 2, \ldots, L_y$. Thus we find

$$
E = -2t \cos\left(\frac{\pi k}{L + 1}\right) \mp t_{1,2} \tag{1.6}
$$

where the minus (plus) sign corresponds to even, $\psi_{n,1} = \psi_{n,2}$, (odd, $\psi_{n,1} = -\psi_{n,2}$) states with the same (opposite) sign for the wave function on each strand. For random on-site disorder, the system is again localized and the localization lengths are known for different energies and disorder values [101].

Iguchi was one of the early authors to suggest that a two-leg ladder model might be a useful starting point [102]. A band gap like behavior was found in [103], which also considered the Coulomb repulsion between different bases. It was further shown that for engineered DNA – modelled as frustration – the band vanishes. The authors of [34] used the two-leg ladder model to study the spatial extent of electronic states in long DNA chains. They found that the extent varies considerably depending on the sequence, but remains rather small. Recently, Caetano and Schulz found very large participation ratios in the two-leg ladder at finite system sizes [104]. They speculated that this might indicate a transition to effectively delocalized states. But this claim is not expected to hold for longer chains [105–107]. The influence of electronic spin and interactions has been studied in [36]. This work concentrates on charge transfer aspects and shows that interaction opens a gap in the electronic states of AT and GC pairs. Further transport properties of Ch22, as well as λ-phage and the histone protein, are investigated in [41] and compared to artificial DNA. It is notable that while the model used in [41] is a two-leg ladder, the rungs of the ladder are now modeling not the π transport channels but rather the charge migration along the sugar-phosphate backbone. This approach is similar to that of [108]. Discrete breather-type solutions caused by environmental effects were studied in a two-leg ladder already in [109]. A Morse potential was used to represent hydrogen bonding. The breathers were found to be pinned by the discrete lattice or trapped in defect regions. A similar model based on the non-linear Schrödinger equation was studied in [110], where the transport of the solitons was assumed to propagate along the sugar-phosphate backbone.

1.4.3 Backbone Effects: The Fishbone Model

This *fishbone model*, shown in Fig. 1.5, retains the central conduction channel in which individual sites represent a base-pair. However, these are now interconnected and further linked to upper and lower sites, representing the backbone. The backbone sites themselves are *not* interconnected along the backbone. Every link between the sites implies the presence of a hopping amplitude. The Hamiltonian for the fishbone model (H_F) is given by

$$H_F = \sum_{i=1}^{L} \sum_{q=\uparrow,\downarrow} \left(-t_i|i\rangle\langle i+1| - t_i^q|i,q\rangle\langle i| + \varepsilon_i|i\rangle\langle i| + \varepsilon_i^q|i,q\rangle\langle i,q| \right) + h.c. \,,$$

$$(1.7)$$

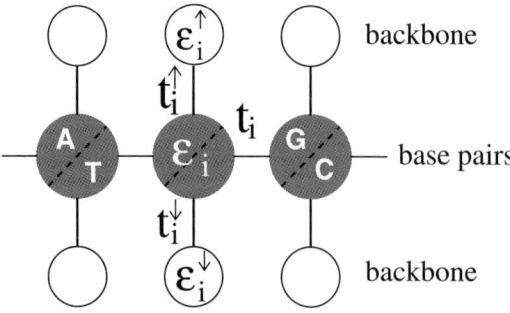

Fig. 1.5. The fishbone model for electronic transport along DNA corresponding to the Hamiltonian given in (1.7). *Lines* denote hopping amplitudes and *circles* give the central (*grey*) nucleobase pairs and backbone (*open*) sites

where t_i is the hopping along the central branch and t_i^q with $q = \uparrow, \downarrow$ gives the hopping from each site on the central branch to the upper and lower backbone respectively. We denote the on-site energy at each site along the central branch by ε_i and, additionally, the on-site energy at the sites of the upper and lower backbone is given by ε_i^q, with $q = \uparrow, \downarrow$. L is the number of sites/bases in the sequence. It is easy to see that the existence of the backbone leads to an effectively renormalized and energy-dependent disorder

$$\left(\epsilon_n - \frac{t^{\uparrow\,2}}{\epsilon_n^{\uparrow} - E} - \frac{t^{\downarrow\,2}}{\epsilon_n^{\downarrow} - E} \right) \tag{1.8}$$

at each base pair n on the π stack. If, just as we have done earlier for the wire model (1.3), we consider the ordered situation $t_i = t$, $t^{\uparrow} = t^{\downarrow}$, $\epsilon_i = \epsilon_i^{\sigma} = 0$ for $\sigma = \uparrow, \downarrow$, we find that the energies are now given by

$$E_{\pm} = -2 \cos \left(\frac{\pi k}{L+1} \right) \pm \sqrt{t^2 \cos^2 \left(\frac{\pi k}{L+1} \right) + 2t^{\uparrow\,2}} \tag{1.9}$$

for $k = 1, 2, \ldots, L$. Hence, there is a highly degenerate state at $E = 0$ corresponding to all the backbone sites and the original single-band of (1.4) splits into two cosine bands such that

$$E \in \left[-t - \sqrt{t^2 + 2t^{\uparrow\,2}}, -t + \sqrt{t^2 + 2t^{\uparrow\,2}} \right] \cup \left[t - \sqrt{t^2 + 2t^{\uparrow\,2}}, t + \sqrt{t^2 + 2t^{\uparrow\,2}} \right].$$
$$\tag{1.10}$$

In [29] it was shown that this model when applied to an artificial sequence of repeated GC base pairs, poly(G)-poly(C) DNA, reproduces the experimental data for current-voltage measurements, when $t_i = 0.37\,\text{eV}$ and $t_i^q = 0.74\,\text{eV}$ are being used. Therefore, we will assume $t_i^q = 2t_i$ and set the energy scale by $t_i \equiv 1$ for hopping between GC pairs. Furthermore, since the energetic differences in the adiabatic electron affinities of the bases are small [111], we choose $\varepsilon_i = 0$ for all i.

The physics of the fishbone model was first discussed for poly(G)-poly(C) in [29]. In fact, the central sites of the fishbone are to model the G nucleotide only, with the effect of the C bases neglected as not so relevant for transport due to their different on-site HOMO/LUMO energies. The model was then independently studied by Zhong [112] for random and natural DNA sequences and he also found an interesting transport enhancing effect in the band gap upon increasing potential disorder. A further study [40] revealed that the extent of electronic states in the two bands of the model can be up to a few dozen base pairs large. Furthermore, upon adding binary disorder, intended to model adhesion of ions from the ionic solution in which DNA strands exist, the band gap closes and the size of initially very well localized band-gap states can be made to increase substantially [113]. This effect was also studied in [38, 39] where the system was coupled to a phonon bath. Here, the band gap was shown to close with increasing temperature and the temperature dependence of the charge transmission near the Fermi energy is exponential.

1.4.4 Backbone in a Ladder

Combining the advantages of the fishbone and the two-channel models, we now model each base as a distinct site where the base pair is then weakly coupled by the hydrogen bonds. The resulting *ladder* model is shown in Fig. 1.6. There are two central branches, linked with one another, with interconnected sites where each represents a complete base and which are additionally linked to the upper and lower backbone sites. The backbone sites as in the fishbone model are not interconnected. In fact, first principle calculations showing that the phosphate molecular orbitals are systematically below the base related ones, do not favor the possible hopping of charge carriers between successive phosphate groups along the backbone [114]. The Hamiltonian for the ladder model is given by

$$
H_{\mathrm{L}} = \sum_{i=1}^{L} \left[\sum_{\tau=1,2} \left(t_{i,\tau} |i,\tau\rangle\langle i+1,\tau| + \varepsilon_{i,\tau} |i,\tau\rangle\langle i,\tau| \right) \right.
$$
$$
+ \sum_{q=\uparrow,\downarrow} \left(t_i^q |i,\tau\rangle\langle i,q(\tau)| + \varepsilon_i^q |i,q\rangle\langle i,q| \right)
$$
$$
\left. + t_{1,2} |i,1\rangle\langle i,2| \right] + h.c. , \tag{1.11}
$$

where as before in (1.5) $t_{i,\tau}$ is the hopping amplitude between sites along each branch $\tau = 1,2$ and $\varepsilon_{i,\tau}$ is the corresponding on-site potential energy. t_i^q and and ε_i^q as in (1.7) give hopping amplitudes and on-site energies at the backbone sites. Also, $q(\tau) = \uparrow, \downarrow$ for $\tau = 1,2$, respectively. The parameter t_{12} represents the hopping between the two central branches as for the two channel model (1.5).

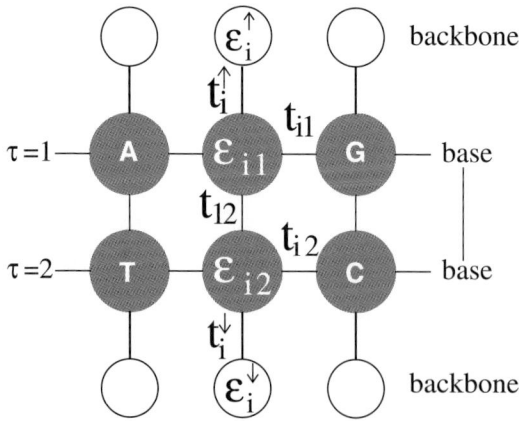

Fig. 1.6. The ladder model for electronic transport along DNA. The model corresponds to the Hamiltonian (1.5) and the reader should compare the figure to Figs. 1.4 and 1.5

For the ordered system with $t_{i,\tau} = t$, $t^\uparrow = t^\downarrow$, $\epsilon_{i,\tau} = \epsilon_i^\sigma = 0$, we find again that the presence of the backbone sites leads to an effective renormalization of on-site energies along the two base pair strands with energy-dependent disorder

$$\epsilon_{n,\tau} - \frac{t^{\sigma 2}}{E - \epsilon_\tau^\sigma} \tag{1.12}$$

and $(\tau, \sigma) = (1, \uparrow)$ or $(2, \downarrow)$. The energies for even states are

$$E^+ = \frac{1}{2} \left\{ -t_{1,2} - 2t \cos\left(\frac{\pi k^+}{L+1}\right) \pm \sqrt{\left[t_{1,2} + 2t \cos\left(\frac{\pi k^+}{L+1}\right)\right]^2 + 4t^{\uparrow 2}} \right\} \tag{1.13}$$

with $k^+ = 1, 2, \ldots, L$. Similarly, the odd states have energies

$$E^- = \frac{1}{2} \left\{ t_{1,2} + 2t \cos\left(\frac{\pi k^-}{L+1}\right) \pm \sqrt{\left[t_{1,2} - 2t \cos\left(\frac{\pi k^-}{L+1}\right)\right]^2 + 4t^{\uparrow 2}} \right\} \tag{1.14}$$

and $k^- = 1, 2, \ldots, L$. Thus we again have two energy bands, with a slightly smaller gap, given as

$$E \in \left[-\left(t + \frac{t_{1,2}}{2}\right) - \sqrt{\left(t + \frac{t_{1,2}}{2}\right)^2 + t^{\uparrow 2}}, \left(t + \frac{t_{1,2}}{2}\right) - \sqrt{\left(t + \frac{t_{1,2}}{2}\right)^2 + t^{\uparrow 2}} \right]$$

$$\cup \left[-\left(t + \frac{t_{1,2}}{2}\right) + \sqrt{\left(t + \frac{t_{1,2}}{2}\right)^2 + t^{\uparrow 2}}, \left(t + \frac{t_{1,2}}{2}\right) + \sqrt{\left(t + \frac{t_{1,2}}{2}\right)^2 + t^{\uparrow 2}} \right].$$

In [40], electronic transport in this model was measured by the *localization length* ξ, which roughly speaking, parametrizes whether an electron is confined to a certain region ξ of the DNA (insulating behavior) or can proceed across the full length L ($\leq \xi$) of the DNA molecule (metallic behavior). Various types of disorder, including random potentials, were employed to account for different real environments. Calculations were performed on poly(dG)-poly(dC), telomeric-DNA, random-ATGC DNA and λ-DNA. The authors find that random and λ-DNA have localization lengths allowing for electron motion among a few dozen base pairs only. An enhancement of the localization lengths similar to that in the fishbone model (1.7) was observed at particular energies for an increasing binary backbone disorder. In [100, 115], the model was used to study differences in different natural and artificial DNA sequences. Specifically, promoter sequences and sequences known to be repetitive from a biological point of view were investigated to see whether there were statistically relevant differences. Using the same sequences as in [89], no support for larger ξ values in regions of coding DNA was found.

1.5 Conclusions

In this chapter, we have aimed at giving a review of current models used for a simplified, tight-binding-based analysis of charge transport in DNA. While the models can be roughly classified according to their geometrical structure, many of the presently available results appear somewhat disjointed and are nearly as widely spread as in the experimental situation. Let us nevertheless attempt to identify some common themes. The vast majority of studies presented here agrees that the transport properties upon including some degree of energetic disorder – be it strictly random or according to some suitable, naturally occurring sequence – tend towards the insulating side. Nevertheless, the size of the electronic states for finite DNA strands might be larger and even exceed the distance between contacts. In such a situation, the experimental results might find finite currents. This finding seems to be largely independent of the set of on-site energies and hopping strengths chosen. Also, most studies agree that there are differences between natural DNA sequences and random DNA with the same ATGC content. However, it is not clear if these differences are due to the special choice of DNA strands and simply become statistically irrelevant when other DNA sequences are considered as well. Thus, a clear correlation between charge transport and a particular DNA sequence or parts thereof is yet to be discovered. We emphasize, however, that if such a correlation were to be found, we would find it useful if it persists across most models reviewed here.

Acknowledgement. It is a pleasure to thank H. Burgert, R. Di Felice, D. Hodgson, R. Gutierrez, D. Porath, R.A. Remer, S. Roche, C.T. Shih, E. B. Starikov, A. Troisi and M.S. Turner for stimulating discussions and collaborations on topics related to this chapter. Free file hosting has been provided by CVSDude.org. G.C. would like to acknowledge support by the the Volkswagen Foundation under grant No. I/78 340, the DFG priority program "SPP 1243", the European Union grant DNAnanoDEVICES under contract No. IST-029192-2, the German Israeli Foundation grant No. 190/2006, and the Hans Vielberth Foundation. E.M. has been supported by the Universidad Complutense de Madrid through project PR27/05-14014-BSCH. R.A.R. thankfully acknowledges support by the EPSRC (EP/C007042/1), the Royal Society and the Leverhulme Trust.

References

1. *Modern Methods for Theoretical Physical Chemistry of Biopolymers*, edited by E.B. Starikov, J.P. Lewis, and S. Tanaka (Elsevier, Amsterdam, 2006).
2. D.D. Eley and D.I. Spivey, Trans. Faraday Soc. **58**, 411 (1962).
3. S.O. Kelley and J.K. Barton, Science **283**, 375 (1999).
4. K. Keren, R.S. Berman, E. Buchstab, U. Sivan, and E. Braun, Science **302**, 1380 (2003).

5. M. Mertig, R. Kirsch, W. Pompe, and H. Engelhardt, Eur. Phys. J. D **9**, 45 (1999).
6. J.H. Reif, T.H. LaBean, and N.C. Seeman, in *DNA '00: Revised Papers from the 6th International Workshop on DNA-Based Computers* (Springer-Verlag, London, UK, 2001), pp. 173–198.
7. C.J. Murphy, M.A. Arkin, Y. Jenkins, N.D. Ghatlia, S. Bossman, N.J. Turro, and J.K. Barton, Science **262**, 1025 (1993).
8. S. Priyadarshy, S.M. Risser, and D.N. Beratan, J. Phys. Chem. **100**, 17678 (1996).
9. E. Meggers, M.E. Michel-Beyerle, and B. Giese, J. Am. Chem. Soc. **120**, 12950 (1998).
10. C.R. Treadway, M.G. Hill, and J.K. Barton, Chemical Physics **281**, 409 (2002).
11. E. Braun, Y. Eichen, U. Sivan, and G. Ben-Yoseph, Nature **391**, 775 (1998).
12. A.J. Storm, J. van Noort, S. de Vries, and C. Dekker, Appl. Phys. Lett. **79**, 3881 (2001).
13. D. Porath, A. Bezryadin, S. De Vries, and C. Dekker, Nature **403**, 635 (2000).
14. H. Cohen, C. Nogues, R. Naaman, and D. Porath, Proc. Nat. Acad. Sci. **102**, 11589 (2005).
15. H.-W. Fink and C. Schönenberger, Nature **398**, 407 (1999).
16. K.-H. Yoo, D.H. Ha, J.-O. Lee, J.W. Park, J. Kim, J.J. Kim, H.-Y. Lee, T. Kawai, and H.Y. Choi, Phys. Rev. Lett. **87**, 198102 (2001).
17. B. Xu, P. Zhang, X. Li, and N. Tao, Nano Lett. **4**, 1105 (2004).
18. A.Y. Kasumov, M. Kociak, S. Guéron, B. Reulet, V.T. Volkov, D. V. Klinov, and H. Bouchiat, Science **291**, 280 (2001).
19. E. Artacho, M. Machado, D. Sanchez-Portal, P. Ordejon, and J.M. Soler, Mol. Phys. **101**, 1587 (2003).
20. A. Calzolari, R. Di Felice, E. Molinari, and A. Garbesi, Appl. Phys. Lett. **80**, 3331 (2002).
21. R. Di Felice, A. Calzolari, and H. Zhang, Nanotechnology **15**, 1256 (2004).
22. F.L. Gervasio, P. Carolini, and M. Parrinello, Phys. Rev. Lett. **89**, 108102 (2002).
23. R.N. Barnett, C.L. Cleveland, A. Joy, U. Landman, and G.B. Schuster, Science **294**, 567 (2001).
24. A. Hübsch, R.G. Endres, D.L. Cox, and R.R.P. Singh, Phys. Rev. Lett. **94**, 178102 (2005).
25. E.B. Starikov, Phil. Mag. Lett. **83**, 699 (2003).
26. E.B. Starikov, Phil. Mag. **85**, 3435 (2005).
27. C. Adessi, S. Walch, and M.P. Anantram, Phys. Rev. B **67**, 081405(R) (2003).
28. H. Mehrez and M.P. Anantram, Phys. Rev. B **71**, 115405 (2005).
29. G. Cuniberti, L. Craco, D. Porath, and C. Dekker, Phys. Rev. B **65**, 241314 (2002).
30. J. Jortner, M. Bixon, T. Langenbacher, and M.E. Michel-Beyerle, Proc. Nat. Acad. Sci. **95**, 12759 (1998).
31. J. Jortner and M. Bixon, Chemical Physics **281**, 393 (2002).
32. S. Roche, D. Bicout, E. Maciá, and E. Kats, Phys. Rev. Lett. **91**, 228101 (2003).
33. S. Roche, Phys. Rev. Lett. **91**, 108101 (2003).
34. M. Unge and S. Stafstrom, Nano Lett. **3**, 1417 (2003).
35. F. Palmero, J.F.R. Archilla, D. Hennig, and F.R. Romero, New J. Phys. **6**, 13 (2004).

36. V.M. Apalkov and T. Chakraborty, Phys. Rev. B **71**, 033102 (2005).
37. V.M. Apalkov and T. Chakraborty, Phys. Rev. B **72**, 161102 (2005).
38. R. Gutierrez, S. Mandal, and G. Cuniberti, Phys. Rev. B **71**, 235116 (2005).
39. R. Gutierrez, S. Mandal, and G. Cuniberti, Nano Lett. **5**, 1093 (2005).
40. D.K. Klotsa, R.A. Römer, and M.S. Turner, Biophys. J. **89**, 2187 (2005).
41. H. Yamada, Phys. Lett. A **332**, 65 (2004).
42. M. R. D'Orsogna and R. Bruinsma, Phys. Rev. Lett. **90**, 078301 (2003).
43. E. Maciá and S. Roche, Nanotechnology **17**, 3002 (2006).
44. E. Maciá, Phys. Rev. B **74**, 245105 (2006).
45. O.R. Davies and J.E. Inglesfield, Phys. Rev. B **69**, 195110 (2004).
46. Y.J. Yan and H. Zhang, J. Theor. Comput. Chem. **1**, 225 (2002).
47. K. Iguchi, Int. J. Mod. Phys. B **18**, 1845 (2004).
48. A. A. Voityuk, J. Jortner, M. Boxin, and N. Rösch, J. Chem. Phys. **114**, 5614 (2001).
49. H. Sugiyama and I. Saito, J. Am. Chem. Soc. **118**, 7063 (1996).
50. H. Zhang, X.-Q. Li, P. Han, X. Y. Yu, and Y. Yan, J. Chem. Phys. **117**, 4578 (2002).
51. X. Yang, X.-B. Wang, E. R. Vorpagel, and L.-S. Wang, Proc. Nat. Acad. Sci. **101**, 17588 (2004).
52. E. Cauet, D. Dehareng, and J. Lievin, J. Phys. Chem. A **110**, 9200 (2006).
53. *Computational Statistical Physics: From Billiards to Monte Carlo*, edited by K. H. Hoffmann and M. Schreiber (Springer-Verlag, Berlin, 2002).
54. B. Kramer and A. MacKinnon, Rep. Prog. Phys. **56**, 1469 (1993).
55. R. A. Römer and M. Schreiber, in *The Anderson Transition and its Ramifications – Localization, Quantum Interference, and Interactions*, Vol. 630 of *Lecture Notes in Physics*, edited by T. Brandes and S. Kettemann (Springer, Berlin, 2003).
56. P.J. Pablo, F. Moreno-Herrero, J. Colchero, J. Gomez Herrero, P. Hererro, P. Baro, A.M. an Ordejon, J.M. Soler, and E. Artacho, Phys. Rev. Lett. **85**, 4992 (2000).
57. B. Alberts, D. Bray, J. Lewis, M. Raff, K. Roberts, and J. Watson, *Molecular Biology of the Cell* (Garland, New York, 1994).
58. A. MacKinnon, J. Phys.: Condens. Matter **13**, L1031 (1980).
59. A. MacKinnon, Z. Phys. B **59**, 385 (1985).
60. R.A. Römer, C. Villagonzalo, and A. MacKinnon, J. Phys. Soc. Japan **72**, 167 (2003), suppl. A.
61. A. Croy, R.A. Römer, and M. Schreiber, in *Parallel Algorithms and Cluster Computing - Implementations, Algorithms, and Applications, Lecture Notes in Computational Science and Engineering*, edited by K. Hoffmann and A. Meyer (Springer, Berlin, 2006).
62. J. D'Amato and H. Pastawski, Phys. Rev. B **41**, 7411 (1990).
63. J.-L. Pichard and G. Sarma, J. Phys. C **14**, L127 (1981).
64. J.-L. Pichard and G. Sarma, J. Phys. C **14**, L617 (1981).
65. A. MacKinnon and B. Kramer, Z. Phys. B **53**, 1 (1983).
66. A. MacKinnon, J. Phys.: Condens. Matter **6**, 2511 (1994).
67. K. Frahm, A. Müller-Groeling, J.L. Pichard, and D. Weinmann, Europhys. Lett. **31**, 169 (1995).
68. R.A. Römer and M. Schreiber, Phys. Rev. Lett. **78**, 4890 (1997).
69. M.L. Ndawana, R.A. Römer, and M. Schreiber, Europhys. Lett. **68**, 678 (2004).

70. M. Büttiker, Y. Imry, and R. Landauer, Phys. Lett. A **96**, 365 (1983).
71. M. Büttiker, Y. Imry, R. Landauer, and S. Pinhas, Phys. Rev. B **31**, 6207 (1985).
72. P.F. Bagwell and T.P. Orlando, Phys. Rev. B **40**, 1456 (1989).
73. E.G. Emberly and G. Kirczenow, Phys. Rev. B **58**, 10911 (1998).
74. E.G. Emberly and G. Kirczenow, J. Phys.: Condens. Matter **11**, 6911 (1999).
75. T. Kostyrko, J. Phys.: Condens. Matter **14**, 4393 (2002).
76. E. Maciá, F. Triozon, and S. Roche, Phys. Rev. B **71**, 113106 (2005).
77. Y. Zhu, C.C. Kaun, and H. Guo, Phys. Rev. B **69**, 245112 (2004).
78. B. Hartzell, B. Melord, D. Asare, H. Chen, J. J. Heremans, and V. Sughomonian, Appl. Phys. Lett. **82**, 4800 (2003).
79. Y. Zhang, R. H. Austin, J. Kraeft, E. C. Cox, and N. P. Ong, Phys. Rev. Lett. **89**, 198102 (2002).
80. P.W. Anderson, Phys. Rev. **109**, 1492 (1958).
81. *The Anderson Transition and its Ramifications – Localization, Quantum Interference, and Interactions*, Vol. 630 of *Lecture Notes in Physics*, edited by T. Brandes and S. Kettemann (Springer, Berlin, 2003).
82. X.-Q. Li and Y. Yan, Appl. Phys. Lett. **79**, 2190 (2001).
83. S. Datta, *Electronic Transport in Mesoscopic Systems* (Cambridge University Press, Cambridge, 1999).
84. Z. Yu and X. Song, Phys. Rev. Lett. **86**, 6018 (2001).
85. W. Zhang and S.E. Ulloa, Phys. Rev. B **69**, 153203 (2004).
86. Y. Asai, J. Phys. Chem. B **107**, 4647 (2003).
87. E.L. Alburquerque, M.S. Vasconcelos, M.L. Lyra, and F.A.B.F. de Moura, Phys. Rev. E **71**, 21910 (2005).
88. C.T. Shih, Phys. Stat. Sol. (b) **243**, 378 (2006).
89. C.T. Shih, Phys. Rev. E **74**, 010903 (2006).
90. E.M. Conwell and S.V. Rakhmanova, Proc. Nat. Acad. Sci. **97**, 4556 (2000).
91. R. Bruinsma, G. Grüner, M.R. D'Orsogna, and J. Rudnick, Phys. Rev. Lett. **85**, 4393 (2000).
92. W. Zhang, A.O. Govorov, and S.E. Ulloa, Phys. Rev. B **66**, 060303 (2002).
93. S. Komineas, G. Kalosakas, and A.R. Bishop, Phys. Rev. E **65**, 061905 (2002).
94. H. Yamada, E.B. Starikov, D. Hennig, and J. F. R. Archilla, Eur. Phys. J. E **17**, 149 (2004).
95. F.C. Grozema, Y.A. Berlin, and L.D.A. Siebbeles, J. Am. Chem. Soc. **122**, 10903 (2000).
96. R.G. Endres, D.L. Cox, R.R.P. Singh, and S.K. Pati, Phys. Rev. Lett. **88**, 166601 (2002).
97. E.I. Kats and V. V. Lebedev, JETP Lett. **75**, 37 (2002).
98. V.D. Lakhno, J. Biol. Phys. **26**, 133 (2000).
99. N. Rösch and A.A. Voityuk, Top. Curr. Chem. **237**, 37 (2004). See Ref. [122].
100. H. Wang, R. Marsh, J.P. Lewis, and R.A. Römer, in *Modern Methods for Theoretical Physical Chemistry of Biopolymers*, edited by E.B. Starikov, J.P. Lewis, and S. Tanaka (Elsevier, Amsterdam, 2006).
101. R.A. Römer and H. Schulz-Baldes, Europhys. Lett. **68**, 247 (2004).
102. K. Iguchi, Int. J. Mod. Phys. B **11**, 2405 (1997).
103. J. Yi, Phys. Rev. B **68**, 193103 (2004).
104. R.A. Caetano and P.A. Schulz, Phys. Rev. Lett. **95**, 126601 (2005).
105. A. Sedrakyan and F. Domínguez-Adame, Phys. Rev. Lett. **96**, 059703 (2006).

106. R.A. Caetano and P.A. Schulz, Phys. Rev. Lett. **96**, 059704 (2006).
107. E. Díaz, A. Sedrakyan, D. Sedrakyan, and F. Domínguez-Adame, Phys. Rev. B **75**, 014201 (2007).
108. K. Iguchi, J. Phys. Soc. Jpn. **70**, 593 (2001).
109. K. Forinash, A.R. Bishop, and P.S. Lomdahl, Phys. Rev. B **43**, 10743 (1991).
110. N.R. Walet and W.J. Zakrzewski, Nonlinearity **18**, 2615 (2005).
111. S.S. Wesolowski, M.L. Leininger, P.N. Pentchev, and H.F. Schaefer III, J. Am. Chem. Soc. **123**, 4023 (2001).
112. J. Zhong, in *Proceedings of the 2003 Nanotechnology Conference, Computational Publications*, edited by M. Laudon and B. Romamowicz (Nano Science and Technology Institute, Cambridge, 2003), Vol. 2, pp. 105–108, (Molecular and Nano Electronics).
113. D.K. Klotsa, R.A. Römer, and M.S. Turner, In Proceedings 27th International Conference on the Physics of Semiconductors(Q5 129), Flagstaff, Arizona 328 (2004).
114. R.G. Endres, D.L. Cox, and R.P. Singh, Rev. Mod. Phys. **76**, 195 (2004).
115. A. Rodriguez, R.A. Römer, and M.S. Turner, phys. stat. sol. (b) **243**, 373 (2005).
116. J.F. Feng and S.J. Xiong, Phys. Rev. E **66**, 021908 (2002).
117. J. Cuevas, E.B. Starikov, J. F. R. Archilla, and D. Henning, Mod. Phys. Lett. B **18**, 1319 (2004).
118. R. Bulla, R. Gutierrez, and G. Cuniberti, *Modern methods for theoretical physical chemistry of biopolymers*, edited by E. Starikow, J. Lewis and S. Tanaka (Elsevier, Amsterdam, 2006).
119. D. Porath, G. Cuniberti, and R. Di Felice, in *Long-Range Charge Transfer in DNA I and II*, Vol. 237 of *Topics in Current Chemistry*, edited by G.B. Schuster (Springer, Berlin, 2004), pp. 183. See Ref. [122].
120. Y. Calev, H. Cohen, G. Cuniberti, A. Nitzan, and D. Porath, Israel Journal of Chemistry **44**, 133 (2004).
121. H. C. D. P. R. Gutirrez, S. Mohapatra and G. Cuniberti, Phys. Rev. B **74**, 235105 (2006).
122. *Long-Range Charge Transfer in DNA I and II*, Vol. 237 of *Topics in Current Chemistry*, edited by G.B. Schuster (Springer, Berlin, 2004).

2 Mechanism and Absolute Rates of Charge Transfer Through DNA

Ferdinand C. Grozema and Laurens D.A. Siebbeles

Opto-Electronic Materials Section, DelftChemTech,
Delft University of Technology, Julianalaan 136, 2628 BL Delft, The Netherlands
l.d.a.siebbeles@tudelft.nl

2.1 Introduction

The mechanism and rates of charge transfer through the DNA double helix have been studied extensively over past decades [1–4]. The possibility of electrical conductivity in DNA was put forward for the first time in 1962 by Eley and Spivey [5], briefly after Watson and Crick described the double helical structure [6]. Eley and Spivey noted that the stack of aromatic base pairs in the interior of the helix shows a striking resemblance to the stacking found in one-dimensional molecular crystals. High mobilities of charge carriers have been demonstrated for aromatic crystals and one-dimensional discotic materials [7]. There are also important differences between DNA and these materials. Natural DNA consists of a non-periodic stack of different nucleobases. This leads to energy variations along the stack that are much larger than for aromatic crystals and discotic materials, where all molecules are the same. Moreover, the structure of DNA in its natural aqueous environment is inherently flexible, and structural variations can be expected to influence the efficiency of charge transport to a large extent.

The interest in charge transfer in DNA took a flight in the early 1990's after a series of papers by Barton and Turro in which they suggested that ultra-fast photoinduced charge transfer takes place over large distances between donors and acceptors that are intercalated in DNA [8–10]. These claims have prompted a wide variety of experimental and theoretical studies into the nature of charge migration through DNA, and the subject is still the source of considerable controversy [3,11,12]. Initially, experiments on charge transfer were mostly interpreted in terms of classical theory for charge transfer from a donor via an intervening (DNA) bridge to an acceptor. In this case the rate of charge transfer depends exponentially on the distance between the donor and the acceptor

$$k_{\mathrm{CT}}(R) = k_0 \exp(-\beta R) , \qquad (2.1)$$

where k_0 is a scaling factor and β is the so-called fall-off parameter which determines the distance dependence. The value of β has often been used to distinguish between different mechanisms for charge transport through DNA. A large β ($\approx 1\,\text{Å}^{-1}$) represents strong distance dependence, characteristic of a single-step tunnelling process in which the charge does not actually become localized on the bridge. A small value for β ($\approx 0.1\,\text{Å}^{-1}$) indicates a weak dependence on distance. This can occur when the energy of the charge at the donor is close to that on the intervening bridge and leads to a considerable charge density on the bridge during the charge transfer process.

In the initial studies of Barton and co-workers, values for β as low as $0.2\,\text{Å}^{-1}$ were reported [13–15], however, several other experimentalists found much higher values [16–20]. Another important development was the experimental work by Giese and co-workers [1, 21, 22]. They found that the rate of charge transfer depends strongly on the base-pair sequence, leading to different distance dependence for different sequences. These experiments and many others, were mostly directed at unravelling the mechanism of charge transport in DNA. The experiments mentioned above all rely on a steady state method to examine the relative rates of charge transfer, either by examining the damage yield at different positions along the base sequence [21, 23–25] or by monitoring the fluorescence quenching [13, 14, 16, 18]. Direct observation of the kinetics of photo-induced charge transfer through DNA can be achieved by femto-second transient absorption spectroscopy. Lewis and co-workers have determined absolute rates for charge transfer in synthetic DNA hairpins by analyzing the kinetics of the decay of the charge separated state involving a donor anion and a positive charge on DNA [26, 27].

In order to understand and describe the experimental results theoretically, we have devised a model based on the tight-binding approximation, that is able to describe all charge transfer mechanisms, ranging from incoherent hopping to single step tunnelling, depending on the base pair sequence [28–30]. The main advantage of this model is that it makes no a-priori assumptions about the charge transport mechanism. In this chapter, we review the model that we use to study charge transport in DNA and show which are the key parameters that determine the charge transport mechanism and the absolute rates for transfer of positive charges through DNA.

2.2 Theoretical Description of Charge Migration in DNA

In general, the rate of charge transfer between two neighboring nucleobases depends on the energy difference between the bases and the charge transfer integral between these bases. This is analogous to electron transfer in other systems. In order to describe the motion of charges along a DNA double helix, a whole sequence of charge transfer reactions along many nucleobases has to be considered. Each of these charge transfer reactions is characterized by a specific charge transfer integral and energy difference, which both depend on the orientation of the bases with respect to each other and the local environment (e.g. the presence of counter ions or the water surrounding the DNA). In order to study the migration of charges through DNA, we use a tight-binding model, which can take these variations along the DNA double helix into account. The wave function of the charge, Ψ, is taken to be a linear combination of basis functions, φ_n, that are localized on each nucleobase, with expansion coefficients c_n

$$\Psi(t) = \sum_{n}^{N} c_n(t)|\varphi_n\rangle .\tag{2.2}$$

In the case of the migration of positive charges through DNA, these basis functions correspond to the highest occupied molecular orbitals (HOMO) on the nucleobases. In a donor-DNA-acceptor system, the charge is initially created on the donor where it is completely localized. This initial condition is achieved by setting the expansion coefficient on the donor site equal to one, and all others to zero. In order to study the motion of the charge along the DNA bridge towards the acceptor, the wavefuncion is propagated in time according to the time-dependent Schrödinger equation

$$i\hbar \frac{\partial \Psi(t)}{\partial t} \hat{H}\Psi(t) .\tag{2.3}$$

In most cases, propagation of the wavefunction leads to a spreading of the charge over the intervening DNA bridge towards the acceptor. In order to ensure that the charge is irreversibly trapped when it arrives on the acceptor, a complex part is added to the diagonal matrix element of the acceptor site (N), $H_{NN} = \varepsilon_{NN} - i\hbar/\tau$. A decay time τ of 100 fs was used. The decay at the acceptor site leads to a decay of the total charge density on the donor-DNA-acceptor system. The survival probability $P(t)$ is the probability that the charge is still present in the system

$$P(t) = \sum_{n}^{N} |c_n(t)|^2 .\tag{2.4}$$

The rate of charge transfer from the donor to the acceptor can be obtained from the decay of the survival probability during time.

The charge transfer integrals and charge transfer energetics of the donor-bridge-acceptor system are described by the Hamiltonian matrix. The diagonal matrix elements of the Hamiltonian correspond to the site-energies, $\varepsilon_{ii} = \langle \varphi_i | \hat{H} | \varphi_i \rangle$, i.e. the energy of a charge carrier when it is localized on a single nucleobase. In the simplest approximation, these site-energies correspond to the ionization potential of a single nucleobase, in the case of transfer of positive charges. However, it is important to note that the site-energies can change considerably depending on the neighboring bases [31]. As an example, the site-energies of two neighboring guanine moieties in a DNA sequence are in general not the same, which leads to different amounts of oxidative damage on both sites [31].

When only nearest neighbor interactions are taken into account, the off-diagonal matrix elements of the Hamiltonian are equal to the electronic coupling, $J_{ij} = \langle \varphi_i | \hat{H} | \varphi_j \rangle$, between the HOMO orbitals on adjacent nucleobases, while all other off-diagonal matrix elements are zero. The Hamiltonian matrix is then given by

$$
H = \begin{pmatrix}
\varepsilon_{11} & J_{12} & 0 & \cdots & & 0 \\
J_{21} & \varepsilon_{22} & & & & \\
0 & & \ddots & & & \\
\vdots & & & \ddots & & \\
0 & & & & \varepsilon_{NN} - \dfrac{i\hbar}{\tau}
\end{pmatrix} .
\tag{2.5}
$$

Both the site-energies and the charge transfer integrals in (2.5) are sensitive to the geometry of the DNA. All changes in the intramolecular degrees of freedom lead to changes in ε and J, and hence in the charge transport properties of DNA. This also implies that the motion of the charge carrier is directly coupled to structural fluctuations in the DNA double helix. These structural fluctuations are described classically in our model as will be discussed below.

2.3 Single Step Tunnelling Through DNA

The controversy on charge transfer through DNA in the early 1990s, kindled by the experiments of Barton and co-workers [8, 10], was centered mostly on the value of the fall-off parameter, β. We have calculated the value of β for DNA sequences studied experimentally by Meggers et al. [21, 25]. In their experiments, these authors site-selectively generated a guanine radical cation G^+ in a DNA double strand with a known base pair sequence. The positive charge was found to migrate to a site consisting of three consecutive GC base pairs. Such a GC triplet is known to act as an acceptor for holes with an ionization energy that is 0.7 eV lower than for a single GC base pair [32]. In the experiments, the transfer of a positive charge through a bridge consisting only of AT base pairs was studied by examining the relative damage yield at the initial GC base pairs and at the GC triplet. The essential parameters for modelling charge transfer through such a sequence are the site-energies and the charge transfer integrals, as discussed in the previous section. In these calculations, we have taken all charge transfer integrals to be the same and equal to 0.11 eV, which was considered a reasonable estimate of the average charge transfer integral based on calculations of orbitals splittings and band structure calculations [33, 34]. In a simplified picture, there are two different site-energies, one corresponding to a GC base pair and another for AT base pairs. The energy difference between these sites was taken from ab initio calculations by Hutter and Clark and was taken to be 0.55 eV [35]. Calculations have been performed for a series of bridges that consist of an increasing number of AT base pairs between the G^+ donor and the GC triplet acceptor, see Scheme 2.1.

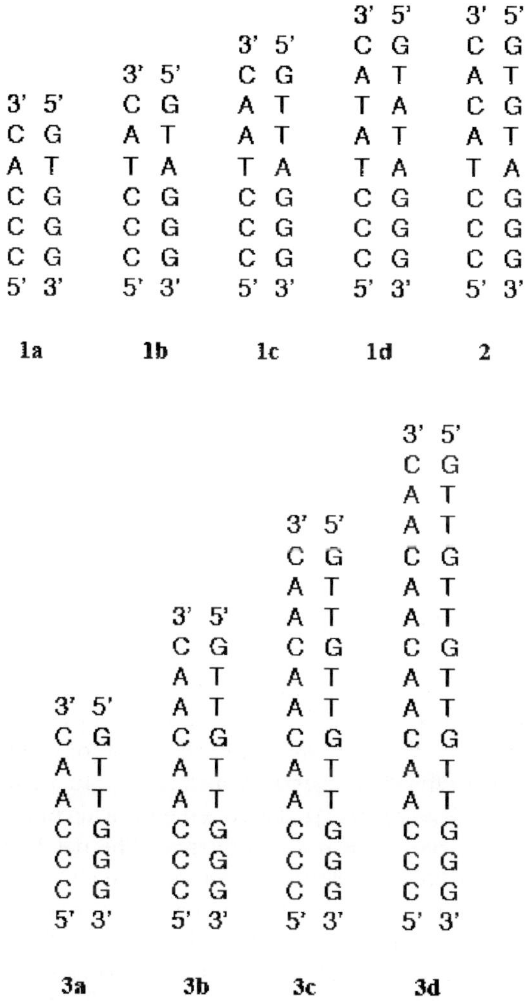

```
                                        3' 5'      3' 5'
                            3' 5'       C  G       C  G
                3' 5'       C  G        A  T       A  T
    3' 5'       C  G        A  T        T  A       C  G
    C  G        A  T        A  T        A  T       A  T
    A  T        T  A        T  A        T  A       T  A
    C  G        C  G        C  G        C  G       C  G
    C  G        C  G        C  G        C  G       C  G
    C  G        C  G        C  G        C  G       C  G
    5' 3'       5' 3'       5' 3'       5' 3'      5' 3'

     1a          1b          1c          1d         2
```

```
                                                    3' 5'
                                                    C  G
                                                    A  T
                                        3' 5'       A  T
                                        C  G        C  G
                                        A  T        A  T
                            3' 5'       A  T        A  T
                            C  G        C  G        C  G
                            A  T        A  T        A  T
                3' 5'       A  T        A  T        A  T
                C  G        C  G        C  G        C  G
    3' 5'       A  T        A  T        A  T        A  T
    C  G        A  T        A  T        A  T        A  T
    A  T        C  G        C  G        C  G        C  G
    A  T        C  G        C  G        C  G        C  G
    C  G        C  G        C  G        C  G        C  G
    C  G        5' 3'       5' 3'       5' 3'       5' 3'
    C  G
    5' 3'

     3a          3b          3c          3d
```

Scheme 2.1.

Figure 2.1a shows the so-called survival fraction (2.4), the fraction of the charge still present on the whole DNA molecule, for different numbers of AT base pairs in the bridge as a function of time. The decay of the survival probability, $P(t)$, is due to charge transfer from the donor to the acceptor where it is trapped irreversibly. The decay of the survival probability can be approximated by an exponential function

$$P(t) = \exp(-k_{CT}(R)t) , \qquad (2.6)$$

where k_{CT} is the effective decay rate. The calculated rates of decay decrease rapidly with increasing number of intervening AT base pairs. This is clearly

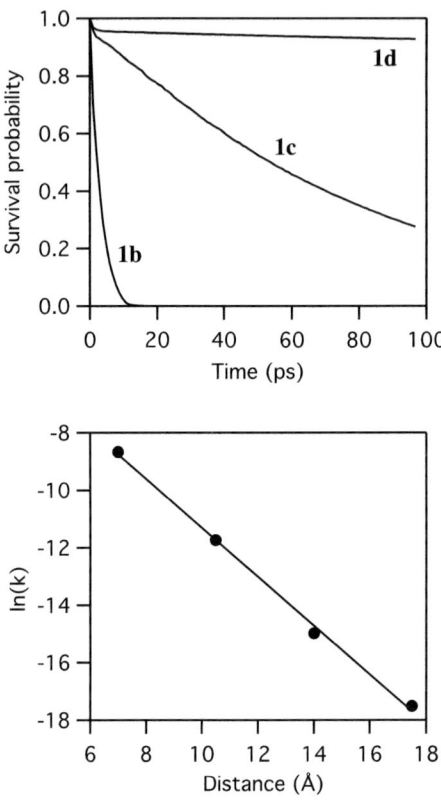

Fig. 2.1. *Top*: Survival probability for sequences **1b–1d** as a function of time. *Bottom*: ln(k) plotted against the donor-acceptor distance for sequences **1a–1d**. The value of β obtained from the linear fit is $0.85\,\text{Å}^{-1}$

visible from Fig. 2.1b where the natural logarithm of k_{CT} is plotted against the donor-acceptor distance. The fall-off parameter obtained from Fig. 2.1b according to (2.1) is $0.85\,\text{Å}^{-1}$. This value is in reasonable agreement with the experimental value of $0.7\,\text{Å}^{-1}$ obtained by Meggers et al. [21]. Such a relatively large fall-off parameter is consistent with a single step process in which the charge tunnels through the AT bridge from the donor GC to the acceptor. This was confirmed by examining the calculated charge density on the bridge and it was found that no significant charge density is present on the bridge during the charge transfer.

The energy difference of $0.55\,\text{eV}$ between the donor and the bridge used here corresponds to G^+ as a donor. In the experimental work, a variety of hole donors have been used. The use of different donors will in general lead to different energy differences between the donor and the bridge. In order to study the effect of this injection barrier on the charge transfer process, the calculations described above have been repeated for a series of energy

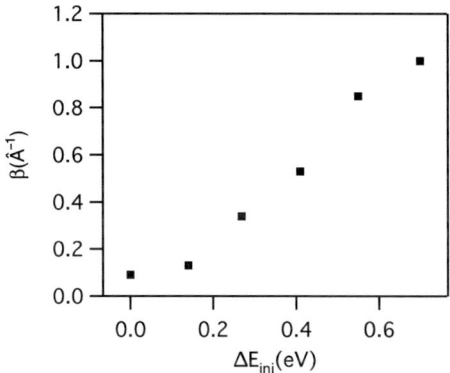

Fig. 2.2. β-value for hole transfer through an AT bridge as a function of the injection energy

differences ranging from 0 to 0.7 eV. The resulting values obtained for β are plotted against the injection barrier in Fig. 2.2.

The fall-off parameter decreases as the injection barrier becomes smaller and attains a limiting value of $0.09\,\text{Å}^{-1}$ for an injection barrier of 0 eV. These calculations show that the value for β can vary widely, depending on the specific type of hole donor used. β-values as low as those found by Barton et al. $(0.2\,\text{Å}^{-1})$ [13, 15] can be obtained for injection barriers close to 0.2 eV. Very high values, such as the one found by Fukui et al. [19] can be obtained for an injection barrier that is higher than 0.7 eV. The results in Fig. 2.2 are also consistent with the work of Lewis et al. who studied hole transfer through DNA segments consisting only of AT base pairs while varying the hole donor [27, 36]. The experimental trend found in these experiments is very similar to that shown in Fig. 2.2 [27]. It can be concluded from these calculations that the specific value of β is not a useful parameter to assess the ability of DNA to transport charge, since β strongly depends on the injection barrier, or equivalently, the hole donor.

2.4 Sequence Dependence

An interesting observation on charge migration through DNA is the large influence of the base pair sequence in the DNA bridge on both the charge transfer rate and its distance dependence. It was found experimentally by Meggers et al. that the charge transfer rate increases dramatically when one of the base pairs in a sequence of 4 AT base pairs is replaced by a GC base pair (sequence **2** in Scheme 2.1) [21].

The calculated survival probability for this sequence is shown in Fig. 2.3. It is evident that the motion of the charge through sequence **2** is almost as fast as the decay on a bridge containing only two AT base pairs (sequence **1b**).

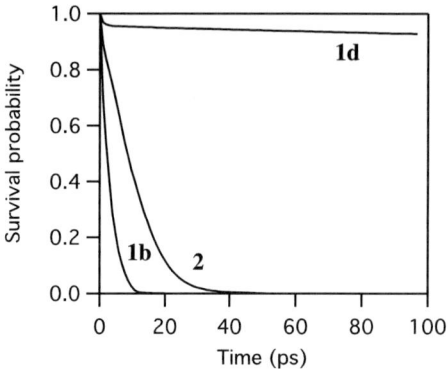

Fig. 2.3. Survival probability for sequences **1b, 1d** and **2** as a function of time

This is in agreement with the experimental findings by Meggers et al. [21]. The observed sequence dependence can be explained by assuming that a hole moves along the bridge by undergoing successive series of 'hops' between G bases [37–39]. These 'hops' are in fact tunnelling steps through regions containing only AT base pairs. The validity of this explanation can be verified by examining the population on the bridge sites. These populations are shown in Fig. 2.4 as a function of time.

It can clearly be seen that the population on the GC site in the bridge is quite large, while the population on the AT sites is always negligible. The hole oscillates back and forth between the donor site and the GC site on the bridge while it slowly leaks through the barrier that is formed by the last two AT base pairs of the bridge, as evident from the overall decay of the full population. This last step is the rate-determining step in the process of charge migration through this particular bridge. Therefore, it is easily understood that the rate of charge migration through this bridge is of the same order of magnitude as that found for a bridge containing only two AT base pairs.

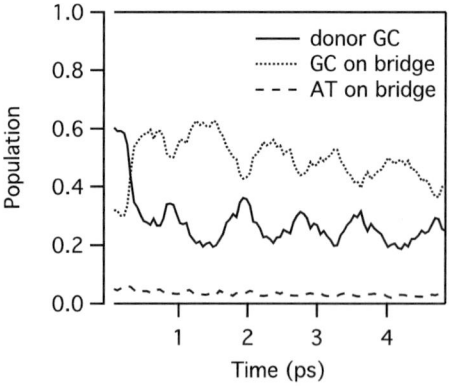

Fig. 2.4. Population on different sites in sequence **2** as a function of time

Giese et al. published an experimental test of this mechanism of hopping between GC base pairs [25]. In this study, a series of DNA bridges with an increasing number of GC base pairs mutually separated by two AT base pairs was considered. These sequences **3a–3d** are shown in Scheme 2.1.

The time evolution of the survival probability obtained from simulations on these DNA sequences are shown in Fig. 2.5.

It is evident that the distance dependence is rather weak. A value for β can be derived again by plotting the logarithm of the effective decay rate [obtained by fitting of (2.6)] against the distance as shown in Fig. 2.6a. The β-value derived from Fig. 2.6a for this series of sequences is $0.09\,\text{Å}^{-1}$. This value agrees nicely with the experimental values of $0.07\,\text{Å}^{-1}$ reported by Giese et al. [25] for this series. The low β indicates that the mechanism of charge migration through this bridge is effectively a process in which the charge effectively hops from GC site to GC site by tunnelling through the intervening sequence of AT base pairs. This was confirmed by examining the population on the AT base pairs which was found to be negligible at all times while a significant amount of charge appeared on the GC sites. As noted in the introduction, in cases where the charge moves by a multi-step hopping mechanism, there is no exponential relation between the length of the chain and the rate. This is also evident if the linear fit is compared to the numerical data points in Fig. 2.6a, where there are considerable deviations from linearity in the data points. In the case of hopping between GC base pairs there is actually a power law relation between the charge transfer rate and the number of hopping steps, N,

$$k_{\text{CT}} \propto N^{-\eta} .$$

Therefore, it is more appropriate to plot the logarithm of the rate against the logarithm of the distance. A much better linear fit is obtained as shown in Fig. 2.6b. The value of the proportionality factor η obtained from the fit is 2.09 which is reasonably close to the experimental value of 1.725.

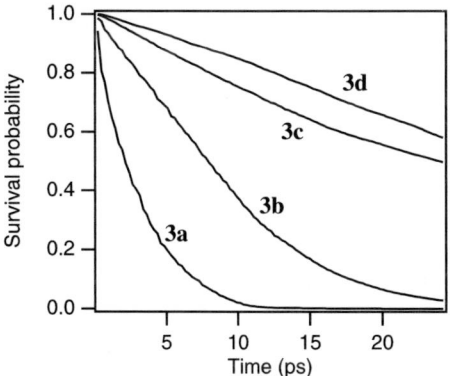

Fig. 2.5. Survival probability for sequences **3a–3d** as a function of time

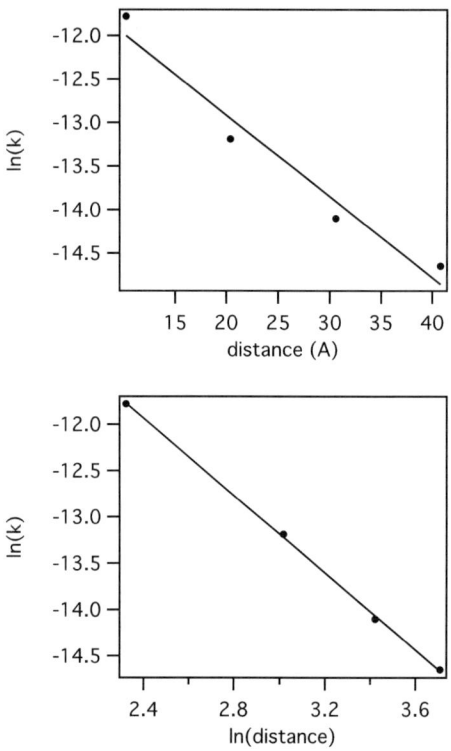

Fig. 2.6. *Top*: $\ln(k)$ plotted against the donor-acceptor distance for sequences **3a–3d**. The value for β obtained from the linear fit is $0.09\,\text{Å}^{-1}$. *Bottom*: $\ln(k)$ plotted against the logarithm of the donor-acceptor distance for sequences **3a–3d**. The value obtained for η [see (2.6)] from the linear fit is 2.09

It is important to note that each 'hop' is in fact a tunnelling step through an AT bridge and therefore this hopping-like transport is quite distinct from thermally-activated hopping over barriers as has been proposed for charge transport in disordered materials. For the systems described in this work, the charge never becomes localized on the AT bridges that separate the GC localization sites. Such hopping onto AT base pairs has been proposed for sequences longer than the ones studied here [40]. In this case, the rate for a single-step-tunnelling process becomes so low that hopping of the charge onto the bridge becomes kinetically favorable.

2.5 Calculation of Accurate Parameters for Hole Transport Calculations

The calculations presented above were all carried out with a very limited set of parameters. The only important parameters used are the site-energy difference between AT and GC base pairs and the charge transfer integral between neighboring units. These were taken to be the same for all combinations of base pairs. Even with these simplifications, the qualitative description of the distance dependence and the sequence dependence of hole transport through DNA is very good as shown above. However, in order to obtain absolute rates for hole transfer from our calculations, a much more detailed description is needed. In an irregular DNA sequence, all site-energies are different, even for the same bases. This is due to the local surroundings of each nucleobase [31, 41]. The same is true for the charge transfer integral; each combination of two bases giving rise to a different value [30, 42–44]. Additionally, in the calculations described above, it was assumed that the spatial overlap between molecular orbitals on neighboring nucleobases is negligible, which is, in general, not the case. The site-energies, charge transfer integrals and spatial overlap integrals can be obtained directly from DFT calculations using the molecular orbitals of the individual nucleobases as a basis set. This fragment orbital approach has been described previously and we will only summarize some of the results obtained for charge transfer parameters in DNA here [30, 45].

The site-energies, which are defined as the diagonal matrix elements of the Kohn-Sham Hamiltonian involving the HOMOs on the nucleobases G, A, C and T, are given in Table 2.1, for all possible combinations of flanking nucleobases at the 5'- and 3'-positions. The effect of the flanking nucleobase

Table 2.1. Site-energies (in eV) for a nucleobase B in 5'-XBY-3' triads (X, B, Y = G, A, C and T)

Y	G	A	T	C
GGY	7.890	8.040	8.290	8.310
AGY	7.900	8.060	8.320	8.341
CGY	7.957	8.115	8.361	8.383
TGY	7.965	8.124	8.380	8.407
GAY	8.343	8.487	8.712	8.716
AAY	8.376	8.558	8.799	8.763
CAY	8.438	8.584	8.793	8.800
TAY	8.434	8.630	8.858	8.810
GTY	9.111	9.308	9.533	9.557
ATY	9.130	9.370	9.586	9.578
CTY	9.268	9.451	9.662	9.701
TTY	9.273	9.499	9.699	9.705
GCY	9.446	9.637	9.870	9.857
ACY	9.441	9.630	9.867	9.851
CCY	9.490	9.667	9.917	9.882
TCY	9.499	9.679	9.925	9.895

at the 5′-position on the site-energies is much less pronounced than the effect of the nucleobase at the 3′-position. The site-energy of G flanked by another G at the 3′-position is considerably lower than in cases where G is flanked by another nucleobase at the 3′-position. For all nucleobases, the site-energy is smaller when G or A are present at the 3′-position than in cases where C or T are present at the 3′-position. It is worth mentioning that the site-energies do not always increase in the order G < A < C < T known for the hierarchy of the vacuum ionization energies of individual nucleobases. For instance, the site-energy of G in 5′-TGC-3′ is higher than the site-energy of A in the sequence 5′-GAG-3′. Similar trends were obtained for the ionization energies of base pair triplets calculated by Voityuk et al. [41].

The charge transfer and spatial overlap integrals for nucleobases within the Watson-Crick base pairs are $J = -0.085$ eV and $S = -0.006$ for G:C, while $J = -0.11$ eV and $S = -0.007$ for A:T. The values of J and S for all other combinations of nucleobases in neighboring base pairs are given in Table 2.2. The data in Table 2.2 were obtained for a geometry with stan-

Table 2.2. Charge transfer integrals, J (in eV), spatial overlap matrix elements, S, and generalized charge transfer integrals, J' (in eV), for nucleobases stacked at a distance of 3.38 Åwith a twist angle of 36°

| | $5'\text{-}B_1B_2\text{-}3'$ | | | $3'\text{-}b_1b_2\text{-}5'$ | | |
	J	S	J'	J	S	J'
GG	0.119	0.008	0.053	0.119	0.008	0.053
AA	−0.038	−0.004	−0.004	−0.038	−0.004	−0.004
CC	0.042	0.002	0.022	0.042	0.002	0.022
TT	0.180	0.012	0.072	0.180	0.012	0.072
GA	−0.186	−0.013	−0.077	−0.013	−0.0003	−0.010
GC	−0.295	−0.020	−0.114	0.026	0.002	0.009
GT	0.334	0.023	0.141	0.044	0.003	0.018
AC	0.091	0.005	0.042	−0.008	−0.001	−0.002
AT	−0.157	−0.010	−0.063	−0.068	−0.004	−0.031
CT	−0.161	−0.011	−0.055	−0.066	−0.004	−0.028

| | $5'\text{-}B_1b_2\text{-}5'$ | | | $3'\text{-}b_1B_2\text{-}3'$ | | |
	J	S	J'	J	S	J'
GG	0.046	0.004	0.012	−0.075	−0.005	−0.032
AA	0.122	0.010	0.031	0.148	0.011	0.049
CC	0.002	0.0001	0.001	0.030	0.002	0.010
TT	0.009	0.001	0.001	0.016	0.001	0.006
GA	−0.048	−0.004	−0.013	−0.037	−0.003	−0.011
GC	0.004	0.0002	0.002	0.059	0.004	0.022
GT	−0.018	−0.001	−0.009	−0.049	−0.003	−0.014
AC	−0.004	−0.0003	−0.001	0.045	0.003	0.017
AT	0.035	0.003	0.007	−0.026	−0.002	−0.007
CT	0.0004	0.001	0.0003	−0.015	−0.002	0.004

dard global helical parameters of B-form DNA [46]. Following the notation of Voityuk [42], the nucleobases B_1 and B_2 (see Scheme 2.2) in one strand involved in intrastrand electronic coupling, are symbolized in Table 2.3 as $5'$-B_1B_2-$3'$. Similarly, the nucleobases b_1 and b_2 coupled within the other strand are denoted as $3'$-b_1b_2-$5'$.

As illustrated in Scheme 2.2, the notations $5'$-B_1b_2-$5'$ and $3'$-b_1B_2-$3'$ stand for partners in interstrand coupling. The values of J and S for identical nucleobases are much larger for GG and TT than for AA and CC. The largest intrastrand charge transfer integral is obtained for $5'$-GT-$3'$. In most cases, the interstrand charge transfer integrals for $5'$-XY-$5'$ and $3'$-XY-$3'$ are smaller than the intrastrand charge transfer integrals involving the same nucleobases. Interestingly, the interstrand charge transfer integrals for two adenines are significantly larger than the intrastrand charge transfer integral, in agreement with the results of Voityuk et al. [42]. It is well-known that dynamic disorder caused by motion of stacked base pairs along different degrees of freedom can strongly affect hole transport in DNA [43,47,48]. It is also expected that the major contribution to this effect is base pair twisting [47]. To include the influence of stack dynamics on hole transport in DNA hairpins the charge transfer and spatial overlap integrals were calculated as a function of the twist angle, θ. As an example, the values of intrastrand charge transfer and spatial overlap integrals for identical nucleobases in neighboring base pairs are given in Fig. 2.7. As follows from the data plotted in Fig. 2.7, the values of J and S indeed exhibit a strong variation with θ, and therefore this effect cannot be ignored in studies of rate processes in DNA.

For each twist angle, the charge transfer integrals discussed above were calculated directly as the off-diagonal matrix elements of the Kohn-Sham Hamiltonian. These values of J can be used in theoretical studies of charge transport in DNA, provided the spatial overlap matrix elements S are explicitly taken into account. This is done in the tight-binding calculations of charge transport discussed in the next section. By contrast, in calculations of electronic couplings for superexchange, the spatial overlap integrals are often assumed to be zero as in the calculations discussed above. If, however, this assumption is not valid, the electronic couplings for superexchange can be calculated using generalized charge transfer integrals [49]

$$J' = J - S(\varepsilon_1 + \varepsilon_2)/2 \tag{2.7}$$

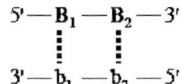   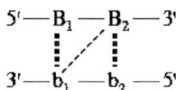

5'—B_1—B_2—3'   5'—B_1—B_2—3'   5'—B_1—B_2—3'

3'—b_1—b_2—5'   3'—b_1—b_2—5'   3'—b_1—b_2—5'

Intrastrand pair B_1-B_2 Interstrand pair B_1-b_2 Interstrand pair b_1-B_2

Scheme 2.2.

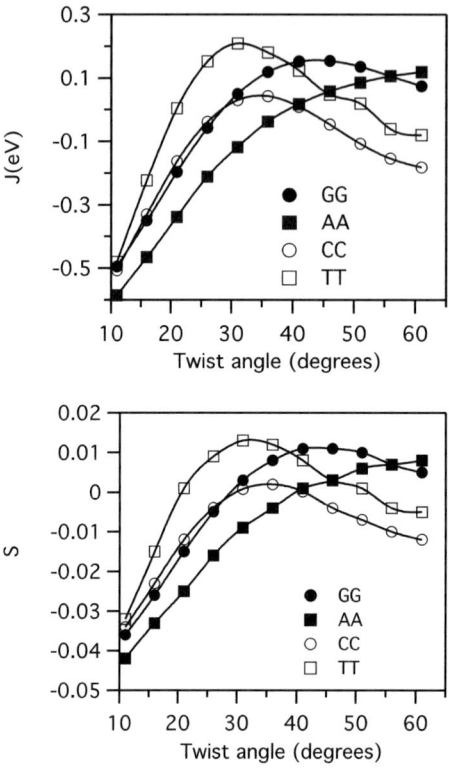

Fig. 2.7. Charge transfer (*top*) and spatial overlap (*bottom*) integrals versus the twist angle between neighboring bases in the same strand

Table 2.3. Experimental rate constants, electronic couplings and reorganization energies in DNA hairpins

Seq.	Rate constants[a] (s^{-1})	Electronic coupling[b] (eV)	Reorganization energies[c] (eV)
	$10^{-7}k_{\mathrm{t}}$	$10^3 V_t$	λ
2b	6.0	8.68	1.00
3b	0.33	2.15	1.46
4c	0.048	0.49	1.09
5b	0.09	0.42	1.00

[a] See Senthilkumar et al. [30].
[b] Calculated from (2.11), (2.12) and (2.13) as explained in the text.
[c] Values needed to reproduce the experimental rate constants using the Marcus equation (2.10).

instead of J. The latter expression reduces to $J' = J - S\varepsilon$ if $\varepsilon_1 = \varepsilon_2 = \varepsilon$. Then the value of J' can be obtained directly from the orbital splitting [49]. The values of J' calculated according to (2.7) are included in Table 2.2. The generalized charge transfer integrals between the nucleobases within a Watson-Crick base pair are $J' = -0.055\,\text{eV}$ for G:C and $J' = -0.047\,\text{eV}$ for A:T.

2.6 Tight-Binding Calculations of Hole Transfer Rates

The rates of hole transfer from the G site nearest to **Sa** to the distal GG doublet in the DNA hairpins shown in Scheme 2.3 were calculated using a quantum mechanical description of the hole combined with a classical description of the twisting motion of the base pairs. Similar to previous studies of charge transfer through DNA [29, 48, 50], the hole was described by the Hamiltonian in (2.5) with site-energies, charge transfer and spatial overlap integrals that are different for all combinations of nucleobases. These parameters also depend on the twist angle between neighboring base pairs as discussed above.

It was found that variations of twist angle and distance between base pairs have a negligible effect on the site-energies. However, the twist angle strongly affects the charge transfer integrals, as shown in Fig. 2.7.

The dynamics of the latter degree of freedom was assumed to be harmonic and was described classically by the Hamiltonian

$$H_{\text{tw}} = \frac{1}{2} \sum_m \left[I_m \theta_m^2 + F_{m,m+1} (\theta_{m+1}(t) - \theta_m(t) - \theta_{m,m+1}^{\text{eq}})^2 \right] , \qquad (2.8)$$

where I_m is the moment of inertia of the m-th base pair, $F_{m,m+1}$ the force constant for twisting and $\theta_{m,m+1}^{\text{eq}}$ the equilibrium twist angle for the base pairs m and $m+1$. Values for the force constants based on molecular dynamics simulations were taken from the work of Lankas et al. [51]. The equilibrium twist angles were taken from the experimental work of Olson et al. [52].

The wave function of the hole is expressed as a time-dependent superposition of the HOMOs on the nucleobases, i.e.

$$\psi(t) = \sum_i c_i(t) \varphi_i . \qquad (2.9)$$

Since initially the hole is localized on the single G site with $i = 1$, the initial condition for the wave function can be written as $c_{i=1}(t = 0) = 1$ and $c_{i \neq 1}(t = 0) = 0$. The initial angular velocities and twist angles were sampled from a Boltzmann distribution at $293\,\text{K}$.

The wave function is propagated during a time step dt taken sufficiently small, so that the twist angles can be considered fixed. The coefficients $c_i(t)$ are obtained numerically by integration of the first-order differential equations that follow from substituting the wave function in (2.9) into the time-dependent Schrödinger equation, which yields $i\hbar S \dfrac{\partial \boldsymbol{c}}{\partial t} = \boldsymbol{H}\boldsymbol{c}$, with $\boldsymbol{c}$ the vector containing the coefficients of the HOMOs in (2.9). In these calculations the overlap matrix S is explicitly taken into account. The twist angles and angular velocities are propagated during the same time step dt by numerically solving the first-order differential equations that follow from the Hamiltonian in (2.8). This procedure is repeated until the decay of the charge is completed.

The rate of hole transfer from the proximal G to the distal GG doublet can be obtained from the probability, $P(t)$, of a positive charge to survive trapping by GG at time t, see (2.4).

Figure 2.8a shows calculated time dependencies of the survival probabilities for the sequences in Scheme 2.3. For sequences with G bases located at the same strand (sequences **4a**, **4b**, and **4c**), hole transfer is seen to be fastest for the stilbene capped hairpin **4a**. The rate k_t of hole transfer is smaller for sequence **4c** than for **4a**. This is due to the longer AT bridge in sequence **4b**, same as described above for sequences **1a–1d**.

For sequences **4a** and **4b**, the distance between the donor and the acceptor is the same, however the hole transfer rate is much higher for **4a**. The difference arises because of the considerably higher site-energy for T in sequence **4b** (9.111 eV) as compared to the site-energy for A in sequence **4a**. This shows that, for these sequences the intrastrand hole transfer is more important than intrastrand processes. For the hairpin **4d** in which G bases are located on different strands, the k_t value is significantly larger than for sequences **4b** and **4c**. This is due to the fact that the *interstrand* charge transfer integral for 5'-AG-5' is comparable to the J' value for *intrastrand* transfer via 5'-AG-3' (see Table 2.2).

Interestingly, the hole transfer rate calculated for sequence **4d** disagrees with the experimental rate (see Table 2.3), which was found to fall into the range between the values obtained for hairpins **4b** and **4c**. This discrepancy is not surprising since the charge transfer rates will be affected by the Coulomb interaction between the $\mathbf{Sa^{-\bullet}}$ anion and the hole generated in the system. The effect of the hole interaction with $\mathbf{Sa^{-\bullet}}$ was taken into account by adding the Coulomb term with the dielectric constant 3.5 [53] to the site-energies of

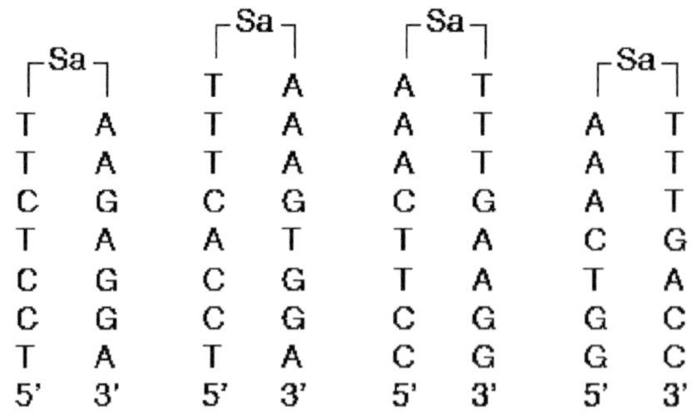

	Sa			Sa			Sa			Sa	
				T	A		A	T			
T	A			T	A		A	T		A	T
T	A			T	A		A	T		A	T
C	G			C	G		C	G		A	T
T	A			A	T		T	A		C	G
C	G			C	G		T	A		T	A
C	G			C	G		C	G		G	C
T	A			T	A		C	G		G	C
5'	3'			5'	3'		5'	3'		5'	3'

4a   **4b**   **4c**   **4d**

Scheme 2.3.

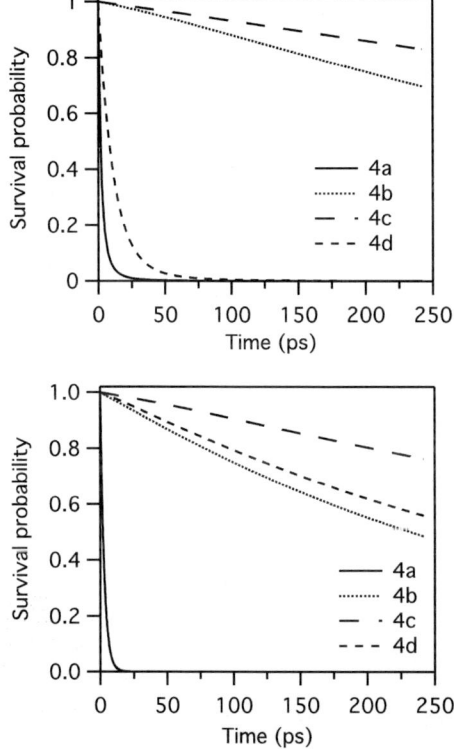

Fig. 2.8. Survival probability for sequences **4a–4d** as a function of time in absence (*top*) and in presence (*bottom*) of the Coulomb potential of Sa.⁻

the nucleobases. The survival probabilities calculated in the case, where the Coulomb interaction is included in the calculations, are shown in Fig. 2.8b. Comparison of the data presented in Fig. 2.8a and b shows that the Coulomb interaction leads to the increase of the rate for hole transfer in sequences **4a, 4b** and **4c**, while for sequence **4d** this rate becomes smaller. As a result, the calculated k_t values increase in the order **4c** < **4d** < **4b** < **4a** in qualitative agreement with the trend observed for the experimental rates.

The effect of the Coulomb interaction on k_t is the direct consequence of changes in energetics of the charge transfer process. In particular, for sequences **4a, 4b** and **4c**, the Coulomb interaction brings the site-energies of the proximal G and the distal GG closer to resonance, thus enhancing the rate of charge transfer. Based on our calculations, the opposite situation is expected to arise for sequence **4d**. As can be seen from the data summarized in Table 2.1, in the absence of the Coulomb interaction between the **Sa**⁻• anion and the hole, the site-energy of the proximal G (8.124 eV) is almost in resonance with that of the G at the 5′-end of the sequence (8.130 eV). The Coulomb interaction decreases the site-energy of the proximal G more than that of the distal GG doublet. As a consequence, the energy gap of 0.14 eV between the G at the 5′-end and the proximal G in sequence **4d** arises. This, in turn leads to a decrease of the rate for hole transfer.

Thus, the tight-binding calculations offer a qualitative explanation of the trend observed for the experimental rates of the forward hole transfer between proximal G and distal GG doublet in hairpins **4a**, **4b**, **4c**, and **4d**. However, the absolute values of the experimental rates are about three orders of magnitude smaller than those obtained from the data in Fig. 2.8b. This can be understood, since it is well known that an excess charge in DNA induces an internal reorganization of nucleobases and an external reorganization of the surrounding water [54–57]. These two processes, which have not been taken into account in the tight-binding calculations considered above, can reduce the rate of hole transfer, as discussed in the next section.

2.7 Effect of Solvent Reorganization Energy on Charge Transfer Rates

According to the standard electron transfer theory [58,59], the charge-induced reorganization is characterized by the so-called total reorganization energy λ. Similar to other electron transfer reactions, λ for hole transfer in DNA can be written as a sum of two terms. These correspond to the contributions to the energetics from an internal reorganization of nucleobases and an external reorganization of the surrounding water.

If temperature T is sufficiently high, so that vibrational modes can be treated classically, the effect of λ on the nonadiabatic charge transfer rate can be described theoretically using the Marcus equation [58, 59]

$$k_{\mathrm{CT}} = \frac{2\pi}{\hbar} \frac{|V_{\mathrm{da}}|^2}{\sqrt{4\pi\lambda kT}} \exp\left(-\frac{(\Delta E_{\mathrm{da}} + \lambda)^2}{4\lambda kT}\right), \qquad (2.10)$$

where k is the Boltzmann constant, V_{da} is the electronic coupling matrix element, and ΔE_{da} is the energetic difference of the hole at the donor and acceptor sites. For superexchange charge transfer through a single bridge of n nucleobases, V_{da} defined by

$$V_{\mathrm{da}} = \frac{J'_{\mathrm{d1}} J'_{na}}{\Delta E_{\mathrm{d},1}} \prod_{k=1}^{n-1} \frac{J'_{k,k+1}}{\Delta E_{\mathrm{d},k+1}}. \qquad (2.11)$$

with J' being the generalized charge transfer integral, defined in (2.7). In (2.11) $\Delta E_{\mathrm{d},i}$ is the energetic difference of the positive charge at the hole donor (single G and GG doublet for forward and backward transfer, respectively) and the i-th bridge site. The difference $\Delta E_{\mathrm{d},i}$ is the sum of two differences. One is the difference between the site-energies of the hole at the donor and i-th bridge site taken from Table 2.1. The other is the difference between the

Coulomb interaction between the **Sa⁻•** anion and the hole on the donor and the i-th bridge site. The dielectric constant was taken equal to 3.5 [53].

Equation (2.10) was used in the present work as a theoretical framework for numerical calculations of rates for hole transfer between the proximal G and the distal GG doublet. Twisting of the base pairs were taken into account by using mean values of the charge transfer integrals

$$\langle|J'_{ij}|^2\rangle = \int_{\theta_{\min}}^{\theta_{\max}} |J'_{ij}(\theta_{ij})|^2 p(\theta_{ij}) \, d\theta_{ij} \; . \tag{2.12}$$

These values were obtained by averaging $J'_{ij}(\theta_{ij})$ over the Boltzmann distribution $p(\theta_{ij})$ of twist angles θ_{ij}

$$p(\Theta_{ij}) = \exp\left(-\frac{F_{ij}(\theta_{ij} - \theta_{ij}^{\mathrm{eq}})^2}{2kT}\right) \Bigg/ \int_{\theta_{\min}}^{\theta_{\max}} \exp\left(-\frac{F_{ij}(\theta_{ij} - \theta_{ij}^{\mathrm{eq}})^2}{2kT}\right) \tag{2.13}$$

calculated using an harmonic potential with experimental force constants F_{ij} from literature [51]. The minimum $\theta_{\min}$ and maximum $\theta_{\max}$ angles in (2.12) were taken equal to 11° and 61°, respectively, which was found to be sufficient for convergence of the results. The effect of the Coulomb interaction between the hole on DNA and the **Sa⁻•** anion on the site-energies was taken into account as described in the previous Section.

It should be noted, that the averaging procedure defined by (2.12) is valid in the limit of slow twisting motion in comparison with hole transport. The averaging in the opposite limit can be done as described Troisi et al. [60]. In most cases, the latter procedure gives results, which differ from the values of $\langle J'^2_{ij}\rangle$ obtained from (2.12) by less than 10%.

The superexchange electronic coupling matrix element for hole transfer calculated for the energetically most favorable pathway between the proximal G and the GG doublet in hairpins **4a–4d** are given in Table 2.3. The superexchange matrix elements for other pathways were found to be significantly smaller. The values of the total reorganization energies λ needed to reproduce the absolute values of the experimental rate constants invoking the semiclassical approach [see (2.10)] are also given in Table 2.3. The λ values for all sequences studied are found to be close to 1 eV, in agreement with earlier results [54–57]. Lebard et al. have calculated a solvent reorganization energy of 0.69 eV within the framework of the molecular-based non-local model of solvent response (NMSR model) for hole transfer in a hairpin similar in structure to sequence **4a** [57]. This estimate together with the internal reorganization energy of 0.65 eV obtained for guanine from DFT calculations [54], gives $\lambda = 1.34$ eV, which does not differ too much from to the total reorganization energy for sequence **4a** in Table 2.1. According to theoretical results reported by LeBard et al. [57], the solvent reorganization energy increases by approximately 0.2 eV when an A:T base pair is added to the bridge between the G primary donor and the GG secondary donor. The same tendency follows from the data on the total reorganization energies for forward hole transfer presented in Table 2.3.

2.8 Implications for the Charge Carrier Mobility in DNA

The reports on the exceptionally efficient charge transfer through DNA have also led to speculations on application of DNA as a wire in nanoscale devices [61,62]. There have even been some direct measurements of charge transport through single DNA molecules positioned between electrodes [63–65]. The usefulness of DNA as a wire in nanoscale electronics depends critically on the mobility of charges along the chain. The similarities between DNA and e.g. discotic liquid crystalline materials mentioned in the introduction might imply that charge carrier mobilities similar to those found in these materials are also achievable in DNA. However, as shown in the previous sections, the charge transport properties of DNA depend strongly on structural variation. The presence of a more or less random sequence of nucleobases gives rise to a strongly disordered energy landscape. The charge moves in this landscape by relative slow tunnelling steps from one GC basepair to another through regions of AT base pairs. But even in regular poly-GC DNA it can be expected that there is considerable disorder. This disorder can strongly affect charge transport since structural variation can lead to considerable variations in the charge transfer integrals between neighboring nucleobases. Moreover, variations in the surroundings, for instance in the solvation shell or the presence of counter ions, can give small variations in the site-energies. Using the same model as used in this work it was shown previously that only small variations in the site-energies and charge transfer integrals are needed to reduce the charge carrier mobility to $0.04\,\mathrm{cm^2/Vs}$ [48]. This estimate disregards the reorganization energy for charge transfer, which was shown to be of the order of $1\,\mathrm{eV}$, as discussed in the previous section. Such high values of the reorganization energy strongly limit the charge carrier mobility. Using a value of $1\,\mathrm{eV}$ for the reorganization energy leads to a mobility of the order of 10^{-4}–$10^{-5}\,\mathrm{cm^2/Vs}$ [30]. These values are close to the experimental results for stacks of AT base pairs [66]. This points to the conclusion that even though charges can migrate over long distances, the mobility is significantly smaller than those found for discotic liquid crystalline materials [7]. The lower mobility is mainly caused by the large reorganization energy.

2.9 Conclusions

We have shown that our tight-binding model can qualitatively describe features relevant for charge transport in DNA. The sequence dependence of charge transport through DNA can already be described using a simplified model containing only two parameters. For a more general description a detailed knowledge of the charge transfer integral and site-energies is necessary. It was also shown that structural fluctuations determine the efficiency of charge transfer through DNA considerably. Therefore it is necessary to account for the structural dependence of the charge transfer integrals also.

To calculate absolute rates for charge transfer through DNA it is very important to include the reorganization energy. This reorganization energy has a value of the order of 1 eV. Due to the large reorganization energy, the mobility of charges in DNA is expected to be considerably lower than for π-stacked discotic liquid crystalline materials.

References

1. B. Giese, Acc. Chem. Res. **33**, 631 (2000).
2. G.B. Schuster, Acc. Chem. Res. **33**, 253 (2000).
3. *Charge transfer in DNA*, edited by H-A.Wagenknecht (Wiley-VCH, Weinheim, 2005).
4. M. Ratner, Nature **397**, 480 (1999).
5. D.D. Eley and D.I. Spivey, Trans. Faraday Soc. **58**, 411 (1962).
6. J.D. Watson and F.H.C. Crick, Nature **171**, 737 (1953).
7. J.M. Warman, M.P. de Haas, G. Dickeri, F.C. Grozema, J. Piris and M.G. Debije, Chem. Mater **16**, 4600 (2004).
8. C.J. Murphy, M.R. Arkin, Y. Jenkins, N.D. Ghatlia, S.H. Bossmann, N.J. Turro and J.K. Barton, Science **262**, 1025 (1993).
9. M.R. Arkin, E.D.A. Stemp, R.E. Holmlin, J.K. Barton, A. Hormann, E.J.C. Olson and P.F. Barbara, Science **273**, 475 (1996).
10. C.J. Murphy, M.R. Arkin, N.D. Ghatlia, S.H. Bossmann, N.J. Turro and J.K. Barton, Proc. Nat. Acad. Sci. USA **91**, 5315 (1994).
11. See, *Long-range charge transfer in DNA*, edited by G.B. Schuster (Springer-Verlag, Berlin, 2004).
12. Y.A. Berlin, I.V. Kurnikov, D. Beratan, M.A. Ratner and A.L. Burin, Top. Curr. Chem. **237**, 1 (2004).
13. S.O. Kelley and J.K. Barton, Chemistry and Biology **5**, 413 (1998).
14. S.O. Kelley and J.K. Barton, Science **283**, 375 (1999).
15. S.O. Kelley, E. Holmlin, E.D.A. Stemp and J.K. Barton, J. Am. Chem. Soc. **119**, 9861 (1997).
16. F.D. Lewis, X. Liu, Y. Wu, S.E. Miller, M.R. Wasielewski, R.L. Letsinger, R. Sanishvili, A. Joachimiak, V. Tereshko and M. Egli, J. Am. Chem. Soc. **121**, 9905 (1999).
17. F.D. Lewis, Y. Zhang, X. Liu, N. Xu and R.L. Letsinger, J. Phys. Chem. B **103**, 2570 (1999).
18. A.M. Brun and A. Harriman, J. Am. Chem. Soc. **114**, 3656 (1992).
19. K. Fukui and K. Tanaka, Angew. Chem. Int. Ed. Eng. **37**, 158 (1998).
20. T.J. Meade and J.F. Kayyem, Angew. Chem. Int. Ed. Eng. **34**, 352 (1995).
21. E. Meggers, M.E. Michel-Beyerle and B. Giese, J. Am. Chem. Soc. **120**, 12950 (1998).
22. B. Giese, Top. Curr. Chem. **236**, 27 (2004).

23. D. Ly, L. Sanni and G.B. Schuster, J. Am. Chem. Soc. **121**, 9400 (1999).
24. P.T. Henderson, D. Jones, G. Hampikian, Y. Kan and G.B. Schuster, Proc. Nat. Acad. Sci. USA **96**, 8353 (1999).
25. B. Giese, S. Wessely, M. Spormann, U. Lindemann, E. Meggers and M.E. Michel-Beyerle, Angew. Chem. Int. Ed. Eng. **38**, 996 (1999).
26. F.D. Lewis, R.L. Letsinger and M.R. Wasielewski, Acc. Chem. Res. **34**, 159 (2001).
27. F.D. Lewis and M.R. Wasielewski in *Charge transfer in DNA*, edited by H.-A. Wagenknecht (Wiley-VCH, Weinheim, 2005).
28. F.C. Grozema, Y.A. Berlin and L.D.A. Siebbeles, Int. J. Quant. Chem. **75**, 1009 (1999).
29. F.C. Grozema, Y.A. Berlin and L.D.A. Siebbeles, J. Am. Chem. Soc. **122**, 10903 (2000).
30. K. Senthilkumar, F.C. Grozema, C. Fonseca Guerra, F.M. Bickelheupt, F.D. Lewis, Y.A. Berlin, M.A. Ratner and L. D. A. Siebbeles, J. Am. Chem. Soc. **127**, 14894 (2005).
31. K. Senthilkumar, F.C. Grozema, C. Fonseca Guerra, F.M. Bickelhaupt and L.D.A. Siebbeles, J. Am. Chem. Soc. **125**, 13658 (2003).
32. I. Saito, T. Nakamura, K. Nakatani, Y. Yoshioka, K. Yamaguchi and H. Sugiyama, J. Am. Chem. Soc. **120**, 12686 (1998).
33. H. Sugiyama and I. Saito, J. Am. Chem. Soc. **118**, 7063 (1996).
34. M.-L. Zhang, M.S. Miao, V.E. Van Doren, J.J. Ladik and J.W. Mintmire, J. Chem. Phys. **111**, 8696 (1999).
35. M. Hutter and T. Clark, J. Am. Chem. Soc. **118**, 7574 (1996).
36. F.D. Lewis, J. Liu, W. Weigel, W. Rettig, I.V. Kurnikov and D.N. Beratan, Proc. Nat. Acad. Sci. USA **99**, 12536 (2002).
37. J. Jortner, M. Bixon, T. Langenbacher and M.E. Michel-Beyerle, Proc. Nat. Acad. Sci. USA **95**, 12759 (1998).
38. M. Bixon, B. Giese, S. Wessely, T. Langenbacher, M.E. Michel-Beyerle and J. Jortner, Proc. Nat. Acad. Sci. USA **96**, 11713 (1999).
39. Y.A. Berlin, A.L. Burin and M.A. Ratner, J. Phys. Chem. A **104**, 443 (1999).
40. B. Giese, J. Amaudrut, A.-K. Kohler, M. Spormann and S. Wessely, Nature **412**, 318 (2001).
41. A.A. Voityuk, J. Jortner, M. Bixon and N. Rösch, Chem. Phys. Lett. **324**, 430 (2000).
42. A.A. Voityuk, J. Jortner, M. Bixon and N. Rösch, J. Chem. Phys. **114**, 5614 (2001).
43. A.A. Voityuk and N. Rösch, J. Chem. Phys. **117**, 5607 (2002).
44. A.A. Voityuk, N. Rösch, M. Bixon and J. Jortner, J. Phys. Chem. B **104**, 9740 (2000).
45. K. Senthilkumar, F.C. Grozema, F.M. Bickelhaupt and L.D.A. Siebbeles, J. Chem. Phys. **119**, 9809 (2003).
46. X.-J. Lu, M.A. El Hassan and C.A. Hunter, J. Mol. Biol. **273**, 681 (1997).
47. A.A. Voityuk, K. Siriwong and N.Rösch, Phys. Chem. Chem. Phys. **3**, 5421 (2001).
48. F.C. Grozema, L.D.A. Siebbeles, Y.A. Berlin and M.A. Ratner, Chem Phys Chem **3**, 536 (2002).
49. M.D. Newton, Chem. Rev. **91**, 767 (1991).
50. Y.A. Berlin, A.L. Burin, L.D.A. Siebbeles and M.A. Ratner, J. Phys. Chem. A. **105**, 5666 (2001).

51. F. Lankas, J. Sponer, J. Langowski and T.E. Cheatham III, Biophys. J. **85**, 2872 (2003).
52. W.K. Olson, A.A. Gorin, Z.-J. Lu, L.M. Hock and V. B. Zhurkin, Proc. Nat. Acad. Sci. USA **95**, 11163 (1998).
53. V. Makarov, B.M. Pettitt and M. Feig, Acc. Chem. Res. **35**, 376 (2002).
54. J. Olofson and S. Larsson, J. Phys. Chem. B **105**, 10398 (2001).
55. H.L. Tavernier and M.D. Fayer, J. Phys. Chem. B **104**, 11541 (2000).
56. G.S.M. Tong, I.V. Kurnikov and D.N. Beratan, J. Phys. Chem. B. **106**, 2381 (2002).
57. D.N. LeBard, M. Lilichenko, D.V. Matyushov, Y.A. Berlin and M.A. Ratner, J. Phys. Chem B. **107**,14509 (2003).
58. M. Bixon and J. Jortner, Adv. Chem. Phys. **106**, 35 (1999).
59. R.A. Marcus and N. Sutin, Biochim. Biophys. Acta **811**, 265 (1985).
60. A. Troisi, A. Nitzan and M.A. Ratner, J. Chem. Phys. **119**, 5782 (2003).
61. S.O. Kelley, N.M. Jackson, M. G. Hill and J. K. Barton, Angew. Chem. Int. Ed. Eng. **38**, 941 (1999).
62. F. Lisdat, B. Ge and F.W. Scheller, Electrochem. Commun. **1**, 65 (1999).
63. D. Porath, A. Bezryadin, S. de Vries and C. Dekker, Nature **403**, 635 (2000).
64. P. J. De Pablo, F. Moreno-Herrero, J. Colchero, J. Gómez Herrero, P. Herrero, A.M. Baró, P Ordejón, J.M. Soler and E. Artacho, Phys. Rev. Lett. **85**, 4992 (2000).
65. A.J. Storm, J. Van Noort, S. de Vries and C. Dekkar, Appl. Phys. Lett. **79**, 3881 (2001).
66. T. Takada, T. Kawai, X. Cai, A. Sugimoto, M. Fujitsuka and T. Majima, J. Am. Chem. Soc. **126**, 1125 (2004).

3 Variable-Range Charge Hopping in DNA

Yuri A. Berlin and Mark A. Ratner

Department of Chemistry, Center for Nanofabrication and Molecular
Self-Assembly, and Materials Research Center, Northwestern University,
2145 Sheridan Road, Evanston, Illinois 60208-3113, USA
berlin@chem.northwestern.edu

3.1 Introduction

The pivotal role of the deoxyribonucleic acid (DNA) in biology is determined by the capability of this molecule for coding, storage and propagation of genetic information. These functions can be performed effectively due to the unique structure of the "molecule of life". More than fifty years ago, Watson and Crick discovered [1] that the double-stranded DNA consists of two intertwined helices with an aromatic π-stack core, where the basis of the pyrimidine deoxynucleotides (thymine, T; cytosine, C) and purine deoxynucleotides (adenine, A; guanine, G) participate in the Watson-Crick base pairing (A:T; C:G). These structural features also determine the self-assembling and the electronic properties, thereby making DNA an attractive object for nanoscience. In particular, the ability of DNA to serve as a medium for the long-range charge transfer has stimulated interest in the possibility to exploit this molecule in nanoscale electronics [2–5], molecular computing [6–9], and in electrochemical biosensoric devices [10–15]. The same property is also shown to be important for developing new methods for detection of the structural changes due to protein binding and the base mismatches [16–23]. For these potential applications, the elucidation of the mechanisms responsible for the charge transport phenomena in DNA turns out to be crucial. This challenging problem is also vital for the current research on the oxidative damage of DNA, which may cause apoptosis, mutations, and cancer [24–28].

A key structural element, which determines the transport properties of DNA, is the array of the π-stacked base pairs. The striking resemblance of the base pair stack to the conductive one-dimensional aromatic crystals prompted the proposal that the interior of the double helix can provide a one-dimensional pathway for charge migration due to the formation of a π-band across different stacked bases [29]. This early mechanistic picture was tested in experiments on the photo-induced oxidation of DNA dating back to the early 1990s (for review see e.g., [30, 31]).

In typical experiments designed to study the transfer of an electronic hole (i.e., a positive charge corresponding to an ionized nucleobase), a donor and an acceptor of these charge carriers are intercalated in the stack of the native base pairs [32–36] or chemically attached to the phosphate-sugar backbone [37–42]. For a properly chosen donor and acceptor species, such chemical modifications enable one to generate holes under irradiation of the sample by light due to the removal of an electron from the nucleobases to the photoexcited donor. The subsequent hole transfer from the donor to the acceptor bridged by the base pair sequence can be probed by measuring, for instance, the quenching of the fluorescence of the donor for the sequences of different lengths [37, 40, 43–45] or the damage yield at certain sites along the sequence [38, 42, 46–50]. Later a similar approach was also utilized in the experiments on the photo-induced reduction of DNA aimed to probe the transport of excess electrons along the stack of the base pairs. In the latter case, the negative charge carriers were generated using the photoexcited donor to inject electrons into DNA or to transfer them directly to the acceptor (for a review see [51]). Some additional information about the motion of the electrons and holes in the interior of the double helix has become available from a series of works on the low-temperature γ-radiolysis of the crystalline DNA, its ice and glassy aqueous solutions [52–59].

Experimental results of the above-mentioned studies allow the conclusion that the ordered π-electron system of the natural DNA bases in a duplex B-form DNA (referred to here simply as DNA) indeed provides an appropriate pathway for the motion of the excess positive and negative charges, once generated on the extended and chemically well-defined stacks of base pairs. Charge carriers can travel along this "π-pathway" for a very large distance [32, 38, 42, 45, 47, 48, 50], which in the case of the holes may be as much as several hundreds of angstroms. However the observed dependence of the charge transfer efficiency on the base pair sequence [47] suggests that the actual behaviors of the generated electrons and holes are more complicated than the band-like picture of the charge motion proposed at the initial stage of investigations [29]. These findings together with a weak distance dependence of the charge transfer deduced from the experiments have triggered a discussion concerning the mechanisms governing the charge propagation along the "π-pathway" and the possible role of the DNA as a molecular wire [60–69].

While this is not intended as a comprehensive review, we detail here the key theoretical results of our own research on the mechanisms governing the charge migration through DNA. Our analysis relies on the concept of the energy landscape for the charge motion along the stack of base pairs considered in Sect. 3.2.1. As will be shown in Sect. 3.2.2, the energetics of the stack suggests that the charge carriers can move through DNA via a series of sequential hops between the nucleobases with the proper energetics. This leads to the formulation of the model of the variable-range hopping [70–76], which is now widely accepted. Although the model has been extremely helpful in explaining the seemingly contradictory observations from different laboratories, there are still many questions that remain unanswered. Some of them are discussed in Sect. 3.3.

3.2 Charge Transfer within a Stack of Base Pairs

Qualitatively, the plausible scenario of the DNA-mediated charge transfer can be inferred from the consideration of the relevant energy landscape. The latter shows how the energy of a charge carrier changes as a hole or an excess electron is consecutively deposited on each nucleobase involved in the formation of the π-pathway between a donor and an acceptor. To construct the simplest landscape possible, one should take into account the energetics of the individual bases and the structural disorder arising from the choice of A, T, G, or C at each substitution base site along the backbone of the DNA helix. These two factors lead to the static energy disorder that determines both the multi-step mechanism of the entire transport process and the mechanisms governing its elementary steps.

3.2.1 Structural Disorder and Disordered Energy Landscape

In the case of the holes, the static energy disorder arises due to the differences between individual nucleobases in the values of the oxidation potential and the ionization energy. To be more specific, the G is known to be the most easily oxidized nucleobase since its *in vitro* oxidation potential is about 0.4 eV lower than that of A and significantly lower than the oxidation potentials of C and T [77, 78]. The same conclusion follows from the comparison of the experimental values of the ionization potentials of the nucleobases in vapors [79–81] and from the computational results [82–92]. Hence, the energy of the hole when residing on the A, C, or T sites is higher than on the G, and therefore the latter base is a more probable place for the positive charge to be localized than the three others. This hierarchy of the hole energies (G < A < C < T) also holds when the stacking interaction between the neighboring nucleobases is taken into account [85,91], although their ionization potentials become lower in comparison to the values found for the individual bases. In particular, the GG doublets and the GGG triplets formed by stacking two or three adjacent G's on the same strand have even smaller ionization potentials compared to that of the single G. As a consequence, the energy of the holes on the GG and GGG molecular units is lower than the energy of G^+ by at least 0.5 and 0.7 eV, respectively [85,91].

Therefore, due to the different energetics of the nucleobases, the structural disorder in the stack of the A:T and G:C pairs gives rise to the static disorder in the energies of the holes residing on the individual bases. Furthermore, a close examination of the resulting energy landscape reveals three main groups of states. The first group consists of states with the lowest energy. They serve as deep hole traps arising when several adjacent G's are stacked on the same strand and form a multiple GG...G unit. The hole states associated with the individual G bases belong to the second group, which is intermediate in energy between the trapped holes and the holes residing on the A, T, and C bases. For this reason, these states will be defined henceforth as "intermediate". Three other native nucleobases A, T, and C are responsible for the formation of the third group comprising of the hole states with the highest energies. Since A, T, and C can be considered as a building block of the bridge connecting two neighboring G bases, all states belonging to the third group will be referred to as "bridging".

Three groups of states discussed above are separated by two energy gaps. At room temperature, the width of both gaps exceeds a typical thermal energy E_{th} given by the product of the Boltzmann constant k_B and temperature T. Therefore in the simplest case considered here, the groups do not overlap and can be considered as isolated.

The main elements of the energy landscape for the transfer of a positive charge mentioned above are shown in Fig. 3.1B, using the fragment of the DNA duplex schematically depicted in Fig. 3.1A as an illustration. In this particular case the doublet $G_7 G_8$ is an example of the relatively deep hole trap, sites G_1 and G_3 exemplify the intermediate states, while T_2, T_4, T_5, and T_6 correspond to the "bridging" states.

Of course, the energy landscape of Fig. 3.1B is oversimplified. In particular, the energies of the holes on each of the single G's are assumed to be equal. Moreover, the hole energies for the T_2, T_4, T_5 and T_6 bridge units are also supposed to be identical for simplicity. Meanwhile, the quantum mechanical calculations [91, 93] suggest that the flanking bases can affect the

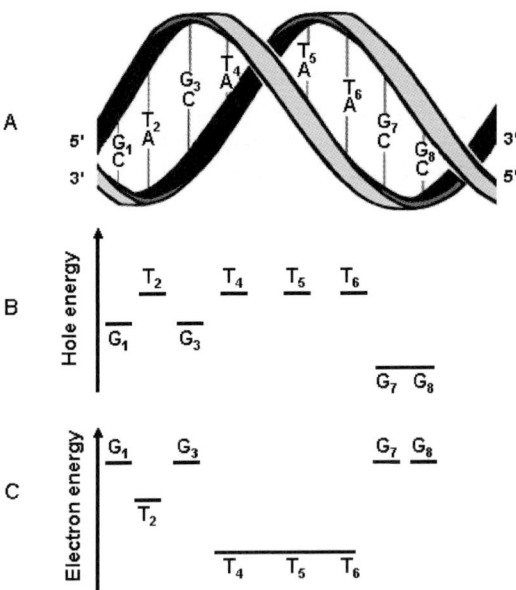

Fig. 3.1. A fragment of the DNA helical structure (**A**) and the energy landscapes for the hole transport (**B**) and the excess electron migration (**C**) along the stack of the base pairs. Both landscapes consist of three groups of states (intermediate, "bridging" and trapping) as explained in the text

energies of holes on the A, T, G, and C bases due to the stacking interactions, thus further increasing the degree of the static energy disorder within the DNA π-stack. For instance, according to theoretical calculations of the ionization potentials for the trimers 5′-XBY-3′ (X,Y,B = A,G,C,T), the energy of a hole on the site T_4 is higher than the energies of T_5^+ and T_6^+ by 0.05 and 0.31 eV, respectively [91]. Levels corresponding to the holes residing on the single G bases can also be shifted relative to each other, if the flanking bases are not identical for all G's. Based on the results of the quantum mechanical studies of the stacking interactions [85,91], one can expect that for the octamer duplex shown in Fig. 3.1A, the energy of the G_3^+ is smaller than the energy of a hole on the terminal base G_1 by about 0.1 eV, but still exceeds the value estimated for a hole on the GG doublet by 0.3 eV. Note, however, that the additional contribution to the energy disorder arising from the stacking interaction does not eliminate the energy gap between the hole intermediate states G^+ and the bridge states A^+, T^+, and C^+. The effect of the stacking interactions also does not reduce dramatically the difference in energies of G^+ and a positive charge on the GG doublet or on the GGG triplet, so that these multiple G units still serve as traps for the positive charges.

Similar to the energetics of the hole transfer considered above, the energy landscape for the motion of the excess electrons along the stack of the base pairs also involves both the intermediate states and the bridging states between them as well as the traps for the negative charges. However, now the intermediate states should be associated with the anions T^- and/or C^-, while the bridging states will correspond to the G and/or A bases (see Fig. 3.1C). This conclusion is derived from the measured redox potentials of the bases, which decrease in the order C $\approx$ T $\gg$ A > G [78,94,95]. Electron affinities calculated both for the individual nucleobases [96] and for the trimers 5′-XBY-3′ [97] exhibit the same trend.

As in the case of the holes, the energy of the electrons residing on each of the four native nucleobases is affected by the stacking interactions, and therefore the stability of the corresponding anion radicals B^- can be considerably influenced by their nearest neighbors located at the same strand. The results of the semi-empirical AM1 calculations show [97] that the most stable state of the triplet tracts 5′-XB⁻Y-3′ are those where all the bases involved, X, B, and Y, are pyrimidines, C or T. Therefore such multiple pyrimidines units as the TTT and CCC triplets are expected to be more favorable places for trapping of excess electrons, as schematically shown in Fig. 3.1C for the particular example of the sequence 5′-GTGTTTGG-3′.

3.2.2 Charge Transport and Its Elementary Steps

The energy landscapes discussed in the previous section suggest two distinct mechanisms of charge transfer between the donor D and the acceptor Ac located at the opposite ends of the nucleobase sequence [70–72,98].

If bridging (and for certain systems also intermediate) states in the landscape are much higher in energy compared to D and Ac, a charge will be transferred via the superexchange mediated tunnelling [2]. Two characteristic features of this coherent quantum mechanism should be mentioned in the context of the charge transport mechanisms in DNA. First, the two-center superexchange charge transfer occurs in a single step and hence does not involve the genuine chemical intermediates. Second, the rate of the whole process k_{CT} rapidly decreases with the donor-acceptor distance R following the familiar exponential law

$$k_{CT} = k_0 \exp(-\beta R) , \tag{3.1}$$

where k_0 is the pre-exponential factor and β is the falloff parameter. In the case of the charge transfer in DNA the β values are theoretically expected to be of the order of $1 \, \text{Å}^{-1}$. For detailed discussion of the reported β values in DNA, see e.g. [70] and [98].

Another situation arises if the intermediate states are comparable in energy with D. Now a charge can be injected from the D site to the proximal base (G in the case of the holes and T or C in the case of the electrons) with the subsequent temporal localization in the corresponding intermediate state of the energy landscape. Thereafter a hole (or an electron) is able either to return back to D or to undergo a transition to the adjacent unoccupied intermediate state through the intervening nucleobases associated with the bridging states (e.g., A or T for the holes and G or C for the electrons). The latter transition represents the first step in a series of consecutive incoherent hopping transitions that allows the charge carriers to move along the stack of the base pairs using the intermediate states as the stepping stones. As a consequence, a charge is able to reach a remote Ac site separated from D by several hundred angstroms, where the hopping transport is terminated by trapping.

Therefore, unlike the coherent single-step superexchange, incoherent hopping in DNA involves several steps, i.e., injection of the charge carrier, their transport along the π-pathway due to successive transitions between the nucleobases with the appropriate energetics, and trapping. In addition, there are several other distinctions between the hopping and the superexchange mechanisms. In particular, the former mechanism implies the formation of reactive chemical intermediates (e.g., G^+ and the pyrimidines anions for the hole and electron transfer, respectively), which propagate from the site of their generation to the distant site of the reaction. By contrast, the single-step superexchange does not include any intermediate active species. The distance dependence of the charge transfer rate for the two mechanisms turn out to be also distinct. For the unbiased hopping on a long one-dimensional regular lattice with $(N+1)$ sites separated by the distance a, this dependence can be approximated by the algebraic function (see e.g. [74,99] and references therein)

$$k_{CT} \propto 1/N \approx 1/(aR) , \tag{3.2}$$

rather than by the exponential law (3.1) typical for the superexchange.

The above-mentioned features of the superexchange and the hopping mechanisms were documented in a number of charge transfer processes observed in DNA. As has been demonstrated experimentally, the single-step superexchange mechanism dominates in the DNA oligomers with short base pair sequences ($R < 20$ Å). Representative examples of such systems and the superexchange-driven reactions are given in Table 3.1. By contrast, experiments with the longer nucleobase sequences ($R > 20$ Å) reveal that the multi-step hopping mechanism prevails. According to the current consensus, the latter mechanism governs a number of processes in different duplexes listed in Table 3.2.

Earlier, we have already emphasized [74,75,98] that there is no dichotomy between the two mechanisms of charge transfer in DNA considered above. On the contrary, each can contribute to the mechanistic picture of the entire process: The superexchange mediated tunneling controls the rate of the short-range (< 20 Å) elementary hops of a charge carrier between neighboring nucleobases that produce the intermediate states in the corresponding energy landscape, while the multi-step hopping is responsible for the long-range migration of the charge along the stack of the nucleobases.

The superexchange mediated tunneling, however, is not the only plausible mechanism for the short-range steps of the hopping motion in DNA: at finite temperatures this mechanism is in competition with classical thermally-induced transitions of the charge carriers between two neighboring "resting" sites (G for holes, C and/or T for the excess electrons). As can be seen from (3.1), the tunneling rate exponentially decreases with the distance separating these two states, while the rate of the thermally-induced transition W_{th} is mainly determined by a thermal population of the bridge. Therefore W_{th} does not vary with the distance, but depends on the energy gap between the intermediate and the bridging states, E_{ib}, in accordance with the Arrhenius law

$$W_{\text{th}} = W_0 \exp[-E_{\text{ib}}/(k_{\text{B}}T)] \,, \tag{3.3}$$

where W_0 is the pre-exponential factor. Due to the distinction in the distance dependencies, a changeover from the superexchange mediated tunneling to the thermally activated regime of the elementary hopping step can be expected as the distance between the neighboring intermediate states becomes equal to a certain critical value [75]. In the case of the hole transfer, the tight-binding model for the elementary step of the hopping motion [75] suggests that a positive charge can be transferred between two neighboring G sites via the superexchange only if the A:T bridge connecting these two sites consists of less than 3–4 base pairs. Since the mean plane-to-plane distance between the base pairs in B-DNA is known to be 3.4 Å, this corresponds to the situation where the AT bridge has the length $R_{\text{AT}} \leq 14$ Å. Otherwise, (i.e., for $R_{\text{AT}} > 14$ Å), the elementary hopping step includes thermal activation of the holes into the tight-binding band followed by their ballistic or hopping motion along the A:T bridge. For the holes, the latter process is known in

Table 3.1. Examples of small-scale DNA-like systems and superexchange driven elementary processes

Process and system	References
Oxidative hole transfer from the first exited singlet ^{1}S* of capped stilbene S to a single G base through the bridge containing up to 4 AT pairs ^{1}S*–(AT bridge)–G$\longrightarrow$S$^-$–(AT bridge)–G$^+$	[39,100]
Charge recombination in stilebene capped DNA hairpines S$^-$(AT bridge)–G$^+$$\longrightarrow$S–(AT bridge)–G	[39,100]
Photo-induced formation of a positive charge on the G site from exited acridine Acr* Acr*–(AT bridge)–G$\longrightarrow$Acr$^-$–(AT bridge)–G$^+$	[35,101]
Trapping of site-selectively generated holes by the GGG triplet G$^+$–(short AT bridge)–(GGG)$\longrightarrow$G–(AT bridge)–(GGG)$^+$	[47,48]
Electron injection in DNA hairpins from a stilbenediether singlet (^{1}Sd*) electron donor to T through the bridge of noncanonical GG base pairs ^{1}Sd*–(GG bridge)–T$\longrightarrow$Sd$^+$–(GG bridge)–T$^-$	[102]
Injection of negative charge in pyrene(Py)-modified duplexes upon excitation due to excess electron transfer from the Py-uracil(U) group to the adjacent C or T bases, e.g. $\cdots - \text{U} - \text{C} - \cdots \longrightarrow -\overset{\bullet-}{\text{U}} - \text{C} - \cdots \longrightarrow \cdots - \text{U} - \text{C}^- - \cdots$ $\qquad\quad | \qquad\qquad\quad | \qquad\qquad\quad |$ $\qquad\text{Py}^* \qquad\qquad \text{Py}^{\bullet+} \qquad\qquad \text{Py}^{\bullet+}$	[51,103]

the literature as A-hopping. Recent experiments (for review, see [100]) provide strong evidences for such thermally-induced transitions through long bridges with the number of A:T pairs $n_{AT} = 4$–10 ($R_{AT} \approx 17$–37 Å) and $n_{AT} = 4$–16 ($R_{AT} \approx 17$–58 Å). Moreover, measurements [111] of the hole transfer efficiency as a function of R_{AT} for the process

$$\text{G}^+ \begin{pmatrix} \text{T} \\ \text{C} \ \text{A} \end{pmatrix}_{n_{AT}} \begin{matrix}(\text{G G G}) \\ \text{C C C}\end{matrix} \rightarrow \text{G} \begin{pmatrix} \text{T} \\ \text{C} \ \text{A} \end{pmatrix}_{n_{AT}} \begin{matrix}(\text{G G G})^+ \\ \text{C C C}\end{matrix} \qquad (n_{AT} = 1\text{–}16)$$

and the analogous theoretical dependence [75] were found to be in good agreement (see Fig. 3.2), thus supporting the theoretical predictions concerning the two competing mechanisms of the elementary hopping step.

Table 3.2. Examples of double helical systems and processes govern by multi-step charge hopping

Process and system	References
Hole migration from the site of its generation to the GGG trap via both intrastrand and interstrand $G^+ \rightarrow G$ ("zigzagging") transitions	[47]

$$G^+ \text{TCAGCT CAGTC TGCA} (\mathbf{GGG}) \quad \text{GTCAGCT CAGTCTGCA} (\mathbf{GGG})^+$$
$$C \ \text{AGTCGAGTCAGACGT C C C} \longrightarrow \text{CAGTCGAGTCAGACGT C C C}$$

Transfer of a positive charge selectively generated at the G site in double helix to the triple G trap along one strand of duplex with bridges of equal length composed of two T bases   [48]

$$G^+ \text{TTGTTG} \ldots \text{TT(GGG)} \longrightarrow \text{GTTGTTG} \ldots \text{TT(GGG)}^+$$

Photo-induced propagation of radical cations in anthraquinone (AQ)-linked duplex DNA oligomers. In the oligomers studied most recently [104] this process proceeds along one of their two strands, which contains $[(T)_l GG]_n$ or $[(A)_m GG]_n$ segments with $l = 1\text{-}5$, $m = 1\text{-}7$ and $n = 4$ or 6, e.g.   [38,42,50,104]

$$AQ * -T(GG)T(GG)T(GG)T(GG)T(GG)T(GG)TATA$$
$$\downarrow$$
$$AQ^- - T(GG)^+T(GG)T(GG)T(GG)T(GG)T(GG)TATA$$
$$\downarrow$$
$$\cdots\cdots\cdots$$
$$\downarrow$$
$$AQ^- - T(GG)T(GG)T(GG)T(GG)T(GG)T(GG)^+TATA$$

Dynamics of holes injected from a capped stilbenedicarboxamide singlet donor (^{1}Sa*) in DNA hairpins with several G sites on one strans, e.g.   [100]

$$^1\text{Sa*} - \text{AAGAGA(GGG)} \rightarrow \text{Sa}^- - \text{AAG}^+\text{AGA(GGG)} \rightarrow \ldots \rightarrow \text{Sa}^- - \text{AAGAGA(GGG)}^+$$

Photo-induced excess electron transfer from an internally conjugated aromatic amine (X) to 5-Bromo-2'-depxyuridine (Y) through the sequence involving up to 5 AT and GC base pairs   [105]

$$X^* - (\text{intervening AT and GC pairs}) - Y \rightarrow X^+ - (\text{intervening AT and GC pairs}) - Y^-$$

Photo-induced excess electron transfer from the excited state of the flavin-capped donor (F*) to the TT dimer in DNA hairpins   [106–109]

$$F^* - (\text{intervening AT and/or GC pairs}) - Y$$
$$\downarrow$$
$$F^+ - (\text{intervening AT and/or GC pairs}) - Y^-$$

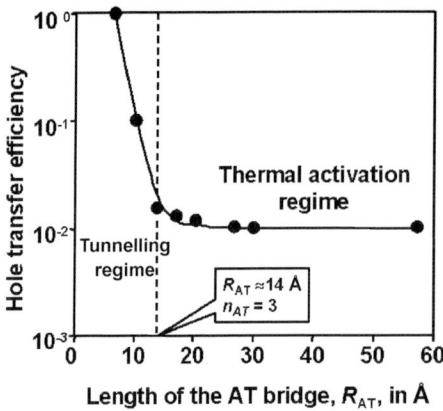

Fig. 3.2. Efficiency of the hole transfer from the site-selectively generated G^+ to the GGG triplet across the A:T bridges of various lengths R_{AT}. The points correspond to the experimental data of Giese et al. [111]. The length dependence calculated for the same system in [75] is shown by the *solid line*. The intersection of the *dotted line* with the *horizontal axis* gives the length of the A:T bridge and the number of A:T pairs, at which the rates of quantum tunnelling and the classical thermally-induced transitions become equal

Hence, a plausible scenario for the entire process of the charge transfer from D to Ac along the sequence of the nucleobases involves variable-range hopping between the intermediate states corresponding to the bases with the appropriate oxidation or the reduction potentials. Short steps made by a moving charge in this multi-step transport process occur due to the coherent superexchange mediated tunneling. In contrast, long steps require thermal activation of the charge carriers needed to overcome the energy gap between the intermediate and the bridging states. Once this thermally-induced transition has completed, electrons or holes can reach the next "resting" site undergoing a ballistic motion or hopping along the pathway provided by the bridging states.

As follows from the detailed kinetic analysis of this scenario [73–75], the model of the variable-range hopping allows quite accurate predictions of both the sequence and distance dependencies for the efficiency of the charge transfer through DNA. This can be illustrated by comparison of the experimental and theoretical data presented in Figs. 3.3 and 3.4.

It is remarkable that our theory reproduces both the values of the charge transfer efficiency and its sequence dependence without invoking any fitting parameters. The only information needed to calculate the efficiency of the charge transfer for the sequence with a given arrangement of A:T and G:C pairs is the values of the relative rates for hopping through the A:T bridges connecting the neighboring Gs. This information is available for the bridges of different lengths from the measurements [47, 48, 111] or can be obtained theoretically [75].

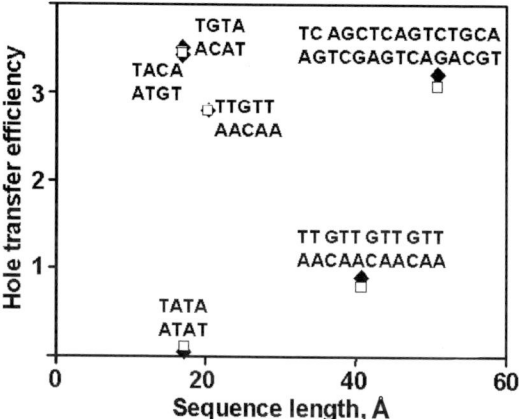

Fig. 3.3. Experimental and theoretical values of the hole transfer efficiency for various sequences of the base pairs. Experimental data shown by *filled diamonds* are taken from [47] and [48]. Theoretical results of the variable-range hopping model (*open squares*) were obtained using (11) of [74]. In both cases the efficiency of the hole transfer is expressed in terms of the damage yield defined as the ratio of the time-independent yields for the reaction of water with $(GGG)^+$ and with the primary G^+ cation. Sequences connecting the primary oxidized G site and the GGG triple are shown for each plotted value of the hole transfer efficiency

Duplex A

3′ –ATGCACCGAAAAGCCAGTGACGTAATCAATTTCCTTACACGCGACTGGTTCCTTGGTTT–5′
5′ –ACGTGGCTTTTCGGTCACTGCATTAGTTAAAGGAATGTGCGCTGACCAAGGAACCAAAG–3′

Duplex B

3′ –ATTTCCGGCATGCGACCAGTACACCAAGTCACCACTGAACCAACGTACCATGCAGGC–5′
5′ –TAAAGGCCGTACGCTGGTCATGTGGTTCAGTGGTGACTTGGTTGCATGGTACGTCCG–3′

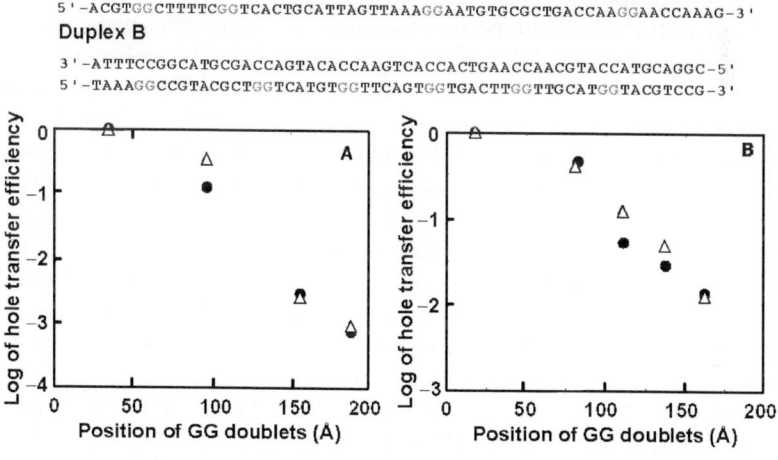

Fig. 3.4. Experimental and theoretical dependencies of the hole transfer efficiency versus the position of the GG doublets in duplexes A (*panel A*) and B (*panel B*). Experimental data shown by the *filled circles* are taken from [42] and [50]. Theoretical values of the hole transfer efficiency (*open triangles*) are the results of our calculations within the variable-range model described in detail in [74]

The model of variable-range hopping also provides reasonable estimations for the distance scale of the propagation of the charge in the DNA duplexes. Based on these estimations, verified by the experiments (see, e.g. [40,50,112]), one can conclude that typically the upper limit for the distance travelled by the charges in DNA is about $200-300$ Å. Charge transfer over such large distances can be accomplished because of the weak distance dependence of the reaction rate (*cf.* (3.2)) offered by the variable-range hopping. In addition, a coexistence of the quantum and the classical steps of the hopping process is also favorable to the long-range migration of the electrons and holes in the interior of the double helix.

3.3 Concluding Remarks

An obvious advantage of the variable-range hopping model is the possibility to estimate both the sequence and the distance dependencies for the efficiency of the charge transfer through the stacks with various combinations of the Watson-Crick base pairs. Information needed for such estimations involves only the data on the relative rates for the hopping steps of different lengths. The significance of the relative, but not absolute, rates for the theoretical analysis of the steady-state experiments follows from the existence of the two competitive channels for the decay of the charge carriers at each step of the transport process. In particular, a hole occupying the G site has two options, namely, it can either be transferred to the nearest-neighbor G nucleobase or undergo the irreversible side reaction with water.

Knowledge of the relative hopping rates, however, is insufficient to decide how fast a hole generated in DNA can be transferred over a certain distance. Theoretical attempts [72,113] to address this kinetic aspect of the problem within the framework of the nonadiabatic electron transfer theory [114] are based on the assumption of charge localization on an individual nucleobase. Although results of recent calculations [115] support this assumption, the problem of competition between the quantum delocalization of charges and their localization due to the vibronic coupling or solvation forces remains an important issue that has not been fully resolved yet. Meanwhile, the solution of this challenging problem is needed for understanding the physical nature of the charge carriers in DNA and the role of solvent in the mechanism of their migration. If a charge is localized on an individual base (for instance, on G in the case of the holes) and hence a non-adiabatic approach to the charge hopping is applicable, the solvent effect can be discussed in terms of the corresponding reorganization energy, λ_S calculated for the DNA oligomers and the hairpins in [116] and [117]. However, if an excess charge is spread over several base pairs, this approach ceases to be valid, and an adiabatic description of the charge motion becomes more appropriate [98]. In this situation, formation of the polarons, is possible due to self-trapping of the charge by a distortion of the base pair stack or by the surroundings water [118]. As a result, the motion of the charge carriers is expected to proceed via the sequential phonon-assisted polaron hopping [42]. This may modify the rates of the elementary hopping steps, but will not require any radical changes in the kinetic equations proposed in our previous publications [73, 74] to describe the variable-range hopping.

The localization/delocalization problem is also closely related to the issue of the static and dynamic disorder in DNA. Since a variety of local conformations and a range of dynamic motions inherent in DNA can affect the energy landscape "seen" by a moving charge, it is expected that different types of disorder will favor temporal localization of the charges. In addition, dynamics of the base pair stack is able to change the electronic coupling, the intra- and interstrand charge transfer and the overlap integrals, thus affecting the absolute rates of the elementary steps of the hopping motion. Indeed, theoretical investigations of the charge transfer rates in DNA hairpins [113] clearly show that the dynamic disorder arising due to the twisting motion of the base pairs have a strong influence on the kinetics of the charge transport.

A particular type of disorder associated with various mismatches in the nucleobase pairing deserves special consideration. Our preliminary results obtained for the hole transport within the framework of the variable hopping model demonstrate that the mismatches are able to change significantly the energies of the G sites. This makes the charge transfer efficiency sensitive to the presence of such "mutations" in the base pair sequences. For example, according to our theoretical findings, supported by the experiment [119], the efficiency of the hole transfer along the sequence containing a single G:T pair instead of a G:C, decreases by a factor of 3 as compared with the "normal" sequence. This provides a route to design the DNA-based nanoscale sensors and again demonstrates the importance of the theoretical studies of charge transport in the double helix for nanoscience.

Acknowledgement. We are grateful to the Chemistry Division of the ONR, the NASA URETI program, to DoD/MURI and DURINT programs for support of the research. We thank many colleagues, particularly A.L. Burin, E.M. Conwell, and J. Jortner for useful discussions.

References

1. J.D. Watson and F.H.C. Crick, Nature **171**, 737 (1953).
2. M.A. Ratner and J. Jortner, *Molecular electronics* (Blackwell, Oxford, 1997).
3. V. Bhalla, R.P. Bajpai and L.M. Bharadwaj, *EMBO Rep.* **4**, 442 (2003).
4. H. Tabata, L.T. Cai, J.H. Gu, S. Tanaka, Y. Otsuka, Y. Sacho, M. Taniguchi and T. Kawai, Synth. Met. **133**, 469 (2003).
5. M. Di Ventra and M. Zwolak, in *Encyclopedia of Nanoscience and Nanotechnology*, edited by H.S. Nalwa American Scientific Publishers, Stevenson Ranch, California, 2004; L. Adleman, Science **266**, 1021 (1994).

6. E. Winfree, in *Proceedings of a DIMACS Workshop*, edited by R.J. Lipton and E.B. Baum American Mathematical Society, Providence, 1996.

7. E. Winfree, J. Biol. Mol. Struct. Dynamics Conversat. **2**, 263 (2000).

8. C. Mao, T. LaBean, J.H. Reif and N.C. Seeman, Nature **407**, 493 (2000).

9. N.C. Seeman, Nature **421**, 427 (2003).

10. G. Hartwich, D.J. Caruana, T. de Lumley-Woodyear, Y.B. Wu, C.N. Campbell and A. Heller, J.Am. Chem. Soc. **21**, 10803 (1999).

11. F. Lisdat, B. Ge, and F.W. Scheller, Electrochem. Commun. **1**, 65 (1999).

12. S.J. Park, A.A. Lazarides, C.A. Mirkin, P.W. Brazis, C.R. Kannewurf and R.L. Letsinger, Angew. Chem. Int. Ed. **39**, 3845 (2000).

13. E.M. Boon, D.M. Ceres, T.G. Drummond, M.G. Hill and J.K. Barton, Nature Biotechnol. **18**, 1096 (2000).

14. C.M. Niemeyer, Angew. Chem. Int. Ed. **40**, 4128 (2001).

15. S.-J. Park, T.A. Taton and C.A. Mirkin, Science **295**, 1503 (2002).

16. A. Marshall and J. Hodgson, Nature Biotechnol. **16**, 27 (1998).

17. S.O. Kelley, N.M. Jackson, M.G. Hill and J.K. Barton, Angew. Chem. Int. Ed. **38**, 941 (1999).

18. F. Lisdat, B. Ge, B. Krause, A. Ehrlich, H. Bienert and F.W. Scheller, Electroanal. **13**, 1225 (2001).

19. J.J. Gooding, Electroanal. **14**, 1149 (2002).

20. S. Kimamoto, H. Nakano, Y. Matsuo, Y. Sugie and K. Yamana, Electrochemistry **70**, 789 (2002).

21. E.M. Boon, J.K. Barton, P.I. Pradeepkumar, J. Isaksson, C. Petit and J. Chattopadhyaya, Angew. Chem. Int. Ed. **41**, 3402 (2002).

22. A. Erdem and M. Ozsoz, Electroanal. **14**, 963 (2002).

23. R.P. Fahlman and D. Sen, J. Am. Chem. Soc. **124**, 4610 (2002).

24. P. O'Neill and E.M. Frieden, Adv. Radiat. Biol. **17**, 53 (1993).

25. B. Armitage, Chem. Rev. **98**, 1171 (1998).

26. C.J. Burrows and J.G. Muller, Chem. Rev. **98**, 1109 (1998).

27. D. Wang, D.A. Kreutzer and J.M. Essigmann, Mutat. Res. **400**, 99 (1998).

28. S. Kawanishi, Y. Hiraku and S. Oikawa, Mutat. Res. **488**, 65 (2001).

29. D.D. Eley and D.I. Spivey, Trans. Faraday Soc. **58**, 411 (1962).

30. U. Diederichsen, Angew. Chem. Int. Ed. **36**, 2317 (1997).

31. M.W. Grinstaff, Angew. Chem. Int. Ed. **38**, 3629 (1999).

32. C.J. Murphy, M.R. Arkin, Y. Jenkins, N.D. Ghatlia, S.H. Bossman, N.J. Turro and J.K. Barton, Science **262**, 1025 (1993).

33. M.R. Arkin, E.D.A. Stemp, R.E. Holmin, J.K. Barton, A. Horman, E.J.C. Olson and P.F. Barbara, Science **273**, 475 (1996).

34. S.O. Kelley, R.E. Holmin, E.D.A. Stemp and J.K. Barton, J. Am. Chem. Soc. **119**, 9861 (1997).

35. K. Fukui and K. Tanaka, Angew. Chem. Int. Ed. Engl. **37**, 158 (1998).

36. C. Wan, T. Fiebig, S.O. Kelley, C.R. Treadway, J.K. Barton and A.H. Zewail, Proc. Natl. Acad. Sci. USA **96**, 6014 (1999).

37. T.J. Meade and J.F. Kayyem, Angew. Chem. Int. Ed. **34**, 352 (1995).

38. S.M. Gasper and G.B. Schuster, J. Am. Chem. Soc. **199**, 12762 (1997).

39. F.D. Lewis, T. Wu, Y. Zhang, R.L. Letsinger, S.R. Greenfield and M.R. Wasielewski, Science **277**, 673 (1997).

40. F.D. Lewis, X. Liu, Y. Wu, S.E. Miller, M.R. Wasielewski, R.L. Letsinger, R. Sanishcili, A. Joachimiak, V. Tereshko and M. Egli, J. Am. Chem. Soc. **121**, 9905 (1999).

41. F.D. Lewis, X. Liu, S.E. Miller and M.R. Wasielewski, J. Am. Chem. Soc. **121**, 9746 (1999).

42. P.T. Henderson, D. Jones, G. Hampikian, Y. Kan and G.B. Schuster, Proc. Natl. Acad. Sci. USA **96**, 8353 (1999).

43. A.M. Brun and A. Harriman, J. Am. Chem. Soc. **116**, 10383 (1994).

44. S.O. Kelley and J.K. Barton, Chem. Biol. **5**, 413 (1998).

45. S.O. Kelley and J.K. Barton, Science **283**, 375 (1999).

46. D.B. Hall, R.E. Holmin and J.K. Barton, Nature **382**, 731 (1996).

47. E. Meggers, M.E. Michel-Beyerle and B. Giese, J. Am. Chem. Soc. **120**, 12950 (1998).

48. B. Giese, S. Wessely, M. Spormann, U. Lindemann, E. Meggers and M.E. Michel-Beyerle, Angew. Chem. Int. Ed. **38**, 996 (1999).

49. K. Nakatani, C. Dohno and I. Saito, J. Am. Chem. Soc. **12**, 10854 (1999).

50. D. Ly, L. Sanii and G.B. Schuster, J. Am. Chem. Soc. **121**, 9400 (1999).

51. H.-A. Wagenknecht, Angew. Chem. Int. Ed. **42**, 2454 (2003).

52. Y. Razskazovskiy, S.G. Swarts, J.M. Falcone, C. Taylor and M.D. Sevilla, J. Phys. Chem. B **101**, 1460 (1997).

53. R.F. Anderson and G.A. Wright, Phys. Chem. Chem. Phys. **1**, 4827 (1999).

54. M.G. Debije, M.T. Milano and W.A. Bernhard, Angew. Chem. Int. Ed. **38**, 2752 (1999).

55. A. Messer, K. Carpenter, K. Forzley, J. Buchanan, S. Yang, Y. Razskazovski, Z. Cai and M.D. Sevilla, J. Phys. Chem. B **104**, 1128 (2000).

56. Z. Cai and M.D. Sevilla, J. Phys. Chem. B **104**, 6942 (2000).

57. Z. Cai, Z. Gy and M.D. Sevilla, J. Phys. Chem. B **104**, 10406 (2000).

58. X. Li, Z. Cai and M.D. Sevilla, J. Phys. Chem. B **105**, 10115 (2001).

59. Z. Cai, X. Li and M.D. Sevilla, J. Phys. Chem B **106**, 2755 (2002).

60. J.M. Warman, M.P. De Haas and A. Rupprecht, Chem. Phys. Lett. **249**, 319 (1996).

61. D.N. Beratan, S. Priyadarshy and S.M. Risser, Chem. Biol. **4**, 3 (1997).

62. E.K. Wilson, Chem. Eng. News **75**, 33 (1997).

63. S. Priyadarshy, S.M. Risser and D.N. Beratan, J. Biol. Inorg. Chem. **3**, 196 (1998).

64. N.J. Turro and J.K. Barton, J. Biol. Inorg. Chem. **3**, 201 (1998).

65. E.S. Krider and T.J. Meade, J. Biol. Inorg. Chem. **3**, 222 (1998).

66. T.L. Netzel, J. Biol. Inorg. Chem. **3**, 210 (1998).

67. E.K. Wilson, Chem. Eng. News **76**, 51 (1998).

68. Y.A. Berlin, A.L. Burin, and M.A. Ratner, Superlatt. and Microstruct. **28**, 241 (2000).

69. C. Wu, Science News **156**, 104 (1999).

70. J. Jortner, M. Bixon, T. Langenbacher and M.E. Michel-Beyerle, Proc. Natl. Acad. Sci. USA **95**, 12759 (1998).

71. M.A. Ratner, Nature **397**, 480 (1999).

72. M. Bixon, B. Giese, S. Wessely, T. Langenbacher, M.E. Michel-Beyerle and J. Jortner, Proc. Natl. Acad. Sci USA **96**, 11713 (1999).

73. Y.A. Berlin, A.L. Burin and M.A. Ratner, J. Phys. Chem. A **104**, 443 (2000).

74. Y.A. Berlin, A.L. Burin and M.A. Ratner, J. Am. Chem. Soc. **123**, 260 (2001).

75. Y.A. Berlin, A.L. Burin and M.A. Ratner Chem. Phys. **275**, 61 (2002).

76. M. Bixon and J. Jortner, Chem. Phys. **281**, 393 (2002).

77. M. Enescu and L. Lindqvist, J. Phys. Chem. **99**, 8405 (1995).

78. C.A.M. Siedel, A. Schultz and M.H.M. Sauer, J. Phys. Chem. **100**, 5541 (1996).

79. C. Lifschitz, E. Bergmann and B. Pullman, Tetrahedron Lett. **46**, 4583 (1967).

80. N.S. Hush and A.S. Cheung, Chem. Phys. Lett. **34** 11 (1975).

81. V.M. Orlov, A.N. Smirnov and Y.M. Varshavshy, Tetrahedron Lett. **48**, 4377 (1976).

82. A.-O. Colson, B. Besler, M.D. Close and M.D. Sevilla, J. Phys. Chem. **96**, 661 (1992).

83. A.-O. Colson, B. Besler and M.D. Sevilla, J. Phys. Chem. **96**, 9787 (1992).

84. M.D. Sevilla, B. Besler and A.-O. Colson, J. Phys. Chem. **99**, 1060 (1995).

85. H. Sugiyama and I. Saito, J. Am. Chem. Soc. **118**, 7063 (1996).

86. M. Hutter and T.J. Clark, J. Am. Chem. Soc. **118**, 7574 (1996).

87. N.S. Kim and P.R. LeBreton, J. Am. Chem. Soc. **118**, 3694 (1996).

88. F. Prat, K.N. Houk and C.S. Foote, J. Am. Chem. Soc. **120**, 845 (1998).

89. I. Saito, T. Nakamura, K. Nakatani, Y. Yoshioka, K. Yamaguchi and H. Sugiyama, J. Am. Chem. Soc. **120**, 12686 (1998).

90. H. Fernando, G.A. Papadantonakis, N.S. Kim and P.R. LeBreton, Proc. Natl. Acad. Sci. USA **95**, 5550 (1998).

91. A.A. Voityuk, J. Jortner, M. Bixon and N. Rösch, Chem. Phys. Lett. **324**, 430 (2000).

92. N. Russo, M. Toscano and A. Grand, J. Comput. Chem. **21**, 1243 (2000).

93. K. Senthilkumar, F.C. Grozema, C.F. Guerra, F.M. Bickelhaupt and L.D.A. Siebbeles, J. Am. Chem. Soc. **125**, 13658 (2003).

94. S. Steenken and S.V. Jovanovic, J. Am. Chem. Soc. **119**, 617 (1997).

95. S. Steenken, J.P. Telo, H.M. Novais and L.P. Candeias, J. Am. Chem. Soc. **114**, 4701 (1992).

96. X. Li, Z. Cai and M.D. Sevilla, J. Phys. Chem. A **106**, 1596 (2002).

97. A.A. Voityuk, M.E. Michel-Beyerle and N. Rösch, Chem. Phys. Lett. **342**, 231 (2001).

98. Y.A. Berlin, I.V. Kurnikov, D. Beratan, M.A. Ratner and A.L. Burin, Top. Curr. Chem. **237**, 1 (2004).

99. Y.A. Berlin and M.A. Ratner, Radiat. Phys. Chem. **74**, 124 (2005).

100. F.D. Lewis, R.L. Letsinger and M.R. Wasielewski, Acc. Chem. Res. **34**, 159 (2001).

101. S. Hess, M. Götz, W.B. Davis and M.E. Michel-Beyerle, J. Am. Chem. Soc. **123**, 10046 (2001).

102. F.D. Lewis, X. Liu, S.E. Miller, R.T. Hayes and M.R. Wasielewski, J. Am. Chem. Soc. **124**, 11280 (2002).

103. N. Amann, E. Pandurski, T. Fiebig and H.-A. Wegenkhecht, Chem. Eur. J. **8**, 4877 (2002).

104. C.-S. Liu, R. Hernandez and G.B. Schuster, J. Am. Chem. Soc. **126**, 2877 (2004).

105. T. Ito and S.E. Rakita, J. Am. Chem. Soc. **125**, 11480 (2003).

106. C. Behrens, M. Ober and T. Carell, Eur. J. Org. Chem. **19**, 3281 (2002).

107. C. Behrens, L.T. Burgdorf, A. Schwögler and T. Carell, Angew. Chem. Int. Ed. **41**, 1763 (2002).

108. C. Behrens and T. Carell, Chem. Commun. **14**, 1632 (2003).

109. S. Breeger, U. Hennecke and T. Carell, J. Am. Chem. Soc. **126**, 1302 (2004).

110. B. Giese, Annu. Rev. Biochem. **71**, 51 (2002).

111. B. Giese, J. Amaudrut, A.-K. Köhler, M. Spermann and S. Wessely, Nature **412**, 318 (2001).
112. M.E. Núñez, D.B. Hall and J.K. Barton, Chem. Biol. **6**, 85 (1999).
113. K. Senthilkumar, F.C. Grozema, C.F. Guerra, F.M. Bickelhaupt, F.D. Lewis, Y.A. Berlin, M.A. Ratner and L.D.A. Siebbeles, J. Am. Chem. Soc. **127**, 14894 (2005).
114. M. Bixon and J. Jortner, Adv. Chem. Phys. **106**, 35 (1999).
115. A.A. Voityuk, J. Chem. Phys. **122**, 204904 (2005).
116. K. Siriwong, A.A. Voityuk, M.D. Newton and N. Rösch, J. Phys. Chem. B **107**, 2595 (2003).
117. D.N. LeBard, M. Lilichenko, D.V. Matyushov, Y.A. Berlin and M.A. Ratner, J.Chem. Phys. B **107**, 14509 (2003).
118. E.M. Conwell, Top. Curr. Chem. **237**, 73 (2004).
119. B. Giese and S. Wessely, Angew. Chem. Int. Ed. **39**, 3490 (2000).

4 Atomistic Models of DNA Charge Transfer

Thorsten Koslowski and Tobias Cramer

Institut für Physikalische Chemie, Universität Freiburg, Albertstraße 23a,
D-79104 Freiburg im Breisgau, Germany
Thorsten.Koslowski@physchem.uni-freiburg.de

4.1 Introduction

With two important conferences held in 2006 on the subject of DNA charge transfer and numerous recent publications, the topic obviously lies in the focus of a large interdisciplinary community, bringing together physicists, chemists, molecular biologists and the nano-engineers [1]. The interest is not purely an academic one, but also motivated by the potential applications in nanodevices and the DNA sensors, including the problem of sequencing the genome at the level of single molecules.

Although the formulation of the question whether DNA can support an electronic charge dates back well into the past century, it was in the two recent decades that the phenomenon has been confirmed on a quantitative and reproducible level. Three types of experiments have contributed to our increasing knowledge of DNA charge transfer: Barton, Giese and Schuster have designed ingenious chemical experiments based upon a photochemically induced charge separation [2, 3]. The excess charge – ususally an electron hole – may propagate along a DNA double strand, until it is trapped at a site that exhibits a particularly low oxidation potential, usually a guanine cluster. At this trap, the consecutive reactions can induce the cleavage of the double strand, thus enabling the analysis of the fragments by means of electrophoresis. In this way, the relative kinetics of the hole propagation can be obtained. Second, in a direct photochemical approach, the charge propagation can be followed on the femtosecond time scale [3]. Third, DNA strands may bridge nanocontacts, thus giving access to the direct measurement of the current-voltage curves [4].

The interpretation of the experiments cited above is far from painting a coherent picture. The photofragmentation studies are usually rationalized within the Marcus theory of charge transfer, with guanines acting as the centers of charge localization. In the DNA double strands, the guanine-cytosine (GC) base pairs are separated by the adenine-thymine (AT) fragments of a variable length, as encoded in the sequence of the strand. Depending on the length of the AT bridges, adenines may contribute virtual energy levels within a superexchange (or tunneling) mechanism, or, for bridges with more than three to four AT pairs, actually act as the stepping stones, giving rise to a diffusion-like hopping transport [5]. This phenomenological hopping model has meanwhile passed the test of time, and its fundamental parameters can be obtained from a set of quantum chemical calculations [6]. The situation is less clear for the nanodevice setups, where the I–V curves are interpreted as either insulating, metallic or semiconducting in roughly equal numbers of experiments [7].

In this chapter, we shall address both the photofragmentation studies in a variety of systems and the transport through the nanowires using the same atomistic Hamiltonian. The results are discussed and – wherever possible – compared with the experimental findings.

4.2 Electronic Structure Model

We assume that the nucleobases with their high-energy frontier molecular orbitals dominate the charge transport properties, and therefore we neglect the deoxyribose and the phosphate units beyond their role of providing a scaffold for the base pairs. We further assume that the σ and π orbitals can be approximately separated, so that theories of the chemical bond appropriate to π electron systems can be applied. In the field of the conductivity phenomena in π systems, the Su-Schrieffer-Heeger (SSH) model [8] has a remarkable record of success. In its standard form, the potential energy of the SSH Hamiltonian reads

$$\hat{H} = \sum_{\langle ij \rangle} \frac{k}{2}(x_i - x_j)^2 - \sum_{\langle ij \rangle} \left[t_0 - \alpha(x_i - x_j) \right] (a_i^\dagger a_j + a_j^\dagger a_i) , \qquad (4.1)$$

where $(x_i - x_j)$ denotes the deviation of the distance between a neighboring pair of atoms from that of a carbon-carbon single bond, k is the corresponding force constant, t_0 is the tight-binding coupling matrix element and α denotes the electron-phonon coupling constant. The angular brackets indicate the restriction of the sum to the distinct pairs of the neighbors, and $a_i^\dagger / a_i$ are the creation/annihilation operators acting on a basis of $2p_z$ atomic orbitals, which is assumed to be orthogonal. The nuclear coordinates and the momenta are treated as the classical quantities. A schematic representation of this model is presented in Fig. 4.1.

The SSH Hamiltonian can be separated into an electronic and a nuclear part using a displaced phonon coordinate [9]. Within a Hartree-like mean-field approximation, the transformed electronic part of the Hamiltonian is given by [9]

$$\hat{H} = -\sum_{\langle ij \rangle} (t_0 + 4U_{\text{SSH}}\bar{n}_{ij})(a_i^\dagger a_j + a_j^\dagger a_i) + 4U_{\text{SSH}} \sum_{\langle ij \rangle} \bar{n}_{ij}^2 , \qquad (4.2)$$

where the $\bar{n}_{ij}$ are the tight-binding bond orders (which are proportional to the displacements $(x_i - x_j)$ of the original SSH Hamiltonian), and we define the so-called off-diagonal Hubbard parameter $U_{\text{SSH}} = \alpha^2/2k$, which is equal

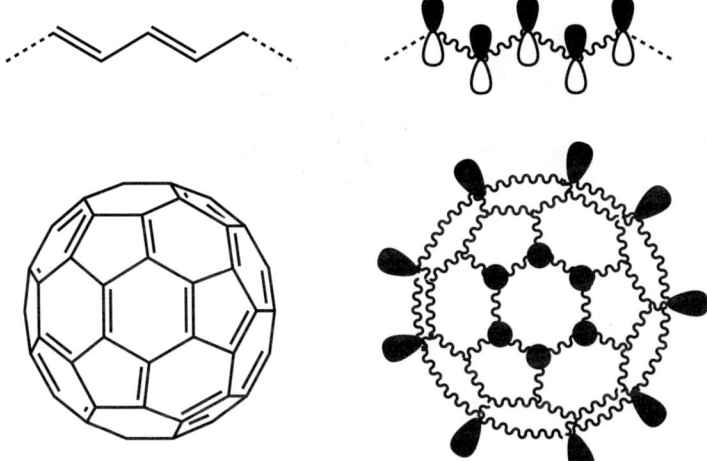

Fig. 4.1. The Lewis picture of the polyacetylene (*top left*) and fullerene (*bottom left*); and a schematic representation of the Su-Schrieffer-Heeger model for these compounds (*right*)

to 0.32 meV in the standard SSH model. We define the operator $n_{ij} = a_i^\dagger a_j$, where $a^\dagger / a$ denote the corresponding creation and annihilation operators. The Hamiltonian (4.2) accounts for the chemical bond within the bases and the inner-sphere contributions to the charge transfer reaction via a possible change in the bond orders or the bond lengths.

In addition to the inner sphere reorganization, a charged species dissolved in a polarizable solvent forces the nearby solvent molecules to reorientate with the thermally averaged dipole moments pointing towards the charge and therefore gives rise to the outer sphere contributions to the charge transfer reaction. The simplest possible model to describe this process stems from classical electrostatics: the charged object is represented by a hard sphere, and the solvent is treated as a polarizable continuum characterized by a dielectric response function $\epsilon(\omega)$, which in the static limit becomes the familiar dielectric constant. Upon polarization, the total energy of the system is lowered; this energy difference constitutes a major part of the solvation energy of the systems consisting of the charged species. The ingredients of the model are shown as a schematic representation in Fig. 4.2. To describe the influence of the solvent polarization effects quantitatively, we apply a straightforward extension of Marcus' treatment of the energetics of the outer-sphere reactions [10] to many-site systems. We write the reorganization energy emerging from an ensemble of the excess charges q_i localized within spheres of radii σ_i as

$$\lambda_{\text{out}} = \frac{e^2}{4\pi\epsilon_0} \left(\frac{1}{\epsilon_\infty} - \frac{1}{\epsilon_s} \right) \left(\sum_i \frac{\Delta z_i^2}{\sigma_i} - \sum_{i<j} \frac{\Delta z_i \Delta z_j}{r_{ij}} \right) . \tag{4.3}$$

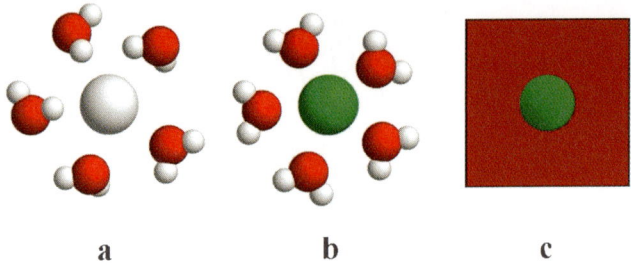

a   **b**   **c**

Fig. 4.2. a–c. The solvent polarization, as treated within the simplest version of the Marcus theory: **a** neutral atom dispersed in a shell of water molecules, **b** charged ion of the same size and the reoriented solvation shell, and **c** representation of the system by a charged hard sphere and a dielectric continuum. The two leftmost figures do not emerge from any computations or simulations, but are only drawn schematically to illustrate the underlying physical chemistry

If the high- and the low-frequency dielectric responses characterized by the constants ϵ_∞ and ϵ_s can be separated and the long-range Coulomb interactions are neglected, this model turns into the interaction term of a spin-free Hubbard Hamiltonian [11]. We replace the charges Δz_i by the corresponding number operators $n_i = n_{ii} = a_i^\dagger a_i$ and arrive at the mean-field expression

$$\hat{H}_{ee} = -U_{ee} \sum_i (n_i - n_{i,0})^2 \simeq -U_{ee} \sum_i [2n_i(\bar{n}_i - \bar{n}_{i,0}) - \bar{n}_i^2 + \bar{n}_{i,0}^2] \,. \quad (4.4)$$

It can be interpreted as a nonretarded reaction field, which extends the linear combination of atomic orbitals (LCAO) approach to systems embedded in a polarizable environment [12]. In this work, we use $U_{ee} = 0.8\,\text{eV}$ [13], while all the other parameters have been obtained by a careful fit to *ab initio* quantum chemical calculations and to the experimental oxidation potentials. Further technical details are given in [13, 14].

As the standard SSH model only describes the hydrocarbons, a chemically specific modification has been introduced to handle the heterocyclic nucleobases, which also contain nitrogen and oxygen atoms. In addition, the interbase tight-binding interactions have been parameterized, and the nucleobases may exhibit an arbitrary mutual orientation. The combined Hamiltonians (4.2) and (4.4) can be solved self-consistently. In this case the $\bar{n}_{ij}$ and $\bar{n}_i$ are computed from a previous self-consistent field (SCF) step.

In the SCF computations, we proceed as follows: Each base is considered to be a potential center of charge localization. Hence, we use the initial conditions that reflect the bond order – or the bond length – distribution of a singly-charged base X and that of the neutral bases for the remaining centers. Here X may be varied. Whenever the initial charge is localized on a guanine, the SCF procedure rapidly converges into the final charge distribution. Thus, in the systems studied here the guanine bases operate as the charge traps and form the centers of localization for a polaron state. As a consequence, the intermediate A–T pairs act as barriers for tunneling.

We will now quantify these findings by accessing a cross section of the potential energy surface relevant to the hole transfer between two guanine bases by a linear synchronous transit (LST) approach [15]. In doing so, we adopt a procedure that contains the elements of an adiabatic charging process [16, 17] and the application of an interpolation method to the charge transfer. The bond orders that characterize each of the minima, which we may denote as X and Y, are given by $\{\bar{n}_{1,1,X}; \bar{n}_{1,2,X}; \dots; \bar{n}_{N-1,N,X}; \bar{n}_{N,N,X}\}$ and $\{\bar{n}_{1,1,Y}, \bar{n}_{1,2,Y}, \dots, \bar{n}_{N-1,N,Y}, \bar{n}_{N,N,Y}\}$, respectively. The interpolation is performed by setting the bond order of the combined Hamiltonians (4.2) and (4.4) as $\bar{n}_{i,j} = \gamma\bar{n}_{i,j,X} + (1 - \gamma)\bar{n}_{i,j,Y}$, where the interpolation parameter γ serves as a reaction coordinate. The charge orders are interpolated in the same way.

A typical energy profile for the hole transfer between two neighboring guanines is shown in Fig. 4.3, from which the parameters relevant to Marcus' theory of charge transfer can be immediately obtained. These are the site-specific reorganization energy λ as the difference between the corresponding ground state minimum and the first excited state, the effective tunnel splitting, t, given as half of the difference between the ground state and the excited state energy at the transition state, i.e. the point of the closest approach between the two curves. In addition, we compute the activation barriers E_A for the forward and the backward reactions, which include the energy differences

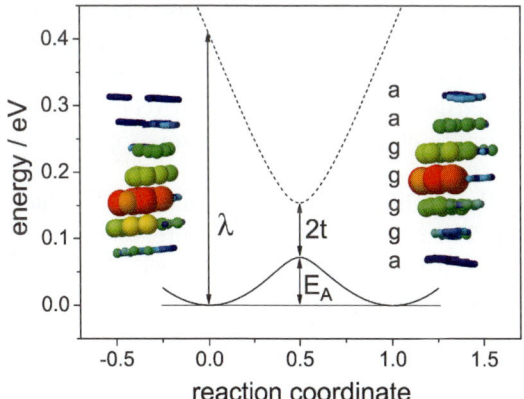

Fig. 4.3. The energy parameters as extracted from a linear synchronous transit approximation to the extended Su-Schrieffer-Heeger model for the guanine-guanine hopping: the reorganization energy λ, the activation barrier E_A, and the effective tunnel splitting t. The insets show the charge distributions at the potential energy minima. The diameters of the spheres are proportional to the logarithm of atomic charges

between the initial and the final states. From these parameters, the hopping rates can be estimated via [10]

$$k_{CT} = \frac{t^2}{\hbar} \sqrt{\frac{\pi}{\lambda k_B T}} \exp\left(-\frac{E_A}{k_B T}\right) .$$ (4.5)

We note that for the high charge transfer rates – as computed for the next-nearest neighbour guanines – the prefactor of the exponential becomes a constant k_0 of the order of 10^{12} to $10^{13}\,\mathrm{s}^{-1}$. We are, however, generally interested in the comparatively slow charge transfer processes, which constitute the bottlenecks of the overall transfer through the DNA assemblies.

4.3 Systems and Numerical Results

4.3.1 Idealized Structures

As a first application of the extended SSH model to DNA charge transfer, we have considered idealized DNA double strand oligomers [13]. As geometries, we have used the models of the A and B forms of the DNA as generated by the NUCLEIC program of Ponder's TINKER suite [18]. In their default form, they correspond to the bulk structure of the DNA fibres with a different water content.

As a general trend, the computed hopping rates decrease by one to two orders of magnitude with each additional A–T or T–A pairs separating the guanines. With three or four of these intervening pairs, k_{CT} becomes as small as $10^5\,\mathrm{s}^{-1}$. Comparing with the experiments, the rate for the sequence $A_5GA_2G_3A_5$ of $2.4 \times 10^8\,\mathrm{s}^{-1}$ is close to the estimate of Bixon et al. [5], ($10^9\,\mathrm{s}^{-1}$). The photochemical experiments, on the other hand, suggest $10^7\,\mathrm{s}^{-1}$ for the hole recombination in the sequence $A_5GAG_3A_6$ [19] in obvious disagreement with the results of Bixon and coworkers [5] and our value of $8 \times 10^{11}\,\mathrm{s}^{-1}$.

For the B form, the decadic logarithm of the inverse half-life – which differs from k_{CT} by a factor of $\ln 2$ – as a function of the intervening A–T pairs is presented in Fig. 4.4. From a linear least-square fit, we obtain the decay parameter of the Marcus-Levich-Jortner relation

$$k_{CT} \propto \exp(-\beta R) ,$$ (4.6)

with $\beta = 1.38 \pm 0.13\,\text{Å}^{-1}$, which is close to the upper boundary of the interval $0.1\,\text{Å}^{-1} < \beta < 1.4\,\text{Å}$ reported in the experiments [20]. For the A form, a similar picture emerges.

The same model provides a potential energy surface relevant to the adenine-adenine hopping that can be thermally populated and which determines the conductivity for a larger number of A–T pairs. From this surface,

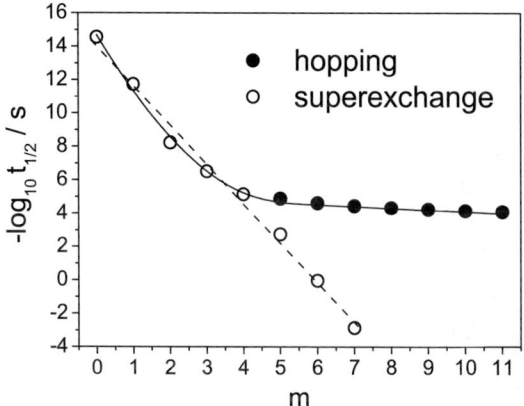

Fig. 4.4. Characteristic time scales for the superexchange (*empty circle*) and the hopping (*full circle*) transport processes between two guanines as a function of the number of intervening adenines, m

a direct hopping to a neighboring guanine or the formation of extended adenine hole states constitute the exit paths. This so-called hopping regime starts to dominate the transport properties if more than three to four A–T pairs separate two guanines.

4.3.2 The Nucleosome

There now exists evidence that charge transfer is operative not only in carefully designed artificial DNA oligomers, but also in structures relevant to living organisms: Núñez et al. have been able to attach a rhodium intercalator to the $5'$ end of a nucleosome core particle and have induced a DNA damage more than 80 Å away from the site of the initial oxidation process [21]. Further evidence suggesting the possibility of *in vivo* charge transfer comes from the long-range oxidative damage in whole nuclei [22].

Chromatine DNA is organized in the nucleosomes, the highly conserved nucleoprotein complexes that contain 145–147 base pairs wrapped around an octameric protein core. In eukaryotic organisms, nucleosomes occur every 200 ± 40 base pairs throughout the genome. The geometry of our model is based on the NCP X-ray crystal structure obtained by Luger et al. [23] with a 2.8 Å resolution. A graphical representation of this complex is shown in Fig. 4.5. The base pairs are organized in the form of a β-helix which is wrapped around the histone octamer in 1.65 turns of a left-handed superhelix. Although the sequence is palindromic, the nucleoprotein complex does not reflect this symmetry.

Within our computations [24], we find that the excess hole charge is transferred along the nucleosome with a superexchange mechanism comparable to the charge transfer in much smaller DNA fragments that are unprotected by

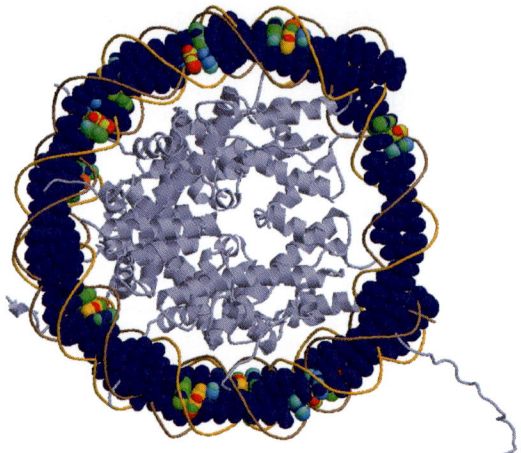

Fig. 4.5. Graphical representation of the nucleosome core complex, after the X-ray structure of Luger et al. [23]. The protein part is shown in *light blue* as a a cartoon model, the DNA backbone is traced in *brown* and *orange* and the DNA bases are displayed as *colored spheres*. The *colors* represent positive charges resulting from a calculation involving twelve polaron states

the proteins. This transport process enables the consecutive cleavage reactions that predominantly occur at the GG clusters. Assuming a cleavage rate of $10^4\,\mathrm{s}^{-1}$, the excess charge can be transferred to the first four GG clusters, but is – according to our calculations – entirely trapped before even a small fraction populates GG5. This finding is in accord with the experiment as the largest fragment found upon irradiation extends only up to GG4.

4.3.3 Three-Way Junctions

Motivated by the recent progress in the design of complex, branched DNA oligomers, the corresponding structural analysis and the first experiments on charge transfer along the DNA junctions [25, 26], we have approached the transport processes within the DNA dendrimers from a theoretical and numerical perspectives [27]. In order to quantitatively predict the hopping rates for large systems, we have used a two-level hierarchy of models.

From the electronic structure models described and cited above, interguanine superexchange hopping rates have been computed, which turn out to reflect the T-shaped anisotropy of the three-way junction by two different rate-limiting reaction coefficients along the branches of the oligomer, $k_{\mathrm{f}} \simeq 10^{11}\,\mathrm{s}^{-1}$ and $k_{\mathrm{s}} \simeq 10^7\,\mathrm{s}^{-1}$.

These rates serve as input parameters to a classical kinetic description of the charge transport within a dendrimer network of three-way junctions, which constitutes the second level of our model. We apply a coarse-grained description that interprets the three-way junctions as vertices connected by bonds with anisotropic hopping rates which are given by the rate-limiting reaction coefficients. A particular realization of this model has been studied in detail: a peripheral charge donor is connected to a central absorbing trap by a chain of bonds enabling fast charge transfer. In turn, this model can be treated with a high accuracy as a linear chain which is separated from the other branches of the dendrimer by contacts that only permit slow transfer or that are disconnected from this branch by the central trap. Up to a hypothetical dendrimer size with generation $g = 20$, the charge transfer along the selected branch is faster than the oxidation and fragmentation processes taking place at the central site. As a potential application for biocompatible drug-delivery based upon a branched charge transfer system, we have considered a target connected to a protective coating by G-rich DNA fragments that is released upon oxidation initiated by a remote peripheral donor.

4.3.4 Dynamic Aspects

Dynamic disorder can be put into realization by collecting snapshots from the molecular dynamics (MD) simulations using the classical AMBER 94 force field, which has been reported as being particularly suitable for this purpose [28]. We find that the effective donor–acceptor couplings show remarkable fluctuations in time [14]. The corresponding charge transfer rate distributions can be described as exponential. Within the accuracy of our data, they are universal with respect to the number of A–T base pairs separating the two guanines, i.e. they can be mapped onto each other by scaling them by their mean value. With a time scale of some ten femtoseconds, the correlation between the rate constants is very short-lived. Any charge transfer (CT) processes slower than $10 - 100\,\text{fs}$ therefore experience the DNA strand as a dynamically averaged object, and for these relatively slow processes, which usually constitute the bottlenecks of the CT in more complex DNA objects, one may replace the CT rate distributions by their average values. A comparison of the calculated values for the CT in model nucleotides to the experimental data based on time resolved spectroscopic measurements, as well as a comparison to the relative photochemical cleavage yields by modelling of a full kinetic scheme, showed that the consideration of the dynamic fluctuations via the MD simulations leads to a significant improvement of the computed results.

4.3.5 Nanoscopic Setups

In the conductivity experiments on the nanoscale, the DNA strands or bundles thereof bridge two electrodes under a large variety of conditions, ranging from the boiling point of liquid Helium to room temperature and at different levels of humidity. An ingenious experiment has recently been reported by Naaman and co-workers [29], who have chemically anchored single strand DNA oligomers at a gold surface and a small number of complementary strands at the tip of an atomic force microscope. In this way, the electronic current as a function of the applied voltage can be measured as reproduced below. Depending on the DNA sequence, currents up to 15 nA have been reported upon the application of ±2 V for a bridge consisting of an estimated upper limit of three strands.

We have applied the extended SSH Hamiltonian to chains with the sequence 5′–CA–TTAATGCTATGCAGAAAATCTTAG–3′ (8GC) and 5′–GC–TGGATGGTATGGAGAAGATGTGCG–3′ (14GC) [31]. To map the many-atom problem to the one where the base pairs are represented by sites, the standard analysis of the potential energy surfaces relevant to the charge transfer has been performed, as described above. To the resulting driving forces, a linearly interpolated electrostatic potential has been added. With the help of the computed CT rates, it is possible to express the kinetics of the entire system via a set of master equations for the base pair populations p_m [30]

$$\dot{p}_m = \sum_n \left[k_{mn} p_n (1 - p_m) - k_{nm} p_m (1 - p_n) \right] . \tag{4.7}$$

We adjust the boundary conditions in such a way that the total charge P populating the bridge is conserved, and we obtain the electronic current I as a function of the applied voltage V and P. Here, I rapidly settles to a plateau with increasing P at a fixed potential.

The numerical results are shown in Figs. 4.6 and 4.7, where the computed currents have been multiplied by a factor of three to take the experimental setup into account, where up to three DNA double strands bind

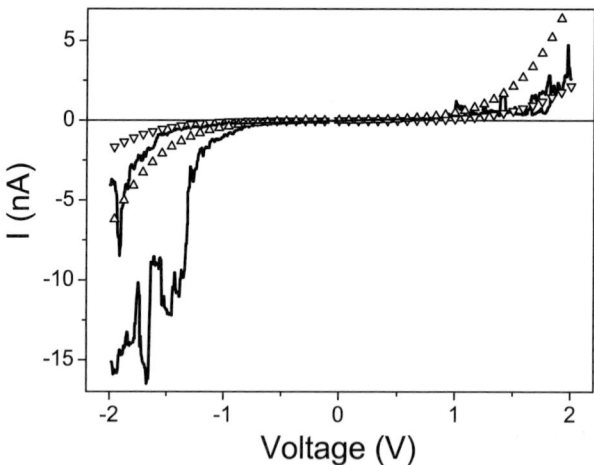

Fig. 4.6. The computed current-voltage curves and the experimental results (*lines*) for the 8GC oligomer. The calculated results are based on the assumption of three DNA double strands bridging the electrodes, each bearing an average charge of $P = 1$ (∇) and $P = 4$ ($\triangle$)

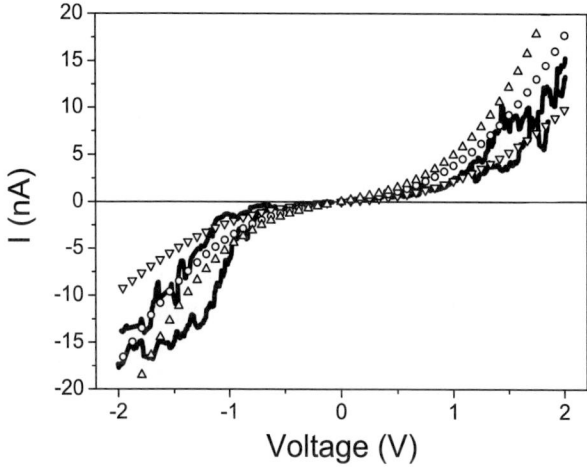

Fig. 4.7. Same as in Fig. 4.6 but for the 14 GC oligomer. Theoretical results are based on the average charges of $P = 1$ (∇), $P - 2$ (o) and $P = 4$ ($\triangle$)

to an AFM tip. Consequently, the absolute values exhibit an uncertainty of considerably less than an order of magnitude, and they provide a reasonable upper limit of the current. Although the model used here is solely based upon the activated hopping processes, a current of the order 10 nA is obtained upon the application of a potential of ± 2 V, a scenario that has hitherto been considered to be compatible only with coherent electronic conduction.

4.4 Conclusions

We have addressed the problem of DNA hole transfer in the electronic ground state from a theoretical and computational perspective. We have extended a standard model for charge transport in doped π-electron systems – the Su-Schrieffer-Heeger Hamiltonian – to account for the presence of the heteroatoms, the polarization effects and the interbase coupling. The model has been parameterized using *ab initio* quantum chemical calculations and the experimental data.

In the presence of G–C Watson-Crick pairs an excess hole charge predominantly localizes on a guanine nucleobase. The bond lengths within the center of the localization are deformed, and a charge reorganization in response to the reaction field part of the Hamiltonian can be computed. Thus, the resulting quasiparticle can be considered as a small polaron. The cross-sections of the potential energy surface relevant to the hole transfer have been computed using a linear synchronous transit approach. They can be analyzed using the Marcus' theory of electron transfer. In this manner, the atomistic model presented here can be mapped onto a much simpler system, which in turn allows the approximate calculation of the energy parameters and the reaction coefficient.

This model does not only reproduce the photochemical fragmentation experiments on the DNA oligomers in solution and on DNA organized in the nucleosome, but also provides a correct order of magnitude estimate for the charge transfer rates through DNA-gold nanojunctions. It reproduces the overall shape of the I–V curves and their dependence on the DNA sequence.

Acknowledgement. It is a pleasure to thank the participants of the Winnipeg symposium on charge migration in DNA – too many to name individually – for fruitful and helpful discussions, and the organizer, T. Chakraborty. Helpful comments by A. Blumen and C.A. Zell are gratefully acknowledged. This work has been supported financially via the DFG SFB 428 program.

References

1. D. Porath, G. Cuniberti, and R. Di Felice, Top. Curr. Chem. **237**, 183 (2004); R.G. Endres, D.L. Cox, and R.R.P. Singh, Rev. Mod. Phys. **76**, 195 (2004).
2. C.-S. Liu, R. Hernandez, and G.B. Schuster, J. Am. Chem. Soc. **126**, 2877 (2003); B. Giese, J. Amaudrut, A.-K. Köhler, M. Spormann, and S. Wessely, Nature **412**, 318 (2001).
3. D.B. Hall, R.E. Holmlin, and J.K. Barton, Nature **382**, 731 (1996); T. Takada, K. Kawai, M. Fujitsuka, and T. Majima, Proc. Natl. Acad. Sci. USA **101**, 14002 (2004); C. Wan, T. Fiebig, S.O. Kelley, C.R. Treadway, J.K. Barton, and A.H. Zewail, Proc. Natl. Acad. Sci. USA **96**, 6014 (1999).
4. H.W. Fink and C. Schönenberger, Nature **398**, 407 (1999); D. Porath, A. Bezryadin, S. de Vries, and C. Dekker, Nature **403**, 635 (2000).
5. M. Bixon, B. Giese, S. Wessely, T. Langenbacher, M.E. Michel-Beyerle, and J. Jortner, Proc. Natl. Acad. Sci. USA **96**, 11713 (1999); B. Giese and M. Spichty, ChemPhysChem **1**, 195 (2000).
6. A.A. Voityuk, Chem. Phys. Lett. **427**, 177 (2006); K. Senthilkumar, F.C. Grozema, C.F. Guerra, F.M. Bickelhaupt, F.D. Lewis, Y.A. Berlin, M.A. Ratner, and L.D.A. Siebbeles, J. Am. Chem. Soc. **127**, 14894 (2005).
7. M. Taniguchi and T. Kawai, Physica E **33**, 1 (2006).
8. W.P. Su, J.R. Schrieffer, and A. Heeger, Phys. Rev. Lett. **41**, 1698 (1979).
9. M. Rateitzak, and T. Koslowski, Chem. Phys. Lett. **377**, 455 (2003).
10. R.A. Marcus, J. Chem. Phys **24**, 966 (1956); R.A. Marcus and N. Sutin, Biochim. Biophys. Acta **811**, 265 (1985).
11. R. Micnas, J. Ranninger, S. Robaskiewicz, Rev. Mod. Phys. **62**, 113 (1990); Shun-Quing Shen, Int. J. Mod. Phys. B **12**, 709 (1998); F. Mancini, M. Marinaro, H. Matsumoto, Int. J. Mod. Phys. B **10**, 1717 (1996); A. Georges, G. Kotliar, W. Krauth, M.J. Rozenberg, Rev. Mod. Phys. **68**, 13 (1996).
12. N. Utz, Th. Koslowski, Chem. Phys. **282**, 389 (2002).
13. T. Cramer, S. Krapf, and T. Koslowski, J. Phys. Chem. B **108**, 11812 (2004).
14. T. Cramer, T. Steinbrecher, A. Labahn, and T. Koslowski, PCCP **7**, 4039 (2005).

15. G. Rauhut, T. Clark, J. Am. Chem. Soc. **115**, 9127 (1993); M. Gröppel, W. Roth, T. Clark, Advanced Materials **7**, 927 (1995).

16. J.-K. Hwang, and A. Warshel, J. Am. Chem. Soc. **109**, 715 (1987).

17. R. A. Kuharski, J. S. Bader, D. Chandler, M. Sprik, M.L. Klein, R.W. Impey, J. Chem. Phys. **89**, 3248 (1988); J. S. Bader, D. Chandler, Chem. Phys. Lett. **157**, 501 (1989); J. S. Bader, R. A. Kuharski, D. Chandler, J. Chem. Phys. **93**, 230 (1990).

18. R.V. Pappu, R. K. Hart, J.W. Ponder, J. Phys. Chem. B **102**, 9725 (1998); M. J. Dudek, K. Ramnarayan, J.W. Ponder, J. Comput. Chem. **19**, 548 (1998); Y. Kong, and J.W. Ponder, J. Chem. Phys. **107**, 481 (1997); J.W. Ponder, F.M. Richards, J. Comput. Chem. **8**, 1016 (1987).

19. F.D. Lewis, X. Liu, S.E. Miller, R.T. Hayes, M.R. Wasielewski, Nature **406**, 51 (2000).

20. B. Giese, Acc. Chem. Res. **33**, 631 (2000).

21. M.E. Núñez, K.T. Noyes and J.K. Barton, Chem. Biol. **9**, 403 (2002).

22. M.E. Núñez, G.P. Holmquist and J.K. Barton, Biochemistry **40**, 12465 (2001).

23. K. Luger, A.W. Mäder, R.K. Richmond, D.F. Sargent and T.J. Richmond, Nature **389**, 251 (1997).

24. T. Cramer, S. Krapf, T. Koslowski, PCCP **6**, 3160 (2004).

25. U. Santhosh, G.B. Schuster, Nucl. Acid Res. **31**, 5692 (2003).

26. D.T. Odom, E.A. Dill, J.K. Barton, Nucl. Acid Res. **29**, 2026 (2001); M.E. Nunez, K.T. Noyes, D.A. Gianolio, L.W. McLaughlin, and J.K. Barton, Biochem. **39**, 6190 (2000).

27. T. Cramer, A. Volta, A. Blumen, T. Koslowski, J. Phys. Chem. B **108**, 16586 (2004).

28. D. A. Case et al. AMBER 8, University of California, San Francisco.

29. H. Cohen, C. Nogues, R. Naaman, and D. Porath, PNAS **102**, 11589 (2004); C. Nogues, S.R. Cohen, S. Daube, N. Apter, and R. Naaman, J. Phys. Chem. B **110**, 8910 (2006).

30. W.F. Pasveer, J. Cottaar, C. Tanase, R. Coehoorn, P.A. Bobbert, P.W.M. Blom, D.M. de Leeuw, and M.A.J. Michels, Phys. Rev. Lett. **94**, 206601 (2005); J. Cottaar and P.A. Bobbert, Phys. Rev. B **74**, 115204 (2006).

31. T. Cramer, S. Krapf, and T. Koslowski, J. Phys. Chem. C, in press.

5 Physics Aspects of Charge Migration Through DNA

Vadim Apalkov[1], Xue-Feng Wang[2], and Tapash Chakraborty[2]

[1] Department of Physics and Astronomy, Georgia State University, Atlanta, Georgia 30303, USA
[2] Department of Physics and Astronomy, University of Manitoba, Winnipeg, Canada R3T 2N2
   tapash@physics.umanitoba.ca

5.1 Introduction

As the scientific activities of the biology, chemistry, and physics communities meet at the nanometer scale, interdisciplinary works are the most efficient avenues to explore many mysteries in science and technology that has been appreciated only in recent decades. One of these mysteries is a deeper understanding and efficient manipulation of charge migration in DNA. Charge migration or the redox process in DNA is directly related to the damage and repair of DNA occurred in the cells of human beings [1]. As we know now, the DNA damage is responsible for many neurological diseases, and plays an important role in aging and many forms of human cancer. On the other hand, molecular electronic devices are believed to be the most promising technology in the near future. DNA has the property of self-assembly and DNA based devices have the advantage of large-scale industrial production. Construction of the artificial DNAs and understanding of charge migration in them then become crucially important [2,3]. Furthermore, DNA sequencing, the process of deciphering the exact order of the 3 billion base pairs that make up the DNA of 24 different chromosomes, has the potential to revolutionize exploration of human biology and medicine. Currently the main concern here is the efficiency of the sequencing process. Study of the transverse charge transport in DNA may result in an efficient tool for rapid DNA sequencing [4,5] as well as fundamental understanding of charge migration across the DNA. In the last decade, charge migration in molecules and DNA has been addressed by many authors which has established a basis for further developments in this field [2,3,6–17]. In this chapter, we investigate some aspects of it from a physical point of view.

5.2 DNA Model for Charge Migration

5.2.1 Molecular Structure

DNA (deoxyribonucleic acid) is the molecule responsible for the storage of genetic information in the cells. The primary structure of DNA as shown in

Fig. 5.1 consists of two chain polymers of the nucleotide units, and is called the DNA duplex. Each nucleotide contains three components: a heterocyclic base, a deoxyribose sugar (pentose), and a phosphate (phosphoric acid). The sugar and phosphate of the successive nucleotide units along each chain are connected in an alternating sequence and form the backbone of the chain. The base of each nucleotide attaches to the sugar on one side and to its counterpart base from the other chain on the other side. The two chains are held together through pairing of their bases by hydrogen bonds. There are four kinds of bases, two purine derivatives, guanine (G) and adenine (A), and two pyrimidine derivatives, cytosine (C) and thymine (T). The pairing occurs only between G and C by three hydrogen bonds or between A and T by two hydrogen bonds, i.e. there are only two kinds of base pairs, (G:C) and (A:T). Along each backbone, the phosphate connects the carbon 5' of one sugar with the carbon 3' of the next sugar (Fig. 5.1a) [18].

The secondary structure of DNA is a double helix with the duplex nucleotide strands twisted around each other. The two strands of the nucleotide polymer in a DNA are oriented in opposite directions, one from carbon 5' to 3' and the other from carbon 3' to 5' [18]. The antiparallel orientation helps to align the hydrogen bond donors and acceptors. Along the double helix, the two strands of the backbone wrap around the stacked base-pair layers. There are three classes of structures, called the B, A, and the Z forms. The form of the B-DNA commonly exists in living beings where the environment is humid. Its helix is about 2 nm in diameter with a vertical distance of about 0.34 nm between layers of the base pairs and about 10 base pairs for each complete turn of the helix (Fig. 5.1b). This is the prototype of DNA for many theoretical works including this work.

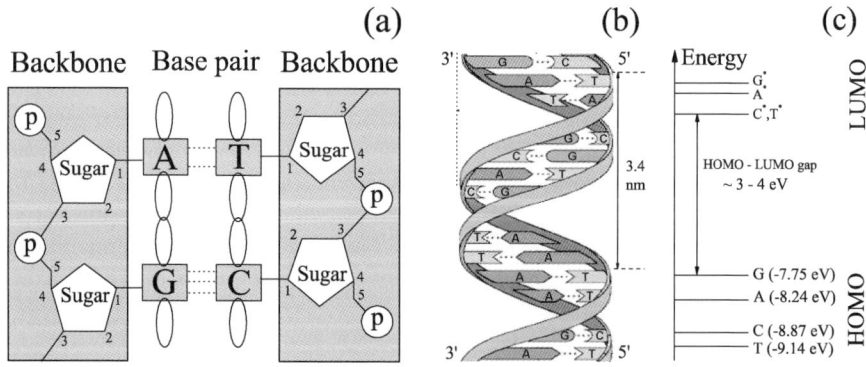

Fig. 5.1. a The primary structure of a DNA duplex with four nucleotides. The elliptical loops show the overlap of the π bonds along the base stacking direction; the *dashed lines* between the (G:C) and (A:T) base pairs are the hydrogen bonds; the numbers around each sugar denote the numbers of the sugar carbons. **b** The DNA double helix. **c** Illustration of the π bond HOMO energies and π* bond LUMO energies of the G, C, A, T bases

From the viewpoint of quantum mechanics, the charge migration in DNA occurs via electronic transitions among states near the chemical potential. The characteristics of charge transfer is then mainly determined by the properties of the highest occupied molecular states (HOMO) and the lowest unoccupied molecular states (LUMO) of the nucleotide in the cases of hole transfer and electron transfer respectively. At zero temperature the chemical potential separates the LUMO from the HOMO. In Fig. 5.1c, the HOMO and LUMO energy levels for the bases G, A, T, and C are also illustrated (approximately). In a DNA with all the bases present, the HOMO-LUMO gap is about $3-4$ eV but the exact value is still an open question [13, 19–21]. The HOMO and LUMO states are mainly composed of molecular states of the π and π^* bonds, i.e. the p_z orbits of the carbon-carbon double bonds, in the purine and pyrimidine bases. The wave function overlap of the π (π^*) bonds between the neighboring bases allows holes (electrons) to jump from one base to another and results in the charge migration along the DNA duplexes.

5.2.2 The Tight-Binding Model

In order to describe the charge migration in DNA quantitatively, both microscopic and macroscopic models have been reported in the literature [7]. In the former case, the system is handled via the first principle; the outer-shell orbits of all atoms in the system and the coupling between them are taken into account explicitly and the transport properties of the system are obtained by the *ab initio* calculations. For the macroscopic models, crucial physical information is extracted from the *ab initio* calculations and are parameterized to simplify the system in the hope of being able to handle bigger systems and also obtain more physical insights than those available from the *ab initio* calculations.

Based on the existing experiments and the *ab initio* results, it has been suggested that the charge migration in DNA is a hole transport via the HOMO states of the bases and the energy gap between the HOMO and LUMO states in each base is about 4 eV [13]. In the zeroth order approximation for a macroscopic model, the system is composed of a series of sites where each site corresponds to a HOMO state of a base. A tight-binding model of the hole transport can then be established with on-site energies for the HOMO energies of the bases and the coupling parameters between the sites for coupling of the HOMO states between the bases.

The on-site energy of each base is then the energy to create a hole in the HOMO state of the base, viz., the ionization energy. The ionization energy is sensitive to the existence of other bases around and also to the environment. This value for the single bases can be calculated by the quantum chemical *ab initio* methods and were confirmed by measurements in the bases' gas phase. The calculated HOMO hole energies for the isolated single bases G, C, T, and A are $E_G = 7.75$, $E_C = 8.87$, $E_T = 9.14$, and $E_A = 8.24$ eV respectively [19, 20, 22, 23]. It is to be noted that these values may depend on the method used [24]. Just as for the on-site energies of the bases, the coupling parameters between different sites (bases) in principle, can also be calculated by the *ab initio* methods. Usually, this effective coupling parameter depends on how the macroscopic model is established. While the intrinsic value comes from the overlap of the π orbit wave functions between the bases, the effective one should be adjusted if other factors (see below) are not explicitly taken into account in the macroscopic tight-binding model. Until now, the calculated coupling parameter from different *ab initio* models are very scattered in a range of $0.01 - 0.4$ eV [19, 25, 26]. Nevertheless, some common qualitative characteristics of the coupling have been extracted from these calculations. Although the purine and pyrimidine bases within each Watson-Crick base pair are strongly coupled by the hydrogen bonds [27], the hydrogen bonds do not participate in the carrier transport because they have a lower energy than the HOMO states. As a result, the interstrand coupling parameter for the HOMO states between them is much weaker than the intrastrand coupling parameter between the neighboring bases along the DNA strands [25, 26]. Because the π bond is highly anisotropic, the coupling parameters are sensitive to the relative position of the two bases in question and a twist of the DNA duplex may modify the coupling parameters significantly [13].

In the above primary picture of the system, we have neglected some other factors which can affect the charge transfer in DNA. In reality, the HOMO states of the bases are not isolated from but are coupled to the other components of the system. First, the HOMO state in a base is coupled to the other outer-shell electronic states with lower or higher energies. The hydrogen bonds, for example, can influence the HOMO states [28]. Second, it is coupled to the inner-shell electronic states and the nuclear states, which introduce the electron-phonon or vibronic coupling [11]. Third, it is coupled to the electronic states in the backbone [29]. Fourth, the charge transfer is affected by the environment, including the static and dynamic screening and random impurities [11]. Fifth, when a finite potential bias is applied over the system, the modification of the potential profile along the DNA duplex and other nonlinear effects will become important [30]. Sixth, spin effects may also be important in some situations [31–33]. Finally, if there is more than one hole present in the DNA, correlation between them may play an important role in the behavior of the charge migration in the system [34].

Different strategies are used to handle these factors. Obviously, a complete and straightforward way would be to take all of these factors into account explicitly. However, this is not very practical because few of these factors are well studied and the corresponding parameters are far from being established. A model with too many uncertain parameters will be confusing and useless. Fortunately, from the mathematical point of view, many characters of these factors can be integrated into the renormalized on-site energy or the coupling parameter. In some cases, even the on-site energy and the coupling parameter can in turn, be represented by each other [35]. Based on this fact, in the numerical calculation of this work, we have used the fixed HOMO energies of the isolated single bases as the on-site energies but leave the effective coupling parameters flexible. However, to which extent and under which condition the parameterization process is valid are still subject of further theoretical and experimental studies.

In the following, we shall discuss only the distance dependence of the charge transfer and neglect the reorganization energetics involved in the charge transfer process, which affects mainly the temperature dependence [11, 16, 36].

5.3 Evaluation of the Electron Transfer Rate in a Chain Model

5.3.1 The One-Dimensional Chain Model

In the simplest single-particle tight-binding model, we assume that the system can be parameterized into a chain model in which an effective on-site energy is used for the HOMO energy of each base pair and an effective coupling parameter between any two nearest neighbor sites [37]. For a homogeneous DNA duplex, such as the Poly(G.C) polymers, this model should be very effective because the charge migration occurs along the purine strand in case of the hole transport.

In the chain model, the Hamiltonian of a N-base-pair DNA as shown in Fig. 5.2a reads

$$\mathcal{H}_{\mathrm{DNA}} = \sum_{n=1}^{N} \varepsilon_n c_n^\dagger c_n - \sum_{n=1}^{N-1} t_{n,n+1}(c_n^\dagger c_{n+1} + c_{n+1}^\dagger c_n)] . \tag{5.1}$$

Here, $c_n^\dagger$ is the creation operator of holes on site n of the DNA chain (for $1 \leq n \leq N$) and $-t_{n,n+1}$ is the coupling parameter between nearest neighbor sites n and $n+1$ [38]. In this tight-binding model, the creation operator $c_n^\dagger$ corresponds to the local electronic state $|n\rangle$ on site n and we assume that all the states are orthogonal to each other, i.e. $\langle m|n\rangle = \delta_{m,n}$ with $\delta_{m,n}$ the Kronecker delta function. In the matrix form, the secular equation then reads $|\hat{\mathcal{H}}_{\mathrm{DNA}} - E\hat{I}| = 0$ with $\hat{I}$ the unit matrix. In a real DNA, the HOMO states, $|\tilde{n}\rangle$, of different bases are not orthogonal to each other and the overlap matrix $\hat{S}$ with elements $S_{m,n} = \langle \tilde{m}|\tilde{n}\rangle$ is not the unit matrix. As a result, the secular equation should be $|\hat{\mathcal{H}}_{\mathrm{DNA}} - E\hat{S}| = 0$. Here an orthogonalization process is assumed to have been done to transform the HOMO molecular states $|\tilde{n}\rangle$ to the on-site states $|n\rangle$ and to construct the site representation of the DNA system [39]. The Hamiltonian is transformed accordingly but the modification to the on-site energy by the orthogonalization process is neglected.

(a)

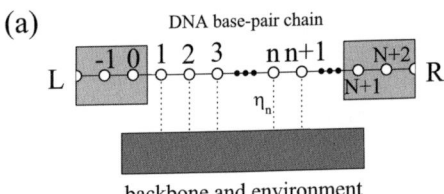

(b)

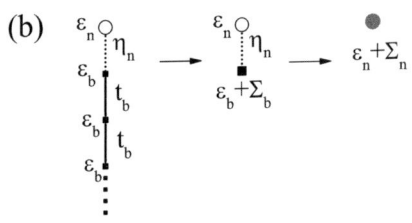

Fig. 5.2. a A schematic illustration of the one-dimensional chain model. The left and right ends of the DNA is connected to the electrode L and R and each site coupled to the dephasing reservoir of the backbone and the environment. **b** The dephasing reservoir for the site n is approximated as a semi-infinite chain and is renormalized as a self energy Σ_n

Any single-particle eigenstate can be expressed by the envelope wave function ψ_n in the site representation as $|\Psi\rangle = \sum_n \psi_n |n\rangle$ and the Schrödinger equation $\mathcal{H}_{\mathrm{DNA}}|\Psi\rangle = E|\Psi\rangle$ reads:

$$
\begin{pmatrix}
\varepsilon_1 - E & -t_{1,2} & \cdots & 0 & 0 & 0 & \cdots & 0 \\
-t_{1,2} & \varepsilon_2 - E & \cdots & 0 & 0 & 0 & \cdots & 0 \\
\cdots & \cdots & \cdots & \cdots & \cdots & \cdots & \cdots & \cdots \\
0 & 0 & \cdots & \varepsilon_{n-1} - E & -t_{n-1,n} & 0 & \cdots & 0 \\
0 & 0 & \cdots & -t_{n-1,n} & \varepsilon_n - E & -t_{n,n+1} & \cdots & 0 \\
0 & 0 & \cdots & 0 & -t_{n,n+1} & \varepsilon_{n+1} - E & \cdots & 0 \\
\cdots & \cdots & \cdots & \cdots & \cdots & \cdots & \cdots & \cdots \\
0 & 0 & \cdots & 0 & 0 & 0 & \cdots & \varepsilon_N - E
\end{pmatrix}
\begin{pmatrix}
\psi_1 \\
\psi_2 \\
\cdots \\
\psi_{n-1} \\
\psi_n \\
\psi_{n+1} \\
\cdots \\
\psi_N
\end{pmatrix}
= 0 .
$$

This is an equation group of N equations having the recursive form

$$-t_{n-1,n}\psi_{n-1} + (\varepsilon_n - E)\psi_n - t_{n,n+1}\psi_{n+1} = 0 , \tag{5.2}$$

and in principle, can be solved exactly for a closed system with a finite N. For a long homogeneous chain with $\varepsilon_n = \varepsilon$ and $t_{n,n+1} = t$, we have a periodic system and a Bloch type of wave function $\psi_n = \psi_0 e^{ikna}$ exists for the system. For the periodic boundary condition, i.e. $|N + n\rangle \equiv |n\rangle$ and $\psi_{N+n} \equiv \psi_n$, the energy $E = \varepsilon_n - t_{n-1,n}\psi_{n-1}/\psi_n - t_{n,n+1}\psi_{n+1}/\psi_n$ in (5.2) becomes

$$E = \varepsilon - 2t \cos(ka) \tag{5.3}$$

with $ka = \ell 2\pi/N$ for integer ℓ. This is an energy band centered at the on-site energy ε with a band width of four times the coupling parameter $4t$. The

Brillouin zone is $-\pi/a \leq k \leq \pi/a$ and the corresponding density of states (DOS) is $Na/(2\pi)(dk/dE) = N/[4\pi t \sin(ka)]$.

5.3.2 The Transfer Matrix Method

In an open system with a source of charge at one end of the DNA molecule and a drain at the other end (Fig. 5.2a), the system becomes infinitely large and the boundary condition for a closed system is no longer valid. The direct diagonalization of the Hamiltonian described above for the closed systems can not be used to find the transport properties. The transfer matrix method [40, 41] and the non-equilibrium Green's function theory [42] have been developed to solve the problem. The transfer matrix method is straightforward but it is not always the convenient choice for the complicated cases that involve taking into account many physical factors. In contrast, the Green's function theory appears more sophisticated for the simple cases but has the technical advantage when many physical factors are to be taken into account. Using simple examples like the homogeneous DNA chain, we can show that they are equivalent [23,42]. In what follows, we have used the transfer matrix method.

One example of the open systems is a DNA duplex that is connected to a circuit via metal electrodes. Here the longitudinal charge migration from the source to the drain through the DNA duplex occurs when a voltage drop is applied between the electrodes. To facilitate the transport calculation in this system, the electrodes can be modelled as semi-infinite periodic one-dimensional tight-binding chain with uniform parameters of the on-site energy ε_e, the band width of $4t_e$, and the Fermi energy ε_e^F measured from ε_e. Furthermore, the contact properties between the DNA duplex and the left (right) electrode are described by the contact parameter t_{de}^L (t_{de}^R). The total Hamiltonian then reads

$$\mathcal{H} = \sum_{n=-\infty}^{\infty} \varepsilon_n c_n^\dagger c_n - \sum_{n=-\infty}^{\infty} t_{n,n+1}(c_n^\dagger c_{n+1} + c_{n+1}^\dagger c_n)] . \tag{5.4}$$

Here the sites for $n \leq 0$ represent the left electrode and for $n \geq N+1$ the right electrode; $t_{0,1} = t_{de}^L$ and $t_{N,N+1} = t_{de}^R$ are the coupling parameters between the electrodes and the DNA chain.

The transport property of this open system is the electronic response of the drain (right) electrode to an injection of charge from the source (left) electrode. If the phase coherence length in the DNA is longer than the DNA length, we can assume a plane wave current injection and calculate the output plane wave function employing the recursion relation of the wave function in DNA (5.2). Rewriting (5.2) with the identity $\psi_n \equiv \psi_n$ in a recursion matrix form

$$\begin{pmatrix} \psi_{n+1} \\ \psi_n \end{pmatrix} = \begin{pmatrix} \dfrac{\varepsilon_n - E}{t_{n,n+1}} & -\dfrac{t_{n-1,n}}{t_{n,n+1}} \\ 1 & 0 \end{pmatrix} \begin{pmatrix} \psi_n \\ \psi_{n-1} \end{pmatrix} , \tag{5.5}$$

we obtain the above 2×2 transfer matrix $\hat{M}_n$ to derive the wave functions on sites $n + 1$ and n from those on sites n and $n - 1$. Note that (5.5) is a general form of the Schrödinger equation for any systems and the eigenstates of a closed system can be derived from it with proper boundary conditions. From (5.5), we can derive the wave function of the whole system once we know the wave functions of any two successive sites. For example, if we know the wave functions of sites 0 and -1 in the left electrode we can derive the wave functions of sites $N + 1$ and N in the right electrode as

$$\begin{pmatrix} \psi_{N+1} \\ \psi_N \end{pmatrix} = \hat{M}_T \begin{pmatrix} \psi_0 \\ \psi_{-1} \end{pmatrix} \tag{5.6}$$

with $\hat{M}_T = \prod_{n=0}^{N+1} \hat{M}_n$ and evaluate the transmission of an electronic wave package from the left to the right electrode. Any propagating wave package can be expanded into a series of plane waves by the Fourier transform, and we can evaluate the transmission of the plane waves to get the overall transport properties. We consider that the hole wave functions in the source electrode has the general propagating form $\psi_n^L = A\, e^{ik_L na} + B e^{-ik_L na}$ $(n \leq 0)$ with A being the incident wave amplitude and B the reflected wave amplitude, and in the drain electrode $\psi_n^R = C e^{ik_R na} + D e^{-ik_R na}$ $(n \geq N + 1)$ with C the transmitted wave amplitude. As the probability of carriers is proportional to the density of states, we choose the normalized incident amplitude $A = 1/\sqrt{|\sin(k_L a)|}$. Here D represents the current injection from the right electrode and does not contribute to the current in this case. Nevertheless, we keep it here for the sake of generality of the formalism.

Substituting the wave function ψ_n^L and ψ_n^R into (5.6), we get

$$\hat{S}_R \begin{pmatrix} D \\ C \end{pmatrix} = \hat{M}_T \hat{S}_L \begin{pmatrix} B \\ A \end{pmatrix} \tag{5.7}$$

with

$$\hat{S}_L = \begin{pmatrix} 1 & 1 \\ e^{ik_L a} & e^{-ik_L a} \end{pmatrix} \tag{5.8}$$

and

$$\hat{S}_R = \begin{pmatrix} e^{-ik_R (N+2)a} & e^{ik_R (N+2)a} \\ e^{-ik_R (N+1)a} & e^{ik_R (N+1)a} \end{pmatrix} . \tag{5.9}$$

We then have the transfer matrix for the amplitude $\hat{M}_A$,

$$\begin{pmatrix} B \\ A \end{pmatrix} = \hat{M}_A \begin{pmatrix} 0 \\ C \end{pmatrix} ; \hat{M}_A = \hat{S}_L^{-1} \hat{M}_T^{-1} \hat{S}_R = \begin{pmatrix} M_{11} & M_{12} \\ M_{21} & M_{22} \end{pmatrix} . \tag{5.10}$$

Using $C = M_{22}^{-1} A$ and the group velocity $v = dE/dk = 2ta \sin ka$ for the tight-binding band, we arrive at the following expression for the transmission

$$T(E) = \frac{|C|^2}{|A|^2} \frac{v_R}{v_L} = \frac{|C|^2 \sin(k_R a)}{|A|^2 \sin(k_L a)} . \tag{5.11}$$

The conductance at an ideal transmission, i.e. 100%, through a quantum one-dimensional channel is the conductance quanta $e^2/h = (25.8\,\mathrm{k\Omega})^{-1}$ and in general, the current through the system is evaluated by the Landauer-Büttiker formula

$$I = \frac{2e}{h} \int_{-\infty}^{\infty} dE\, T(E)[f(E - \mu_{\mathrm{L}}) - f(E - \mu_{\mathrm{R}})] . \tag{5.12}$$

Here, $f(E - \mu_{\mathrm{X}}) = 1/\exp[(E - \mu_{\mathrm{X}})/k_{\mathrm{B}}T_{\mathrm{e}}]$ is the Fermi function with the chemical potential μ_{X} of the electrode X for $X = \mathrm{L}$ or R and T_{e} is the environment temperature.

5.4 Charge Migration Through DNA

5.4.1 Charge Transfer Measurement

Both the chemical and physical techniques have been successfully used to measure the charge transfer rate in DNA. Usually, the conductance is measured indirectly in the chemical techniques such as the fluorescence quenching and the poly(G) trap methods. In the former case, a fluorescent molecule complex is inserted into the DNA and its fluorescence spectrum is measured after it is excited. The time-dependent fluorescent quenching is used to determine the transfer rate of the excited electrons in the fluorescent molecule [43]. In the latter case, a charge is injected into a single G base optically or electrically and the trapping rate at a double GG or a triple GGG trap is measured by water cleavage of the DNA strand [44]. In recent years, the conductance of DNA was also measured directly by physically connecting it to a circuit [3, 45, 46]. However, the conductance of the DNA extracted from different measurements appears to be very different [3, 45] and a consistent explanation of these results requires a systematic study and understanding of the mechanism of the charge transfer in different systems used in the experiments [47]. A reliable conclusion from any specific measurement depends on the understanding of the corresponding mechanism for the system in question, including the understanding of the charge transfer process in the DNA itself, the boundary condition or the contact effect, and participation of the environment.

5.4.2 Charge Transfer Via the DNA Molecule

In a long DNA complex, several mechanisms have been known to contribute to the charge transfer. Generally, we are considering a hole initially introduced to the DNA by the oxidation of a G base [43, 44]. Because the HOMO state of the G base has a lower ionization energy than the other bases, the G bases in DNA can work as charge stops and charge may hop back and forth from one G base to another during its long-distance migration [48–50]. In a poly(G:C) DNA, an energy band forms in this periodic system and a charge can transport quickly through this band from one end to the other.

In DNA with a mixed (A:T) and (G:C) base pairs [51, 52], the (A:T) base pairs work as barriers for charges with energy of the G-base being the HOMO energy. A charge can tunnel from one G base to the next G base through the (A:T) barrier bridge between them. If the temperature is high enough, the charge can also gain thermal energy and oxidize the A bases so the charge can migrate through the DNA by thermal hopping. The role of the (A:T) bases in charge migration is a major focus of recent activities in the field.

Conventionally, a DNA is treated as a polymer chain of the base pairs when the charge transport is considered. Since the HOMO state in a (G:C) base pair is located at the G base and in an (A:T) base pair at the base A, charge is believed to migrate along a channel composed of the G and A bases. For the tunneling mechanism [53–57] the (A:T) base pairs located between the (G:C) base pairs are the tunneling potential barriers and the tunneling current should decay exponentially with the number of the (A:T) base pairs that are in the middle. For the thermal hopping mechanism [58–60], on the other hand, once an A base is oxidized with the help of the thermal energy, the other A bases can be easily oxidized. As a result, the hole can transport freely through the poly(A) channel and reach to the other end of the DNA without much resistance. In this case, the total resistance comes mainly from the first hop from the G base to an A base and the transfer rate is almost independent of the number of the (A:T) base pairs in the bridge.

In the above DNA chain model, each base pair is a unit. This picture is natural when discussing the entire electronic energy of the system stored in both inner and outer shell electronic states. The two paired bases in each base pair is strongly coupled with each other by the hydrogen bonds of the sigma orbits. In contrast, two stacked bases along the DNA chain are more weakly coupled by the overlap of the π orbits similar to the coupling between stacked graphene layers in a graphite. However, as far as the longitudinal charge transfer is concerned, the contribution of the σ orbits to the HOMO state and the participation of electrons in the σ orbits in the charge transfer process are negligible. This is due to the fact that the σ orbits have a much lower energy than that of the π orbits in the DNA bases. Accordingly, the charge transfer between the bases in a base pair is determined by the overlap of the π orbits between them. The *ab initio* calculations have shown that the overlap of the HOMO states between the two neighboring bases in the same strand is much stronger than that in different strands [25].

Based on the above considerations, the geometry of DNA (in a form of ladder network of the bases) may play an important role in the charge transfer and in some cases, it is more accurate to view a conducting DNA as two paired base strands rather than a chain of stacked base pairs. In that respect a DNA duplex is modelled as a ladder network [61–64] instead of a chain of HOMO sites. When the geometry of the DNA molecule is taken into account, a charge can tunnel or thermally hop through different one-dimensional channels in the two-dimensional network.

5.4.3 Contact Effect

The contact property between the electrodes and the DNA duplex depends on the material of the electrode, the geometry of the contact, and the environment, and is by itself an active field of research in both physics and chemistry [65, 66]. How a DNA duplex makes contacts to the charge source or drain determines the efficiency of the charge injection and affects the measured results in the experiments. Unfortunately, in many cases, the details of the contact especially between the metal and DNA in direct transport measurements are not very clear. In the tight-binding model, an effective contact parameter is used to phenomenologically describe the contact. When a fixed voltage is applied between the source and the drain [30], the contact may significantly modify the potential profile across the DNA and the on-site energies vary accordingly. In the quantum tunneling-transport process, the charge injection efficiency is not a linear function of the contact parameter because of the phase interference. It has been shown that when a periodic DNA base chain with a uniform nearest-neighboring coupling parameter t is connected to a metal electrode of band width $4t_e$, an optimal injection is achieved at $t_{de} = \sqrt{t_d \times t_e}$ in the linear transport regime [22].

5.4.4 Dephasing Effect

It has been widely accepted that an electron in the HOMO state of a base in DNA can not only interact with the electrons in other bases but also with the background including the backbone and the environment, as illustrated by the dotted lines in Fig. 5.2a. The electron or the hole can jump out of the base to the background through the overlap of the outer-shell atomic orbits and through the electron-phonon interaction (coupling with the inner-shell atomic orbits and the nuclear states). As a simple approximation, the effect of the background on site n can be integrated into a self energy $\Sigma_n = \Sigma_n^R + i\Sigma_n^I$, in which the real part offers the energy correction to the on-site energy and the imaginary part to the dephasing effect. Just as for the electrodes, we can model the background as a semi-infinite one-dimensional tight-binding chain as depicted schematically in Fig. 5.2b. The effective Hamiltonian for the coupling between the site n and the background reservoir can be simplified as a 2×2 matrix for a two-level system [42]

$$\hat{\mathcal{H}}_n^{dph} = \begin{pmatrix} \varepsilon_n & -\eta_n \\ -\eta_n & \varepsilon_b + \Sigma_b \end{pmatrix} , \tag{5.13}$$

where η_n is the coupling parameter between the site n and the background, ε_b is the on-site energy of the first site in the semi-infinite chain of the background, and the self energy Σ_b represents the effect from the other sites of the dephasing chain. We have to keep in mind that here again we are concerned with the response of the system to a wave package of any energy E rather than the eigenstates of a closed system and the Schrödinger equation $\mathcal{H}_n^{\mathrm{dph}}|\varphi\rangle = E|\varphi\rangle$ requires $(\varepsilon_n - E)(\varepsilon_{\mathrm{b}} - \Sigma_{\mathrm{b}} - E) = \eta_n^2$. The self energy on the site n in the DNA chain due to the dephasing reservoir, $\Sigma_n = E - \varepsilon_n$, for a carrier of energy E then reads

$$\Sigma_n = \frac{\eta_n^2}{(E - \varepsilon_{\mathrm{b}} - \Sigma_{\mathrm{b}})} \; . \tag{5.14}$$

Similarly, Σ_{b} can be obtained self consistently using this expression by replacing Σ_n with Σ_{b} and η_n with t_{b}, the nearest neighbor coupling parameter for the dephasing chain. We have $\Sigma_{\mathrm{b}} = (E - \varepsilon_{\mathrm{b}})/2 + \mathrm{i}[t_{\mathrm{b}}^2 - (E - \varepsilon_{\mathrm{b}})^2/4]^{1/2}$ [67,68].

5.5 Understanding the Weak Distance Dependence

The possibility of charge transfer in DNA has been proposed soon after the atomistic DNA structure was established [8,9]. However, quantitatively the distance dependence of charge transfer in DNA was measured systematically only several decades later [69]. It has been well established that the charge can transfer from the donor to the acceptor through the intermediate bridges (molecular clusters) of higher energy via the electronic superexchange interaction along a molecular chain. In this picture, the bridges work as a tunneling barrier and the donor and acceptor are treated as charge traps. With the help of the electron-phonon (vibronic) interaction, the tunneling can occur when the donor and the acceptor states are not degenerate. A perturbation theory based on the tight-binding model has been developed to describe this charge transfer mechanism as early as the first proposal of charge transfer in DNA [53]. Exponential decay of the transfer rate with the distance was predicted. This strong exponential distance dependence in the molecular chain has been confirmed by many experiments and also by different theoretical formalisms in the years that followed. However, many measurements have also shown a weak distance dependence of charge transfer in the DNA [43,44,69–72]. This means that charges may transfer along a very long distance and indicate the possible importance of the thermally-induced hopping between the G and G's bases [58–60]. In order to clearly identify the regimes for the validity of the different mechanisms, Giese et al., carried out a systematic measurement of the charge transfer between the (G:C) and (G:C)$_3$ charge traps over (T:A)$_{\mathcal{M}}$ bridges of higher energy [44]. That experiment demonstrated an exponential decay rate versus the distance for $\mathcal{M} \leq 3$ and an almost flat distance dependence for $\mathcal{M} > 3$ with a crossover around $\mathcal{M}_{\mathrm{c}} = 3$ as shown in Fig. 5.3.

Naturally, this flat distance dependence was connected qualitatively to the thermally-induced hopping mechanism. Based on this proposal, many

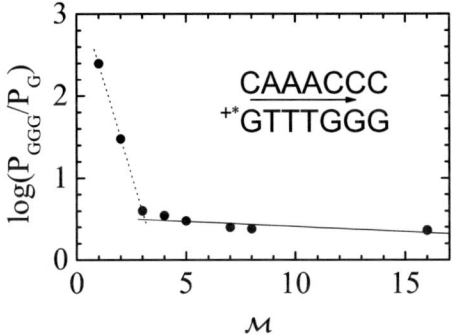

Fig. 5.3. Logarithm of the efficiency of the charge transfer (*filled circles*), measured by the ratios of the irradiation products P_{GGG}/P_G in [44], from the (G:C) to triple (G:C) base pairs, plotted as a function of the number $\mathcal{M}$ of (A:T) base pairs in a $(G:C)(A:T)_{\mathcal{M}}(G:C)_3$ DNA duplex. The linear fit to the results for $1 \leq \mathcal{M} \leq 3$ is $\log(P_{GGG}/P_G) = 3.3\text{--}0.9\mathcal{M}$ as shown by the *dotted line* and for $4 \leq \mathcal{M} \leq 16$ is $\log(P_{GGG}/P_G) = 0.5\text{--}0.01\mathcal{M}$ as shown by the *solid line*

authors have contributed to a quantitative explanation of the experimental result. Berlin, Burin, and Ratner [37] used a tight-binding (Hückel) model for a one-dimensional chain with each base pair as a site and derived the crossover number when the tunneling transfer rate is equal to the transfer rate by the activation process to the (T:A) tight-binding band. For a barrier height of 0.46 eV (between the site energies of (G:C) and (T:A) or (A:T) base pairs) and the coupling parameter in the range from 0.1 eV to 0.4 eV, the crossover number is found to lie between 3 and 4, in agreement with the experiment. Bixon and Jortner [73, 74] emphasized the difference between the intra- and interstrand couplings for different base sequences and proposed that the charge transfer occurs along a dominant path. They applied the kinetic-quantum mechanical model for the thermally-induced hopping process and fitted the experimental result. However, they noticed some inconsistency in the theory and concluded that the theory does not explain the result. Renger and Marcus [75] proposed that the flat distance dependence can be explained by integrating the variable-range hopping concept into the kinetic model. In addition, describing the system with the Su-Schrieffer-Heeger model and the Hubbard Hamiltonian, Cramer, Krapf, and Koslowski [76] obtained the energy potential surface in the atomistic level for evaluation of the transfer rate based on the Marcus theory. They also explained the crossover from the exponential to the flat distance dependence in the picture of tunneling and thermal hopping transition. Basko and Conwell [77] emphasized the importance of phonons in the process and proposed that the formation of polarons [78–82] in the system is key for the flat distance dependence.

It is known [39] that the quantum interference can play an important role in a system with multi-tunneling channels and a resistance ladder network has very different properties than a series of resistance. In this work, we explore the transport properties of a DNA duplex that is treated as a tunneling ladder network [83].

5.6 Electron Tunneling Through Multi-Path Barriers

5.6.1 The Double-Stranded Model

In order to take into account the duplex geometry of the primary structure, we consider a DNA duplex chain of N Watson-Crick base pairs connected to four semi-infinite one-dimensional electrodes with one for each end of the first and the second strand as illustrated in Fig. 5.4. The four electrodes are assumed independent of each other since in many cases the charges are injected into one base and trapped in another bases at the ends. The tight-binding Hamiltonian of the system is

$$\mathcal{H} = \sum_{n=-\infty}^{\infty} [\varepsilon_n c_n^\dagger c_n - t_{n,n+1}(c_n^\dagger c_{n+1} + c_{n+1}^\dagger c_n)]$$
$$+ \sum_{n=-\infty}^{\infty} [u_n d_n^\dagger d_n - h_{n,n+1}(d_n^\dagger d_{n+1} + d_{n+1}^\dagger d_n)]$$
$$- \sum_{n=1}^{N} \lambda_n (c_n^\dagger d_n + d_n^\dagger c_n) .$$

Here $c_n^\dagger$ $(d_n^\dagger)$ is the creation operator of holes in the first (second) strand on site n of the DNA chain (for $1 \leq n \leq N$), the left electrodes ($n \leq 0$), and the right electrodes ($n \geq N + 1$). The coupling parameter of the first (second) strand $t_{n,n+1}$ $(h_{n,n+1})$ is equal to the intra-strand coupling parameter t_d

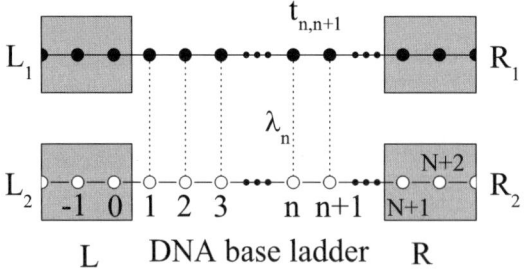

Fig. 5.4. Schematic illustration of the two-stranded model. The first strand (*filled circle*) has a DNA base sequence G(T)$_M$GGG and the second strand (*empty circle*) a sequence C(A)$_M$CCC. The four *gray areas* indicate the four virtual electrodes connected to the DNA chain. Current is injected into the first strand through the left electrode L_1 and measured at the right electrode R_1

between neighboring sites n and $n+1$ of the DNA for $1 \leq n \leq N-1$, one-fourth of the conduction band-width in the electrodes t_e for $n \leq -1$ and $n \geq N+1$, and the coupling strength t_{de} between the electrodes and the DNA strands for $n = 0$ and $n = N$. The inter-strand coupling between the sites in the same Watson-Crick base pair is described by λ_n.

In the site representation, the Schrödinger equation is an equation group with two inequivalent form of equations

$$t_{n-1,n}\psi_{n-1} + (E - \varepsilon_n)\psi_n + \lambda_n\phi_n + t_{n,n+1}\psi_{n+1} = 0$$
$$h_{n-1,n}\phi_{n-1} + (E - u_n)\phi_n + \lambda_n\psi_n + h_{n,n+1}\phi_{n+1} = 0 ,$$

where ψ_n (ϕ_n) is the wave function of the first (second) strand on site n. The wave functions of the sites $n+1$ and n are related to those of the sites n and $n-1$ by a 4×4 transfer matrix $\hat{M}_n$,

$$\begin{pmatrix} \psi_{n+1} \\ \phi_{n+1} \\ \psi_n \\ \phi_n \end{pmatrix} = \hat{M}_n \begin{pmatrix} \psi_n \\ \phi_n \\ \psi_{n-1} \\ \phi_{n-1} \end{pmatrix} , \tag{5.15}$$

with

$$\hat{M}_n = \begin{bmatrix} \dfrac{(\varepsilon_n - E)}{t_{n,n+1}} & \dfrac{-\lambda_n}{t_{n,n+1}} & -\dfrac{t_{n-1,n}}{t_{n,n+1}} & 0 \\ \dfrac{-\lambda_n}{h_{n,n+1}} & \dfrac{(\varepsilon_n - E)}{h_{n,n+1}} & 0 & -\dfrac{h_{n-1,n}}{h_{n,n+1}} \\ 1 & 0 & 0 & 0 \\ 0 & 1 & 0 & 0 \end{bmatrix} .$$

Assuming the plane wave injection and transmission in the electrodes and following the same process as described in (5.6)–(5.12) for the one-dimensional chain model, we can evaluate the charge transfer rate to any electrode from one injection electrode in the double-stranded model.

5.6.2 Charge Transfer Through a (G:C)(T:A)$_\mathcal{M}$(G:C)$_3$ DNA

We now apply the double-stranded model to describe the intra-molecular hole transfer along the DNA duplex chain (G:C)(T:A)$_\mathcal{M}$(G:C)$_3$ measured by Giese et al. and described in Sect. 5.5. To minimize the contact effect introduced by the virtual electrodes we introduced to facilitate the calculation, we assume a strong coupling (of coupling parameter $t_{0,1} = t_{N,N+1} = h_{0,1} = h_{N,N+1} = t_{de} \geq 1.5\,\mathrm{eV}$) between the electrodes and the sites at the ends of the DNA strands, and choose a band width ($4t_e$) in the electrodes such that the optimal injection condition $t_d \times t_e = t_{de}^2$ [22] is satisfied. The result is found to be independent of the choice of the value of t_{de} once it is much larger than the coupling parameter between the sites inside the DNA. In this case, the added electrodes do not become a bottleneck of the system for the charge transfer and the calculated result predominantly reflects the properties of the DNA duplex.

To evaluate the transfer rate or current of a charge (hole) from the donor at the left-end site to the acceptor at the right-end site of the first strand, we need to know the chemical potential at each end. In the experiment of [44], a hole was injected to the left-end site. This means that the left chemical potential is approximately the on-site energy of this site while the right one is less. During the charge transfer process, the hole may retain the same energy if no inelastic scattering occurs or loose energy via the electron-phonon scattering or other inelastic collisions [17]. Real electron-phonon scattering at the donor and acceptor contributes to the reorganization energy and affects the temperature dependence of charge transfer [11, 36, 84] while the virtual electron-phonon interaction may affect the electronic coupling between different sites [85]. In the case of strong electron-phonon coupling, the charge may be dressed by the phonon cloud and transforms into a quasiparticle, the polaron [78, 79]. Here we assume that the virtual phonon effect and the polaron effect can be simplified as an adjustable to the coupling parameter between sites. Since we are concerned with distance dependence of the transfer, we do not deal with the inelastic scattering mechanisms at the donor and the acceptor sites explicitly but analyze two limiting situations, between which the real charge transfer process would occur. Since our results for the distance dependence of the transfer rate from the two limits converge (see below), we conclude that our results are reliable.

In the first limit, we assume that there is no inelastic scattering involved and the hole energy is conserved during the transfer process. The transfer rate is proportional to the conductance of the system at equilibrium. For a small electric potential difference $k_B T_e / e$, the current is

$$I = \frac{2e}{h} \int_{-\infty}^{\infty} dE \, T(E)[1 - f(E - \mu)] f(E - \mu) \,, \qquad (5.16)$$

with μ equal to the on-site energy of site 1 in the first strand and $T_e = 300 \, \text{K}$.

In the second limit, we assume that the hole can lose energy freely during the charge transfer process before or after the tunneling, and the transfer rate is proportional to the total current via all channels of energies below the hole's initial energy. This corresponds to an infinitely low chemical potential at the right electrode and the current is

$$I = \frac{2e}{h} \int_{-\infty}^{\infty} dE T(E) f(E - \mu) \,. \qquad (5.17)$$

We now calculate the distance dependence of the transfer rate using (5.16) and (5.17) in a DNA duplex, where the first strand has the base sequence $G(T)_{\mathcal{M}}GGG$ as in the experiment of [44]. For the sake of simplicity and to focus on the geometry effect, a uniform intra-strand hopping parameter

$t_{n,n+1} = h_{n,n+1} = t_d$ $(1 \leq n \leq N - 1)$ and a uniform inter-strand hopping parameter $\lambda_n = \lambda_d$ $(1 \leq n \leq N)$ between any two neighboring bases in the DNA are used.

First we switch off the inter-strand coupling and calculate the dependence of the current I on $\mathcal{M}$ as shown in Fig. 5.5a, for different values of the intra-strand coupling parameter t_d. We find an exponential dependence of the current

$$I = I^{\mathcal{M}} \propto e^{-\beta \mathcal{M} a} \tag{5.18}$$

when t_d is much smaller than the bridge barrier height $E_T - E_G$. We then extract the values of β for different t_d and plot in Fig. 5.5c as β versus $\ln(t_d)$ calculated via (5.17). The curves are almost linear, very similar to the results of (5.16), and converge to the approximate formula

$$\beta = \frac{2}{a} \left| \ln \frac{t_d}{E_T - E_G} \right|. \tag{5.19}$$

This is the well-known one-dimensional superexchange result in the literature and has been derived in many different ways [37, 39, 53, 73].

In the next step, we fix t_d and switch on the inter-strand coupling by varying λ_d. The result is displayed in Fig. 5.5b where we choose $t_d = 0.5$ eV and plot I versus $\mathcal{M}$ for a series of λ_d. Note that the charge transfer occurs via π-electrons and generally $\lambda_d < t_d$ [25]. For finite λ_d, the current drops

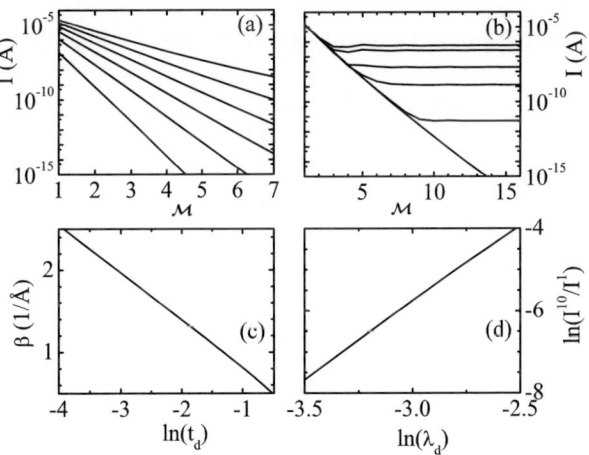

Fig. 5.5. a Current I versus $\mathcal{M}$ for $t_d = 0.1, 0.2, 0.3, 0.4, 0.5, 0.6$ eV (from *lower* to *upper curves*) for zero inter-strand coupling. The displayed results are from (5.17) and identical results are obtained from (5.16) in all the *panels*. **b** Same as in **a** at fixed $t_d = 0.5$ eV but for $\lambda_d = 0, 5, 20, 40, 80, 100$ meV corresponding to curves counted from the *bottom*. **c** The β value calculated from the slope of the lines in **a** versus $\ln t_d$. **d** $\ln(I^{10}/I^1)$, where $I^{\mathcal{M}}$ is the current for a chain with $\mathcal{M}$ (A:T) base pairs, versus $\ln \lambda_d$. The unit of t_d and λ_d is eV and $t_{de} = 1.5$ eV [83]

exponentially with increasing $\mathcal{M}$ for small $\mathcal{M}$ and then becomes almost flat with oscillations around a limiting current I^∞ for large $\mathcal{M}$. The crossover number $\mathcal{M}_c$ depends on the strength of the inter-strand coupling parameter. The weaker the inter-strand coupling is, the bigger the $\mathcal{M}_c$. The dependence of I^∞ on λ_d is approximately illustrated in Fig. 5.5d, where the normalized current I^{10}/I^1 of the DNA chain at $\mathcal{M} = 10$ is plotted versus $\ln(\lambda_d)$. Again, two almost identical straight lines are found corresponding to the two limiting situations based on (5.16) and (5.17) and can be approximately expressed as

$$\ln(I^{10}/I^1) = 5.7 + 3.9\ln(\lambda_d) . \tag{5.20}$$

From (5.18)–(5.20), we estimate the ratio of inter- and intra-strand coupling from the crossover number $\mathcal{M}_c$. Since the environment can change λ_d/t_d, we predict that the transition number may vary and be different from 3 when the experimental environment changes.

Calculating the current I before and after adding a (T:A) base pair at site n with zero or nonzero inter-strand coupling λ_n, we find that the distance-dependence crossover has a topological origin, e.g., from the one-dimensional chain charge transport to a partly two-dimensional ladder network. When a new (T:A) base pair is inserted into the DNA chain, a new superexchange channel is opened through its inter-strand coupling and the corresponding contribution compensates the loss of charge transfer rate that would incur because of an extra barrier to the existing channels [83].

In Fig. 5.6, we fit the $\mathcal{M}$ dependence of the charge transfer rate observed in [44] using intra- and inter-strand coupling parameters $t_d = 0.52\,\text{eV}$ and $\lambda_d = 0.07\,\text{eV}$ respectively. (5.17) is employed in the calculation. The agree-

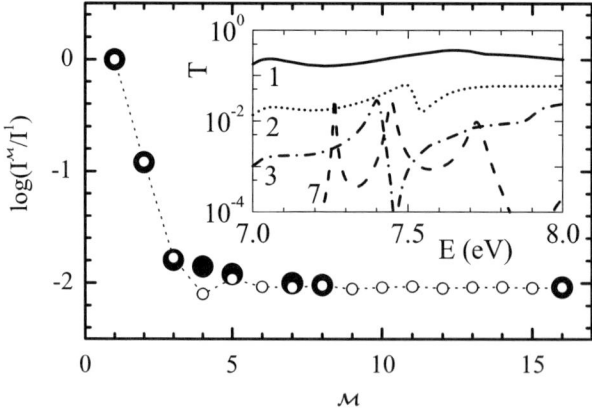

Fig. 5.6. Normalized transfer rate measured in [44] (*filled circle*) and theoretical fit using this model (*open circle*), $\log(I^{\mathcal{M}}/I^1)$, are plotted as functions of $\mathcal{M}$ (T:A) base pairs between the (G:C) and the triple (G:C) base pairs. *Inset:* The corresponding transmission T versus energy E for $\mathcal{M} = 1$ (*solid line*), 2 (*dotted line*), 3 (*dot-dashed line*), and 7 (*dashed line*). Here $t_{de} = 3\,\text{eV}$ is used [83]

ment between the experimental and theoretical results are very good except for a small oscillation in the theoretical result near $\mathcal{M}_c$. This oscillation results in the deviation of the empty circle from the filled circle at $\mathcal{M} = 4$. When (5.16) is used, similar result is obtained but with a stronger oscillation. The oscillations reflect the fact that we have treated the system as a coherent system by neglecting the dephasing effect from the environment and the relaxation process from phonons.

To get a clear picture of the process, we plot as inset in Fig. 5.6, the transmission T as a function of the hole energy E for systems with $\mathcal{M} = 1, 2, 3$, and 7 in an energy range near and below the G base HOMO energy E_G. In the T spectrum, each peak represents a transport channel and there are more fine structures or peaks when more base pairs are added to the system. When $\mathcal{M}$ varies from 1 to 3, the one-dimensional chain transport dominates and only one principal transmission peak is important. The principal peak shifts when $\mathcal{M}$ varies due to the shift of energy of the channel; its height drops rapidly leading to an exponential decrease of charge transfer rate. If we add more (T:A) base pairs to the DNA duplex, the principal T peak drops to a level comparable to that of other peaks and results in a crossover from the one-dimensional chain transport to a two-dimensional network transport. In the absence of any inelastic scattering the charge transfer rate versus $\mathcal{M}$ oscillates as a result of the energy shift of the transport channels and the energy conservation of the charge. With the assistance of the phonon, however, the charge can use channels of energy different from its initial energy and phonons may play an important role in assisting the charge transfer.

In the above analysis, we also neglect the dephasing effect and the interstrand coupling between two neighbor base pairs. The dephasing effect exists in a real system and can help damp the oscillation of the current observed in Figs. 5.5 and 5.6. It has been shown by the *ab initio* calculation that the interstrand coupling between two nearest neighbor base pairs is also important. This coupling can be easily integrated into the transfer matrix by adding non-zero elements. This coupling has similar effects on charge tunneling as λ_n and similar curves as in Fig. 5.5b are obtained if replace λ_n by it.

5.7 Transverse Tunneling Current

In this section, we provide a detailed review of charge transport in singlestranded DNA in the direction *perpendicular* to the backbone axis [6, 86]. As pointed out in [4], this approach might be useful in providing a low-cost, but rapid DNA sequencing. We also discuss the conditions for formation of bipolarons in DNA, and possible experimental manifestation of bipolarons in transverse tunneling experiments.

5.7.1 Rate Equation

The longitudinal transport along the DNA molecule is determined by the properties of the whole DNA, which consists of many basic elements, viz.

the base pairs. In this sense longitudinal charge transfer is the tool to probe the properties of the whole molecule. In what follows, we address another problem, namely, how to study the local properties of the DNA molecule. We discuss here the method to extract the local energy characteristics of the molecule. The only way to measure the local characteristics of an extended DNA molecule is to study the transport in the direction perpendicular to the backbone axis, i.e. the transverse tunneling current [6, 86]. Since the most important local parameters of DNA are the characteristics of electron and hole traps, below we consider only transverse tunneling through electron or hole DNA traps. In this case, if the electrodes have a relatively small width, the tunneling occurs through a single DNA base pair. The linear (unstructured) tunneling conductance then depends on the particular type of base pair [6]. This fact can be used to discover the sequence of DNA by scanning it with conducting probes. We demonstrate below that not only the linear conductance of the tunneling current but also the *structure* of the I–V curves can provide important information on the properties of the DNA, in particular, about the trapping spots. This is because the tunneling current through the system is determined by its density of states (DOS). For a finite system, the DOS has peaks corresponding to discrete energy levels. These peaks will result in a staircase structure of the tunneling current as a function of the applied voltage whenever the Fermi levels align with a new state of the system and thereby open an additional channel for tunneling. Therefore, from the staircase structure of the I–V curve one can learn about the energy spectra of the system. For DNA the trapping spots consist of a finite number of base pairs. Hopping between the base pairs within the traps determines the energy spectra of the spots. In addition to the energy scale due to hopping, there is also an energy scale due to the hole-phonon (or electron-phonon) interaction. Finally, for DNA trapping spots, the I–V dependence has two types of staircase structure; one due to the hopping and the other due to the phonons.

The tunneling transport through a single molecule or a quantum dot with electron-phonon coupling has been extensively studied in the literature [87–90]. The main outcome of these works is the staircase structure of the I–V curves due to the phonon sidebands. The heights of the steps in this structure depend on the strength of the electron-phonon interactions, temperature, and on the equilibrium condition of the electron-phonon system. These studies were mainly restricted to a molecule with a single-electron energy level, although a general approach to a many-level system is also formulated [89]. The DNA trap can be considered as a system of a few molecules (base pairs) with the hopping between them and the electron-phonon coupling. Below, we consider only the hole traps and the tunneling current of holes, but the analysis is also valid for electron traps and electron transport. Whether it is a hole transport or electron transport depends on the gate potential, i.e. on the position of the chemical potential at the zero source-drain voltage V_{sd}.

At first, we study a single-hole transport through the DNA molecule and disregard the effects related to the Coulomb blockade [91] or to a double occupancy of the DNA traps, assuming that the repulsion between the holes is strong enough. In the next section we will analyze the possibility for formation of a bound state of two holes in a trap due to the bipolaronic effect and discuss the manifestation of such a bound state in the transverse tunneling transport.

For a single hole in the trap, the Hamiltonian of the DNA trap and the electrodes consists of three parts: (i) the DNA trap Hamiltonian which includes the tight-binding hole part with hopping between the nearest base pairs (one-dimensional chain model discussed in Section 5.2.1) and the Holstein's phonon Hamiltonian with diagonal hole-phonon interaction [78], (ii) the Hamiltonian of the two leads, left (L) and right (R), and (iii) the Hamiltonian corresponding to the tunneling between the leads and DNA traps

$$\mathcal{H} = \mathcal{H}_{\text{trap}} + \mathcal{H}_{\text{leads}} + \mathcal{H}_{\text{t}} , \tag{5.21}$$

with

$$\mathcal{H}_{\text{trap}} = \sum_{n=1}^{N_t} \varepsilon\, c_n^\dagger c_n - t \sum_i \left[c_n^\dagger c_{n+1} + h.c. \right] +$$
$$+ \hbar\omega \sum_n b_i^\dagger b_n + \chi \sum_n c_n^\dagger c_n \left(b_n^\dagger + b_n \right) , \tag{5.22}$$

$$\mathcal{H}_{\text{leads}} = \sum_{k,\alpha=\text{L,R}} \varepsilon_k\, d_{\alpha,k}^\dagger d_{\alpha,k}, \tag{5.23}$$

$$\mathcal{H}_{\text{t}} = -t_0 \sum_{\alpha=\text{L,R},k} \left[c_{n_0}^\dagger d_{\alpha,k} + h.c. \right] , \tag{5.24}$$

where c_i is the annihilation operator of the hole on site (base pair) n, ε is the on-site energy of the hole in the trap (same for all base pairs within the trap and is determined by the gate voltage or doping of DNA), b_i is the annihilation operator of a phonon on site i, t is the hopping integral between the nearest base pairs, ω is the phonon frequency, χ is the hole-phonon coupling constant, and $d_{\alpha,k}$ is the annihilation operator of a hole in the lead $\alpha = \text{L, R}$ with momentum k. The index $n = 1, \ldots, N_t$ in (5.22) labels the sites (base pairs) in the trap and N_t is their total number. Tunneling from the leads to the trap occurs only to the site n_0 with the tunneling amplitude t_0. In the hole-phonon part of the DNA Hamiltonian $\mathcal{H}_{\text{trap}}$, we include only the optical phonons [92] with diagonal hole-phonon interaction.

We describe the process of tunneling through the trap as a sequential tunneling [93]. In the weak lead-trap coupling regime, the tunneling Hamiltonian $\mathcal{H}_{\text{t}}$ can be considered as a perturbation which introduces the transitions between the states of the trap Hamiltonian, $\mathcal{H}_{\text{trap}}$. We denote the eigenstates of the trap Hamiltonian without coupling to the leads as $|0, m\rangle$ with energy $E_{0,m}$ for the trap without any holes, and $|1, p\rangle$ with the energy $E_{1,p}$ for a trap with a single hole. In the weak lead-trap coupling limit, the master equation for the density matrix of the trap reduces to the rate equation [88] for probability $P_{0,m}$ to occupy the state $|0, m\rangle$ and probability $P_{1,p}$ to occupy the state $|1, p\rangle$,

$$
\frac{dP_{1,p}}{dt} = \sum_{m,\alpha=L,R} W_{\alpha,mp}^{0\to1} P_{0,m} - \sum_{m,\alpha=L,R} W_{\alpha,pm}^{1\to0} P_{1,p} -
$$
$$
- \frac{1}{\tau} \left[P_{1,p} - P_{1,p}^{eq} \sum_{n'} P_{1,p'} \right] ,
\tag{5.25}
$$

$$
\frac{dP_{0,m}}{dt} = \sum_{n,\alpha=L,R} W_{\alpha,pm}^{1\to0} P_{1,p} - \sum_{n,\alpha=L,R} W_{\alpha,mp}^{0\to1} P_{0,m} -
$$
$$
- \frac{1}{\tau} \left[P_{0,m} - P_{0,m}^{eq} \sum_{m'} P_{0,m'} \right] .
\tag{5.26}
$$

In the above equations the distributions $P_{1,p}^{eq}$ and $P_{0,m}^{eq}$ are the corresponding equilibrium distributions with temperature T,

$$
P_{1,p}^{eq} = \exp\left(-E_{1,p}/kT\right) / \sum_{p'} \exp\left(-E_{1,p'}/kT\right)
$$

and

$$
P_{0,m}^{eq} = \exp\left(-E_{0,m}/kT\right) / \sum_{m'} \exp\left(-E_{0,m'}/kT\right) .
$$

Here τ is the relaxation time which is assumed to be the same with or without a hole in the trap. The transition rate $W_{\alpha,pm}^{1\to0}$ is the rate of hole tunneling from the state $|1, p\rangle$ of the trap to the $\alpha = L, R$ lead leaving the trap in the state $|0, m\rangle$. Similarly, the rate $W_{\alpha,mp}^{0\to1}$ is the rate of hole tunneling from the lead α to the state $|1, p\rangle$ of the trap, while originally the trap was in the state $|0, m\rangle$. These rates can be found from Fermi's golden rule

$$
W_{\alpha,pm}^{1\to0} = \Gamma_0 f_\alpha \left(E_{1,p} - E_{0,m}\right) \left| \langle 0, m | c_{n_0} | 1, p \rangle \right|^2 ,
\tag{5.27}
$$
$$
W_{\alpha,mp}^{0\to1} = \Gamma_0 \left[1 - f_\alpha \left(E_{1,p} - E_{0,m}\right)\right] \left| \langle 0, m | c_{n_0} | 1, p \rangle \right|^2 ,
\tag{5.28}
$$

where $\Gamma_0 = 2\pi t_0 \rho / \hbar$ and ρ is the density of states in the leads, which is assumed to be the same in "L" and "R" leads, and $f_\alpha(\epsilon)$ is the Fermi distribution function of the lead α with a chemical potential μ_α. The rate equations (5.25)–(5.26) also assume that the temperature is high enough, i.e. $kT \gg \Gamma$. This means that during the tunneling, the hole state loses its coherence, so the system can be characterized only by the diagonal elements of the density matrix, i.e. by the probabilities to occupy the states in the trap.

For the stationary case, the time derivatives of $P_{1,p}$ and $P_{0,m}$ are zero and (5.25)–(5.26) become a system of linear equations with the normalization condition $\sum_p P_{1,p} + \sum_m P_{0,m} = 1$. The corresponding stationary current can be calculated as

$$I = \sum_{p,m} \left[P_{0,m} W_{L,mp}^{0 \to 1} - P_{1,p} W_{L,pm}^{1 \to 0} \right] . \tag{5.29}$$

The procedure of finding the I–V dependence is the following: At first we calculate the energy spectra and wave functions of the hole-phonon trap system. Then at a given bias and the gate voltages, we calculate the tunneling rates (5.27)–(5.28). As a last step, we solve the linear system of equations (5.25)–(5.26) to find the probabilities $P_{0,m}$ and $P_{1,p}$ and the tunneling current (5.29).

5.7.2 Single-Particle Tunneling

There are few general remarks we can make in relation to the system of equations (5.25)–(5.29). Since the tunneling occurs only through a single base pair, the I–V characteristics should also depend on the position of the base pair through which the tunneling current is measured. This dependence can be illustrated for the hole system without the hole-phonon interactions. For such a system we have only the hopping of the hole within the finite trap system with a finite number of sites (base pairs). The corresponding hopping Hamiltonian of the trap takes the form

$$\mathcal{H}_{\text{trap}} = \sum_{n=1}^{N_t} \varepsilon\, c_n^\dagger c_n - t \sum_i \left[c_n^\dagger c_{n+1} + h.c. \right] . \tag{5.30}$$

Assuming zero boundary conditions at the ends of the trap, i.e. deep trap approximation, we can easily find the hole wave functions within the trap as

$$\Psi_K(n) = \sin\left(\frac{\pi K}{N_{t+1}} n \right) , \tag{5.31}$$

where $K = 1, \ldots, N_t$. It is easy to check that the functions $\Psi_K(n)$ satisfy the boundary conditions $\Psi_K(n = 0) = \Psi_K(n = N_{t+1}) = 0$. The energy corresponding to the state Ψ_K is $\epsilon_K = -2t \cos(\pi K/(N_{t+1}))$. Therefore, for a finite trap with N_t sites there are N_t energy levels within the trap. Generally, if we measure the transverse tunneling current then we should expect N_t steps in the I–V dependence, where each step corresponds to a single energy level. This is not the case when the tunneling occurs through a single base pair because then the contribution to the tunneling current of the Kth state will be proportional to $\sin^2 (\pi K n_0/(N_{t+1}))$. If this coefficient is zero then there is no contribution of the corresponding state and the step related to this state will be suppressed. For example, for $N_t = 2$ and 4 and for any positions of the tunneling site, n_0, there are always N_t steps in the I–V dependence. A more interesting structure is expected for $N_t = 3$ and 5. It is easy to see that for

$N_t = 3$ there are three steps for $n_0 = 1$ and two steps for $n_0 = 2$ since in the last case the contribution from $K = 2$ will be suppressed. Similarly, we can find that for $N_t = 5$ and for $n_0 = 1$ there are 5 steps in the I–V dependence, while for $n_0 = 2$ and 3 there are 4 and 3 steps, respectively. The position of all the steps in the I–V dependence will determine the energy structure of the trap. If we take into account the hole-phonon interaction then an additional scale in the energy spectra of the trap system and an additional structure in the I–V dependence due to the phonons should be expected.

The energy spectra of the hole-phonon quantum system, described by the Hamiltonian $\mathcal{H}_{trap}$ can be found only numerically. To find the eigenfunctions and eigenvalues of a single-hole trap system we make the system finite by introducing limitations on the total number of phonons [94] $\sum_i n_{ph,i} \leq 10$, where $n_{ph,i}$ is the number of phonons on site i. The energy spectra of a trap system without a hole can be easily found. In this case the Hamiltonian H_{trap} is just the Hamiltonian of free phonons at each site of the trap, so the energy of the trap system is just the sum of the energy of all phonons present in the system.

After we derive the energy spectra of the DNA Hamiltonian (5.22) without holes and with a single hole in the trap we solve the system of linear equations (5.25)–(5.26) for a given bias voltage to find the probabilities $P_{1,p}$ and $P_{0,m}$. Then we substitute this solution into (5.29) to find the stationary tunneling current under a given bias voltage. We have calculated the current (5.29) as a function of V_{sd} for different values of on-site energy, ϵ, which can be changed by the gate voltage or by doping. By varying V_{sd} we are keeping the on-site energy ϵ the same and vary the chemical potentials of the leads as $\mu_L = V_{sd}/2$ and $\mu_R = -V_{sd}/2$.

There are five dimensionless parameters which characterize the I–V dependence: the nonadiabaticity parameter [95] $\gamma = \hbar\omega/t$, with a typical value of ~ 0.01–0.5 for DNA, the canonical hole-phonon coupling constant [95] $\lambda = \chi^2/(2\hbar\omega t)$ which is ~ 0.2–1 for DNA, dimensionless bias voltage V_{sd}/t, on-site energy ϵ/t, and the ratio of the relaxation time and the tunneling time $\tau\Gamma_0$.

The calculations have been performed for $N_t = 2$ and $N_t = 3$, i.e., for 2 and 3 base pairs in the trap. The example of such a system could be the guanine hole traps: GG and GGG spots surrounded by adenines. In all the calculations we kept the ratio of the relaxation and the tunneling time equal to 1 ($\tau\Gamma_0 = 1$), i.e., the hole-phonon system in the trap is not in the equilibrium. Different values of $\tau\Gamma_0$, ranging from $\tau\Gamma_0 \ll 1$ (equilibrium case) to $\tau\Gamma_0 \gg 1$ (nonequilibrium case) do not modify qualitatively the behavior of the I–V curve. The phonon steps in the I–V dependence can be seen only when the temperature is less than the phonon frequency. In our calculations the temperature is equal to $0.01t$.

In Fig. 5.7, our results are shown for two base pairs (sites) in the trap. The tunneling occurs through one of the sites, $n_0 = 1$. For an uncoupled hole-phonon system the I–V dependence has two steps corresponding to two

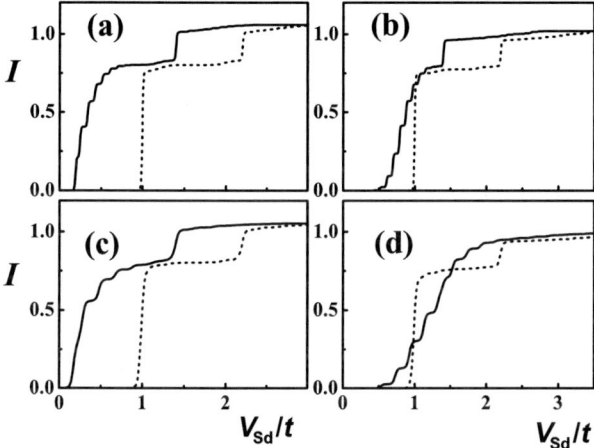

Fig. 5.7. a–d. Current vs the source-drain voltage shown for two base pairs in the trap ($N_t = 2$) and different values of phonon frequency and hole-phonon interaction strength: **a** $\gamma = 0.1$, $\lambda - 0.5$; **b** $\gamma = 0.1$, $\lambda = 1.0$; **c** $\gamma = 0.2$, $\lambda = 0.5$; **d** $\gamma = 0.2$, $\lambda = 1.0$. *Solid line* corresponds to $\epsilon = 1.3t$ while the *dashed line* is for $\epsilon = 2.7$

single-hole energy levels. The distance between the steps is $\delta V_{sd} = 2t$. For a small hole-phonon coupling constant $\lambda = 0.5$ (Fig. 5.7a,c), the additional structures of width $\delta V_{sd} \simeq \hbar\omega$ due to the phonon sidebands appear only at the first step and the second step can still be clearly distinguished. At the same time for a large gate voltage (large on-site energy ϵ), the phonon steps are suppressed and the I–V structure becomes similar to that of a zero-coupling strength, which is shown in Fig. 5.7a,c by dashed lines. For a strong hole-phonon interaction ($\lambda = 1$), the phonon steps suppress the steps due to inter-site hopping within the trap (Fig. 5.7b,d). This suppression becomes stronger for a larger non-parabolicity γ, which is illustrated in Fig. 5.7b,d by a solid line for $\gamma = 0.1$ and $\gamma = 0.2$. With increasing gate voltage the phonon steps disappear and the I–V dependence shows a clear two-step structure.

The origin of such a suppression of the phonon steps can be understood by considering the case of a very short relaxation time. If the relaxation time is much smaller than the tunneling time, then before tunneling the hole-phonon system is in equilibrium. At low temperatures this means that the system will be at the ground state, i.e., without any phonons if there are no holes in the trap and in the polaronic state when there is one hole. The tunneling current through the trap can be considered as a two-step process: the tunneling from the left contact into the trap and the tunneling from the trap into the right contact. Since we consider only the states with no more than one hole in the trap then the tunneling from the left contact will be tunneling to the state without any phonons. The final state after the tunneling will be the state of a single hole-phonon system. The condition of the tunneling is that the energy of the hole in the left contact is equal to the energy of a single hole-phonon system, which can be either in the ground state or in the excited states. Since the tunneling rate is proportional to the overlap of the hole-phonon trap state, the state without any phonons and the hole in the n_0 site (see (5.28)), the tunneling occurs only into two states of the coupled hole-phonon system. These states are polaronic states of the hole-phonon system originated from a single hole state of the trap system without hole-phonon coupling. Therefore, the tunneling from the left contact will probe only a single-hole state, i.e. it results into a two-step structure due to hole-hopping in the I–V dependence.

Tunneling to the right contact is from the ground state of the coupled hole-phonon system. After the tunneling the trap system is just the phonon system, which can be either in the ground state or in the excited states. The energy conservation during the tunneling (see (5.27)) means that the energy of the ground state of the hole-phonon system is equal to the energy of the hole in the right contact plus the energy of the phonon state in the trap. Therefore, the tunneling from the trap into the right contact should produce the phonon steps in the I–V dependence.

Finally, tunneling from the left contact into the trap results in steps in the I–V dependence due to hole-hopping between the sites of the trap, while the tunneling from the trap into the right contact produces the steps due to the phonons. If the gate voltage or the on-site energy is increased, then when the tunneling from the left contact into the traps is allowed, the chemical potential of the right contact will be low enough. This means that after the tunneling, the trap system can be left in the state with many phonons. In terms of the phonon steps this means that the steps will be suppressed.

From Fig. 5.7, we can conclude that for typical parameters of the DNA structure the hopping integral between the sites within the DNA traps and phonon frequency which determine the energetics of the hole-phonon trap system, can be found from the dependence of the tunneling current on V_{sd}. From a small gate voltage, the phonon frequency can be found from the I–V curve, while for a larger gate voltage, the hopping integral can be obtained.

The I–V curve should show even richer structure for a larger number of sites in the trap. In Fig. 5.8, the current as a function of the bias voltage is shown for $N_t = 3$ sites. In this case, the tunneling is possible through the sites $n_0 = 1$ and $n_0 = 2$. For an uncoupled hole-phonon system, the I–V curve shows three steps for $n_0 = 1$ (dotted line in Fig. 5.8a), and two steps for $n_0 = 2$ (dotted line in Fig. 5.8b). This means that for $n_0 = 2$, only two states have non-zero amplitude at $n = 2$ and they contribute to the tunneling current. The finite hole-phonon coupling results in two effects: the phonon steps in the I–V dependence similar to a two-site trap (Fig. 5.7), and the polaronic effect which redistributes the hole density along the trap and increases or decreases the tunneling current.

For a small hole-phonon coupling ($\lambda = 0.5$), the phonon steps are seen only at the first hopping step (Fig. 5.8a,b (solid lines)). The separation between the steps is the phonon frequency. Similar to Fig. 5.7, an increase of the gate

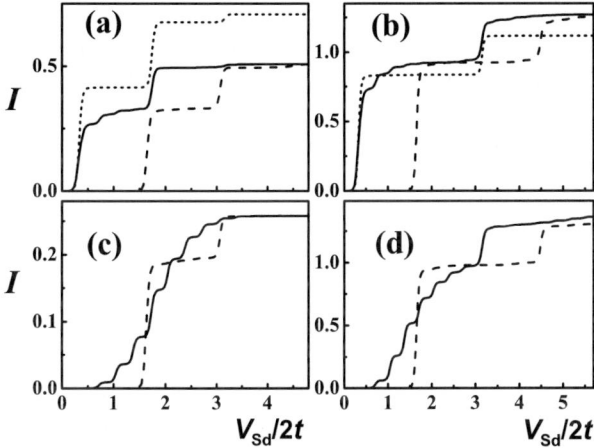

Fig. 5.8. a–d. Current vs the source-drain voltage shown for three base pairs in the trap ($N_t = 3$) and $\gamma = 0.2$, but for different tunneling points n_0 and for different values of hole phonon interaction strength: **a** $n_0 = 1$: $\lambda = 0.5$ and $\epsilon = 1.7t$ (*solid line*), $\lambda = 0.5$ and $\epsilon = 3.0t$ (*dashed line*), $\lambda = 0$ and $\epsilon = 1.7t$ (*dotted line*); **b** $i_0 = 2$: $\lambda = 0.5$ and $\epsilon = 1.7t$ (*solid line*), $\lambda = 0.5$ and $\epsilon = 3.0t$ (*dashed line*), and $\lambda = 0$, $\epsilon = 1.7t$ (*dotted line*); **c** $i_0 = 1$: $\lambda = 1$ and $\epsilon = 1.7t$ (*solid line*), $\epsilon = 3.0t$ (*dashed line*), **d** $i_0 = 2$: $\lambda = 1$ and $\epsilon = 1.7t$ (*solid line*), $\epsilon = 3.0t$ (*dashed line*)

voltage (the on-site energy) suppresses the phonon steps and the I–V curve becomes similar in structure to the uncoupled case (Fig. 5.8a,b, dashed lines). The main difference between the coupled and the uncoupled systems is the different amplitude of the steps. This difference is due to the redistribution of the hole within the trap due to the interaction with the phonons. This results in a suppression of the tunneling current when the tunneling occurs through the site $n_0 = 1$ and enhancement of the tunneling current for $n_0 = 2$ (Fig. 5.8a,b). Therefore, the interaction with the phonons or the polaronic effect increases the probability for the hole to occupy the site $n = 2$.

For a larger hole-phonon coupling ($\lambda = 1$), the steps due to the hole-hopping almost completely disappear for $n_0 = 1$ (Fig. 5.8c, solid line), but some structure is still visible for $n_0 = 2$ (Fig. 5.8d, solid line). As we mentioned above, with increasing gate voltage the phonon steps should be suppressed and the I–V structure should clearly show the steps due to the hole hopping between the sites of the trap. This behavior is illustrated in Fig. 5.8c,d by dashed lines.

To illustrate the polaronic effects due to the hole-phonon coupling which is clearly seen in Fig. 5.8, we have calculated the density of holes and the average number of phonons within the traps for different strengths of the hole-phonon interaction. The results are shown in Fig. 5.9. With increasing hole-phonon interaction, the hole states become more localized at the center of the trap (see Fig. 5.9), i.e. we observe the polaronic effect in the trap system.

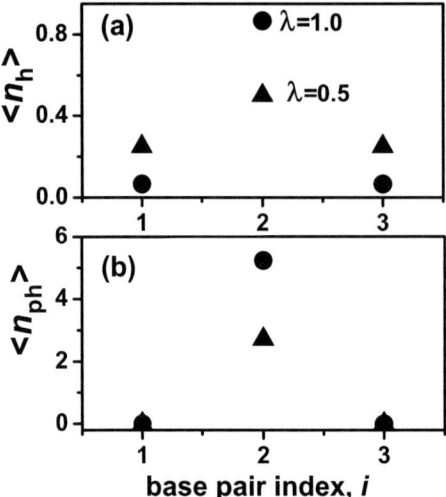

Fig. 5.9. a,b. The average number of holes (**a**) and the average number of phonons (**b**) for a single-hole system in a GGG trap, shown as a function of the base index. *Dots* and *triangles* corresponds to the hole-phonon interaction strength $\lambda = 1.0$ and 0.5, respectively

Localization of the hole at the center of the trap results in an increase of the tunneling current for tunneling through the central site of the trap $n_0 = 2$ (see Fig. 5.8b,d), and a decrease of the tunneling current through $n_0 = 1$ (see Fig. 5.8a,c). In addition to the changes to the tunneling current, the polaronic effect also modifies the structure of the I–V curve. This can be seen in Fig. 5.8a, where increasing the hole-phonon interaction, the third step due to the hole hopping disappears. Therefore, in the hole-phonon coupled system with $N_t = 3$, only two steps due to the hole hopping can be clearly seen in the I–V dependence at a large gate voltage.

5.7.3 Bipolaron Formation in a DNA Molecule

In the previous subsections, we disregarded the Coulomb blockade, the effects related to the hole-hole interactions. Therefore, we assumed the trap system can be occupied only by a single hole. This assumption is valid as long as the repulsion between the holes is quite strong. The specific feature of the hole-phonon system is that the repulsion between the holes can be strongly suppressed in such systems and might even result in an effective attraction between the holes. If there is an attraction between the holes then the trap with the two holes should have a lower energy than the trap with a single hole. Such energetics should modify the I–V dependence of the transverse tunneling current. To study this problem, we first analyze the condition of the trapping of two holes by the trapping spots, such as a GG, GGG, or

GGGG. The formation of the bound state of two holes trapped by the G-sites is analogous to bipolaron formation in the homogeneous one-dimensional system [96].

To write the Hamiltonian of a many-hole system within a trap, we need to add to the single-particle Hamiltonian (5.22) the term which describes the Coulomb interaction between the holes. Therefore, now the Hamiltonian of the hole system within a trap consists of three parts: (i) the tight-binding Hamiltonian which includes the hole-hopping between the nearest base pairs and the on-site energies of a hole, (ii) the hole-hole interaction Hamiltonian, and (iii) the Holstein's phonon Hamiltonian with the diagonal hole-phonon interaction [78]

$$\mathcal{H}_{\text{trap}} = \mathcal{H}_{\text{t}} + \mathcal{H}_{\text{i}} + \mathcal{H}_{\text{ph}} , \tag{5.32}$$

with

$$\mathcal{H}_{\text{t}} = \sum_{i,\sigma} \epsilon_i c_{i,\sigma}^{\dagger} c_{i,\sigma} - t \sum_{i,\sigma} \left[c_{i,\sigma}^{\dagger} c_{i+1,\sigma} + h.c. \right] , \tag{5.33}$$

$$\mathcal{H}_{\text{i}} = \sum_{i,j,\sigma} V_{i,j} n_{i,\sigma} n_{j,-\sigma} + \sum_{i,j \neq i,\sigma} V_{i,j} n_{i,\sigma} n_{j,\sigma} , \tag{5.34}$$

$$\mathcal{H}_{\text{ph}} = \hbar\omega \sum_i b_i^{\dagger} b_i + \chi \sum_{i,\sigma} c_{i,\sigma}^{\dagger} c_{i,\sigma} \left(b_i^{\dagger} + b_i \right) , \tag{5.35}$$

where $c_{i,\sigma}$ is the annihilation operator of a hole with spin σ on site i, and $n_{i,\sigma} = c_{i,\sigma}^{\dagger} c_{i,\sigma}$. The Hamiltonian (5.21) without the phonon part $\mathcal{H}_{\text{ph}}$ was studied for a homogeneous system in [34].

In the tight-binding Hamiltonian (5.33), we assume that the site i can be either an adenine or a guanine. We then take the on-site energy of the hole at adenine (A) site as the zero energy, i.e. $\epsilon_A = 0$, and the on-site energy of the hole at the guanine (G) site to be negative, $\epsilon_G = -\Delta_{\text{GA}} < 0$. In the interaction Hamiltonian $\mathcal{H}_{\text{i}}$, we take into account only the Hartree interaction between the holes. The first term in (5.34) describes the repulsion between the two holes with different spin. The holes can then occupy the same site. The second term in (5.34) corresponds to the repulsion between the two holes with the same spin. To get the basic idea about the typical range of the interaction parameters resulting in the formation of a bound state of two holes within the region of the G-trap, we introduce a single-parameter interaction potential of the form

$$V_{i,j} = V_0 \left[(i-j)^2 + 1 \right]^{\frac{1}{2}} , \tag{5.36}$$

where V_0 is the on-site repulsion between the two holes. The form of the interaction potential, $V_{i,j}$, (5.36) takes into account the finite spreading of the hole on-site wave function. This spreading is about the distance between the nearest base pairs. Although the actual dependence of the interaction potential on the separation between the holes is more complicated [97] than (5.36), this difference is not important for our analysis since only the on-site interaction plays the main role in the formation of the bound state of two holes [98].

Similar to the analysis in the previous subsection, we include in the hole-phonon Hamiltonian only the optical phonons with diagonal hole-phonon interaction, and do not take into account the acoustic phonons which results in non-diagonal hole-phonon interaction [99, 100], i.e. modify the tunneling integral. In (5.33)–(5.35), we also assumed that the hopping integral t, the phonon frequency ω, and the hole-phonon coupling constant χ, do not depend on the specific type of the base pairs (A or G).

The form of the total Hamiltonian, (5.32)–(5.35), leads to four dimensionless parameters which characterize the system: the nonadiabaticity parameter [95] $\gamma = \hbar\omega/t$, the canonical hole-phonon coupling constant [95] $\lambda = \chi^2/(2\hbar\omega t)$, dimensionless hole-hole interaction strength V_0/t, and the dimensionless difference between on-site energies of G and A, $\delta_{\mathrm{GA}} = \Delta_{\mathrm{GA}}/t$.

We determine the eigenfunctions and eigenvectors of the hole-phonon system numerically by exactly diagonalizing the Hamiltonian (5.32)–(5.35) for a finite size system consisting of six base pairs (sites). We also introduce limitations on the total number of phonons [94], $\sum_i n_{\mathrm{ph},i} \leq N_{\mathrm{max}}$. To compare the energy spectrum of the systems with different number of holes, we keep the maximum number of phonons *per hole* the same for all the systems. Therefore, for the two-hole system the maximum number of phonons is $N_{\mathrm{max}} = 16$ and for the one-hole system $N_{\mathrm{max}} = 8$.

Our finite size system contains six sites which are originally adenines. We then introduce the G-traps with a different number N_{G} of guanines, G, GG, GGG, and GGGG, in the middle of the system. For example the system with two guanines is AAGGAA. For different traps we calculate the energy of the ground state of the systems with one and two holes. There are different ways to define the bound state of two holes within the trap. The first one is based on the analysis of the hole density distribution within the trap. When the two holes occupy the same site then we can tell that this is the bound state of the two-hole system. The second one is based on energetics of the two-hole system. Denoting the corresponding energies of the hole system as $E_{1,N_{\mathrm{G}}}$ (for the one-hole system with N_{G} guanines) and $E_{2,N_{\mathrm{G}}}$ (for the two-hole system with N_{G} guanines), we can write the energetic condition that the trap with N_{G} guanines will accommodate two holes as

$$E_{2,N_{\mathrm{G}}} < E_{1,N_{\mathrm{G}}} + E_{1,1} \tag{5.37}$$

or

$$E_{2,N_{\mathrm{G}}} < E_{1,N_{\mathrm{G}}} + E_{1,N_{\mathrm{G}}} . \tag{5.38}$$

The meaning of the first condition (5.37) is as follows [101]: If the two holes are injected initially into the single guanine traps ($N_{\mathrm{G}} = 1$) of the DNA and then one of the holes is trapped by the N_{G} trap, then the condition (5.37) means that the second hole will also be trapped by the same N_{G} trap. This condition corresponds to the experimental realization of the injection of the holes into the DNA molecule.

The second condition (5.38) is relevant to the transverse tunneling experiments. This condition actually means that if the first hole tunnels into the trap system then the second hole can also tunnel into the same trap system, i.e. the energy of a two-hole system is less than twice the energy of a single-hole system.

Since we are interested in the transverse tunneling current, we concentrate below on the condition (5.38) for the bound state of two holes. The condition (5.38) will determine the critical value of the hole-hole interaction strength, V_0^{cr}. That means for $V_0 < V_0^{cr}$ two holes will be trapped by the same trap with N_G guanines. For $V_0 > V_0^{cr}$ such a trapping is energetically unfavorable.

For our investigation of the system (5.21)–(5.24), we consider the following typical DNA parameters: $0.1\,eV < t < 0.3\,eV$, $0.1\,eV < \Delta_{GA} < 0.5$ [102, 103], $0.05\,eV < \hbar\omega < 0.1\,eV$. For the dimensionless canonical hole-phonon coupling constant we have taken the value $\lambda = 1$. For this coupling constant, the size of the polaron is about 2–3 base pairs. Our calculations show that the critical value V_0^{cr} is very small ($V_0^{cr} \approx 0.1\,eV$) when two holes have the same spin and they can not occupy the same site. This small value of V_0^{cr} also illustrates the fact that the phonon mediated attraction between the holes is largest when the holes occupy the same site. Therefore, in what follows we shall consider only the case of two holes with opposite spin.

In Fig. 5.10, the ground state energy of a single hole is plotted as a function of the hole-phonon coupling constant, λ, for different types of traps. For $\lambda \approx 1$, the difference between the bound state of a hole in G and GG traps is about 0.03 eV, which is smaller than the value (0.05 eV) obtained in [104]. The size of the polaron in our calculations is 2–3 base pairs depending on the values of t and ω.

Following the condition (5.38), we need to compare the energy of a single-hole system with the energy of a two-hole system. In Fig. 5.11, the ground state energy E_{2,N_G} of two holes bound in a single trap is plotted for $N_G = 3$

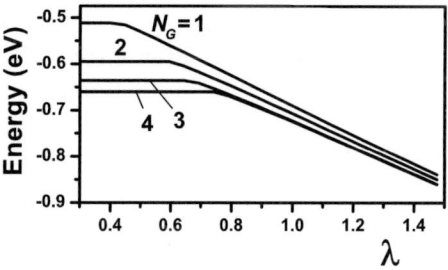

Fig. 5.10. The ground state energy of a single hole in a trap containing N_G guanines is shown as a function of the hole-phonon coupling constant, λ at $t = 0.2\,eV$ and $\Delta_{GA} = 0.3\,eV$. The numbers next to the lines are the number of guanines in the trap

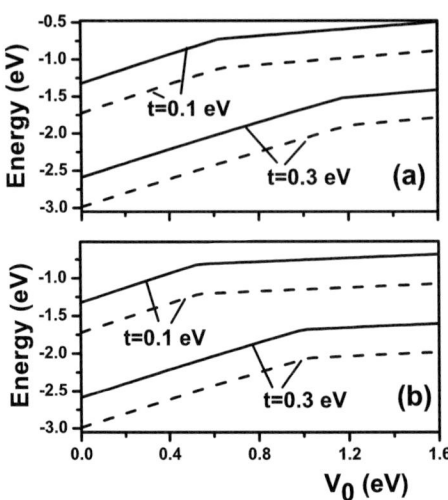

Fig. 5.11. a,b. Ground state energy of two holes in the trap containing $N_G = 3$ guanines (a) and $N_G = 4$ guanines (b) as a function of the inter-hole interaction strength, V_0, for $\Delta_{GA} = 0.3\,\text{eV}$ (*solid line*) and $\Delta_{GA} = 0.5\,\text{eV}$ (*dashed line*)

(Fig. 5.11a) and for $N_G = 4$ (Fig. 5.11b) as a function of the hole-hole interaction strength for different values of Δ_{GA} and t. Here we notice that at some critical value V_0^b of the hole-hole interaction strength, there is a change of slope in the $E_{2,N_G}(V_0)$ dependence. This critical value corresponds to the condition that the two holes are bound in the G-traps, forming a bipolaron. The bound state in this case means that the holes are at the same site of the trap. The illustration of this fact is given in Fig. 5.12. In Fig. 5.12a the average number of holes, $\langle n_h \rangle = \langle n_{i,\sigma} \rangle + \langle n_{i,-\sigma} \rangle$, is shown as a function of the base pair index for a GGGG trap and two different values of the hole-hole interaction strength, V_0. It is clearly seen that for $V_0 = 0.8\,\text{eV} < V_0^b$, the two holes are almost at the same G sites, while at $V_0 = 1.2\,\text{eV} > V_0^b$ the holes are away from each other. The corresponding distribution of the average number of phonons $\langle n_{ph} \rangle$, is shown in Fig. 5.12b.

Another critical value of V_0 is introduced by the equation (5.38). The competition between $2E_{1,N_G}$ and E_{2,N_G} is illustrated in Fig. 5.13. Comparing the energies $2E_{1,N_G}$ and E_{2,N_G} for $\lambda = 1$ and different values of t, Δ_{GA}, and ω, one can determine V_0^{cr}. The result is summarized in Table 5.1 for the GGGG trap. The corresponding results for the GGG trap gives about $0.1\,\text{eV}$ smaller values for V_0^{cr}. The dimensionless parameters, γ, δ_{GA}, and V_0^{cr}/t, are also given in Table 5.1. From these data we can conclude that within the present range of parameters the dependence of V_0^{cr} on Δ_{GA} is weak, and V_0^{cr}/t depends mainly on γ. This dependence can be approximated by a linear function as

$$V_0^{cr} \approx 2.3\gamma t + 1.6t \approx 2.33\hbar\omega + 1.6t \ . \tag{5.39}$$

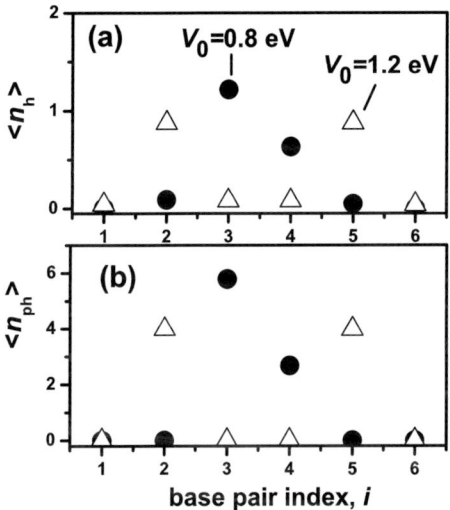

Fig. 5.12. a,b. The average number of holes (**a**) and the average number of phonons (**b**) for a two hole system in a GGGG trap are shown as a function of the base index. The tunneling integral is $t = 0.3\,\text{eV}$ and the hole-phonon coupling is $\lambda = 1$. *Dots* and *triangles* corresponds to inter-hole interaction strength $V_0 = 0.8\,\text{eV}$ and $1.2\,\text{eV}$ respectively

Table 5.1. Calculated values of V_0^{cr} for various values of the dimensionless DNA parameters

t (eV)	$\hbar\omega$ (eV)	Δ_{GA} (eV)	V_0^{cr} (eV)	γ	δ_{GA}	V_0^{cr}/t
0.1	0.1	0.1	0.38	1.00	1.00	3.8
0.1	0.1	0.3	0.41	1.00	3.00	4.1
0.1	0.1	0.5	0.43	1.00	5.00	4.3
0.1	0.05	0.1	0.32	0.50	1.00	3.2
0.2	0.1	0.1	0.48	0.50	0.50	2.4
0.2	0.1	0.3	0.56	0.50	1.50	2.8
0.2	0.1	0.5	0.58	0.50	2.50	2.9
0.3	0.1	0.1	0.78	0.33	0.33	2.6
0.3	0.1	0.3	0.75	0.33	1.00	2.5
0.3	0.1	0.5	0.80	0.33	1.67	2.7
0.3	0.05	0.3	0.54	0.17	1.00	1.8
0.3	0.05	0.5	0.58	0.17	1.67	1.9

The condition (5.37) of formation of the bound state of two holes within the guanine traps gives the higher [101] critical values of the on-site hole-hole repulsion potential, V_0^{cr}, by approximately $0.3\,\text{eV}$.

We see from these data that depending on the parameters of DNA, the critical hole-hole interaction strength V_0^{cr} can range from $0.3\,\text{eV}$ to $0.8\,\text{eV}$.

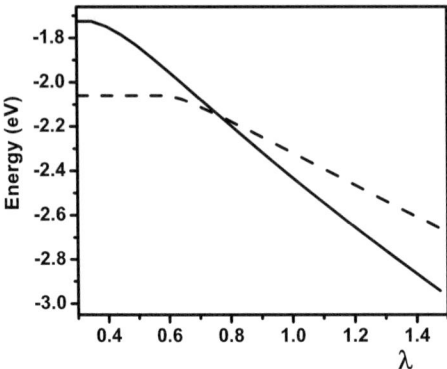

Fig. 5.13. Energies $2E_{1,4}$ and $E_{2,4}$ of a two-hole system are shown as a function of hole-phonon coupling, λ, by *dashed* and *solid lines*, respectively. Tunneling integral is $t = 0.3\,\mathrm{eV}$ and $\Delta_{\mathrm{GA}} = 0.3\,\mathrm{eV}$

Numerical analysis of the electron correlations in different types of DNA [97] shows that the hole-hole interaction strength is around $0.9\,\mathrm{eV}$ for A-DNA and $1.5\,\mathrm{eV}$ for B-DNA. Additional suppression of the inter-hole interaction by a factor of ≈ 0.7 [77] can occur for DNA in solution, when hole-hole interaction is screened by polar solvent molecules and mobile counterions. Under this condition trapping of two holes by GGG and GGGG traps would be possible. Formation of the bound state of two holes at the G-traps requires also a strong hole-phonon interaction, which should overcome the hole-hole Coulomb repulsion. In our calculations, the hole-phonon coupling constant was $\lambda = 1$ which is larger than the experimentally reported $\lambda \approx 0.2$ in Ref. [105]. Hence, experimental observation of the two-hole bound state should give an additional estimate for the strength of hole-phonon interaction.

5.7.4 Pair Tunneling

Experimental observation of the bipolaron formation within the DNA traps should provide additional information about the internal parameters of the DNA molecule, such as the hole-hole repulsion strength, the hole-phonon coupling constant and others. In this section, we discuss the possible manifestation of the bipolaron in the transverse tunneling experiments. We show below that the presence of a bound state of two holes results in the specific I–V dependence of the tunneling current.

Formation of the bound state of two holes within the guanine trap means that the energy of the two holes in the trap is less than the energy of a single hole. This fact results in a modification of the I–V dependence of the tunneling current. Indeed, since there is energy conservation during the tunneling the energy of the hole in the contact should be equal to the energy of the hole in the trap. This means that if the bipolaron has a lower energy than a single hole, then the tunneling of two holes [106] simultaneously becomes more energetically favorable than the tunneling of a single hole.

To illustrate the manifestation of pair tunneling in the transverse tunneling experiments, we consider below a simple model. In this model, we assume that there is no phonon in the system, but there is an effective attraction between the holes, i.e. the energy of two holes is less then the energy of a single hole in the trap. We will also concentrate only on the competition between the contributions of the two-hole states and a single-hole state into the tunneling current and assume that the trap system has a single two-hole level and a single one-hole level. In this case the corresponding rate equations take the form

$$\frac{dP_0}{dt} = -\left(W_{0\to1} + W_{0\to2}\right) P_0 + W_{1\to0} P_1 + W_{2\to0} P_2$$

$$\frac{dP_1}{dt} = W_{0\to1} P_0 - \left(W_{1\to0} + W_{1\to2}\right) P_1 + W_{2\to1} P_2 \qquad (5.40)$$

$$\frac{dP_2}{dt} = W_{0\to2} P_0 + W_{1\to2} P_1 - \left(W_{2\to1} \mid W_{2\to0}\right) P_2 \, ,$$

where P_0, P_1, and P_2 are the probability that there are no holes, one hole, and two holes in the trap, respectively. These probabilities should also satisfy the normalization condition $P_0 + P_1 + P_2 = 1$. The transition rates in the system of linear equations (5.40) can be written as

$$W_{1\to0} = \Gamma_1 \left[1 - f_L(E_1) + 1 - f_R(E_1)\right]$$
$$W_{0\to1} = \Gamma_1 \left[f_L(E_1) + f_R(E_1)\right]$$
$$W_{1\to2} = \Gamma_1 \left[f_L(E_2 - E_1) + f_R(E_2 - E1)\right]$$
$$W_{2\to1} = \Gamma_1 \left[1 - f_L(E_2 - E_1) + 1 - f_R(E_2 - E1)\right]$$
$$W_{2\to0} = \Gamma_2 \sum_{i,j} \{[1 - f_L(\epsilon_i)] [1 - f_L(\epsilon_j)]$$
$$+ \left[1 - f_R(\epsilon_i)\right] \left[1 - f_R(\epsilon_j)\right]$$
$$+ \left[1 - f_L(\epsilon_i)\right] \left[1 - f_R(\epsilon_j)\right]\} \, \delta \left(\epsilon_i + \epsilon_j - E_2\right)$$
$$W_{0\to2} = \Gamma_2 \sum_{i,j} \left\{f_L(\epsilon_i) f_L(\epsilon_j) + \right.$$
$$\left. f_R(\epsilon_i) f_R(\epsilon_j) f_L(\epsilon_i) f_R(\epsilon_j)\right\} \delta \left(\epsilon_i + \epsilon_j - E_2\right) \, ,$$

where Γ_1 and Γ_2 are the tunneling rates for one- and two-hole tunneling, and E_1 and E_2 are the energies of one hole and two hole systems, respectively. Here the energy E_2 takes into account the interaction between the holes and can be expressed in terms of the bound energy, Δ_B, of two holes as $E_2 = 2E_1 - \Delta_B$. When the ground state of the two hole system is a bipolaron then $\Delta_B > 0$, otherwise $\Delta_B < 0$.

We can see from the expression for the transition rates that the specific feature of the pair tunneling is the presence of the sum over the hole states in the contacts. This sum should modify the dependence of the transition rates on the chemical potentials of the contacts and, correspondingly, on the bias voltage. If we assume that the density of states of the hole in the contacts is constant, then the transition rate corresponding to the tunneling of a single hole from the contact into the trap will not depend on the chemical potential. The transition rate corresponding to the tunneling of two holes from the same contact into the trap will be proportional to the chemical potential, i.e., $\propto (2\mu_s - E_2)$, while the transition rate of the two hole tunneling from different contacts will not depend on the chemical potential of the contacts. The dependence of the transition rates on the chemical potential and the bias voltage results in the special structure of I–V characteristics for the systems with pair tunneling.

Under the given bias and gate voltages, the stationary solution of the linear system of equations (5.40) can be found and the corresponding tunneling current can be calculated from the following expression

$$I = [W_{R,1\to 0} - W_{R,1\to 2}] P_1 + 2W_{R,2\to 0}P_2 - [W_{R,0\to 1} + W_{R,0\to 2}] P_0 . \quad (5.41)$$

The I–V characteristics has been found for different gate voltages and different ratios Γ_2/Γ_1. Since the tunneling rate Γ_2 corresponds to the pair tunneling it is smaller than Γ_1. Below we assume that at zero bias and zero gate voltage, the chemical potentials of the contacts coincide with the energy level of a single hole in the trap, i.e. E_1. In this case it is convenient to measure the bias voltage and the gate voltage in the units of $\Delta_B = 2E_1 - E_2$. The results of the calculations are shown in Figs. 5.14 and 5.15 for $\Delta_B > 0$ and $\Delta_B < 0$, respectively. For a positive Δ_B, i.e. when the bound state of

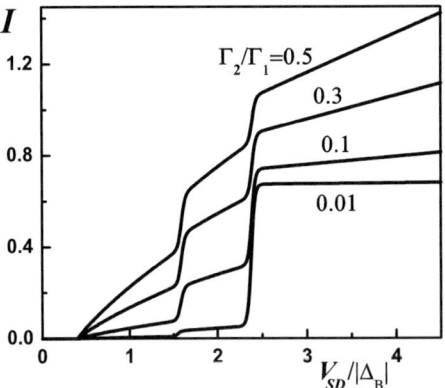

Fig. 5.14. Current vs bias voltage is shown for $\Delta_B > 0$ and for different values of ratio Γ_2/Γ_1. The gate voltage is $0.2\Delta_B$ and the temperature is $0.01\Delta_B$. The ground state of the trap is the bound state of two holes

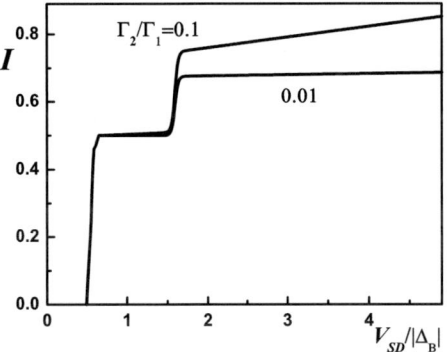

Fig. 5.15. Current vs bias voltage is shown for $\Delta_{\mathrm{B}} < 0$ and for different values of ratio Γ_2/Γ_1. The gate voltage is $0.2\Delta_{\mathrm{B}}$ and the temperature is $0.01\Delta_{\mathrm{B}}$. The ground state of the trap is a single-hole state

two holes has the lower energy than a single-hole state, the I–V structure has a clear linear dependence on the bias voltage within the whole region of the parameters (Fig. 5.14). There are two steps in the I–V dependence, where each step corresponds to the opening of an additional channel for tunneling. In what follows, we analyze these channels in more detail. The variation of Γ_2/Γ_1 has a strong effect on the I–V curve. With decreasing Γ_2/Γ_1, the contribution of the pair tunneling becomes suppressed and at a very small Γ_2/Γ_1 only a single step due to a single-hole tunneling can be seen in the I–V dependence.

The behavior of the I–V dependence becomes very different at $\Delta_{\mathrm{B}} < 0$ (Fig. 5.15). In this case we also have two steps. Now the first step is due to a single-hole tunneling, while the second step is due to a combination of pair tunneling and a single-hole tunneling. At a high bias voltage, i.e. within a second step, we can see the linear dependence in the I–V characteristics, which is a specific feature of the pair tunneling of the holes. With a decrease of the ratio Γ_2/Γ_1, the pair tunneling and correspondingly, the linear dependence becomes suppressed, but still the structure has the two steps. Now the second step is entirely due to a single-hole tunneling and is a manifestation of the Coulomb blockade.

The origin of the different steps at $\Delta_{\mathrm{B}} > 0$ is analyzed in Fig. 5.16, where the occupations of the levels of the traps are shown as a function of bias voltage. The gate voltage is $V_g = 0.2\Delta_{\mathrm{B}}$ and at the zero bias voltage, the trap is occupied by two holes. With an increase of the bias voltage, tunneling to the left contact becomes allowed and we can see the linear dependence of the tunneling current on the bias voltage. The trap is partially occupied by two holes, and only the pair-tunneling contributes to the tunneling current. With an additional increase of the bias voltage, the Fermi level of the right contact becomes equal to the energy of a single-hole state in the trap and the tunneling of a single hole becomes energetically allowed. These results in

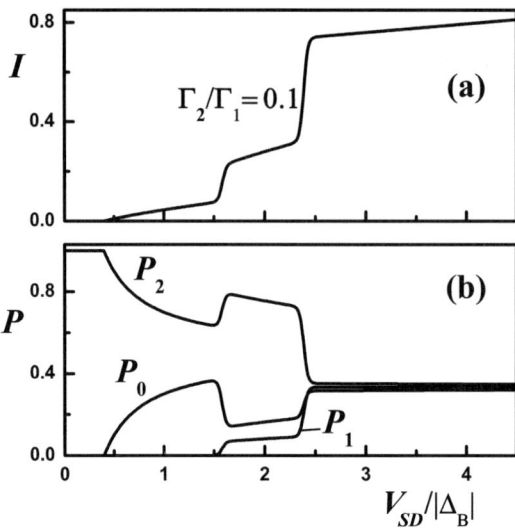

Fig. 5.16. a Current vs the bias voltage, shown for $\Delta_B > 0$, $\Gamma_2/\Gamma_1 = 0.1$, and gate voltage $V_g = 0.2\Delta_B$. **b** The corresponding probabilities, P_0, P_1, and P_2, are shown as a function of the bias voltage. The ground state of an isolated trap is the bound state of two holes

the first step of the I–V curve. Within this region we have the pair-tunneling and a single-hole tunneling to the empty trap or from the trap occupied by a single hole. The origin of the second step at a higher bias voltage is the opening of an additional channel for tunneling: the single-hole tunneling

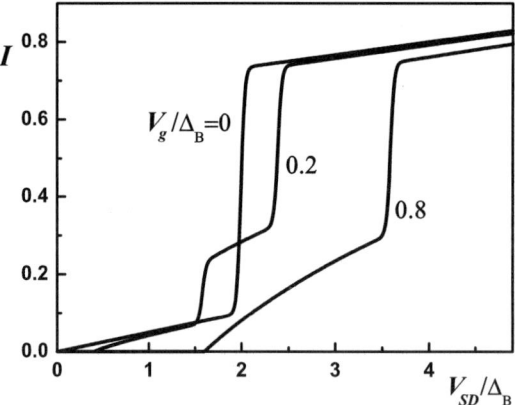

Fig. 5.17. Current vs bias voltage is shown for $\Delta_B > 0$ and for different values of gate voltage V_g. The ratio of the pair tunneling rate and a single-hole tunneling rate is $\Gamma_2/\Gamma_1 = 0.1$. The temperature is $0.01\Delta_B$. The ground state of the trap is the bound state of two holes

from the two-hole state in the trap. Finally, the shape of the I–V curve is determined by the following three channels of tunneling: pair tunneling, a single-hole tunneling to the empty trap or from a single-hole state of the trap, and a single-hole tunneling from the two-hole state of the trap or to a single-hole state of the trap. The opening of different channels depends on the gate voltage. Therefore, by variation of the gate voltage we can modify the I–V structure of the tunneling current through the DNA trap. In Fig. 5.17, different possible I–V dependencies are shown at different gate voltages. At small and high gate voltages there is only a single step in the I–V curve, while at an intermediate gate voltage there are two steps.

5.8 Summary

Many experimental measurements of charge migration along DNA have been carried out in the last decade, especially after the direct measurement of DNA conductance became available. The apparently diverse conclusions extracted from different experiments have made it imperative to initiate systematical and comprehensive theoretical efforts for fundamental understanding of the underlying mechanisms for the charge transfer in DNA. One focus of these efforts is the mechanistic understanding of the observed weak distance dependence of charge transfer along a DNA with the specific sequence: $(G:C)(T:A)_\mathcal{M}(G:C)_3$. Previously, thermally-induced hopping mechanism was invoked to explain it by many authors. In this mechanism, strong dephasing effects is assumed to introduce phase incoherence in the spatial scale of a nanometer. We have proposed that the phase coherence is maintained in the nano-scale of distance in DNA but the two-stranded geometry plays an important role in the weak distance dependence. In other words, the distance dependence is a geometrical characteristic of the quantum transport rather than a trivial property of the classical transport. Within this framework, a quantitative analysis based on the multichannel superexchange mechanism successfully explains the main feature of the experimental result and makes some predictions for future experiments. In the existing experiment, a critical number $\mathcal{M}_c = 3$ is observed when the crossover from strong to weak distance dependence occurs. For the multichannel superexchange mechanism, this crossover number depends on the ratio of the intra- to interstrand coupling parameter in DNA. A crossover number different from three may be observed in other experiments.

   The experimental analysis of the transverse transport through a DNA trap can provide additional information about the parameters of the DNA molecule. The dependence of the tunneling current on the applied bias voltage has a staircase structure. The shape of the structure can be changed by applying the gate voltage to the trap. If the repulsion between the holes within the trap is strong then the main mechanism of tunneling is a single-hole tunneling. In this case, the staircase structure of the I–V dependence has two types of steps: the first one is due to the hole hopping between the sites of the trap, while the second one is due to hole-optical phonon interactions, i.e. the phonon sidebands. At a small gate voltage, both types of steps are present in the I–V dependence and the phonon frequency can be extracted from the I–V curve. At a large gate voltage, the phonon steps become strongly suppressed and the steps due to hole hopping can be clearly seen in the I–V dependence. In this case the width of the steps gives the value of the hopping integral between the sites of the trap.

The transverse tunneling measurements can also be used to analyze the possibility for formation of the bipolaron, i.e. the bound state of the two holes (polarons), within the DNA trap. If the bound state of two holes has a lower energy than a single-hole state then the main contribution to the tunneling current at a low bias voltage comes from the pair tunneling of two holes. This tunneling process results in a specific dependence of the tunneling rate on the bias voltage. As a result of this dependence the I–V curve in the case of a bipolaron formation can be distinguished from the I–V curve corresponding to a single-hole tunneling.

Acknowledgement. This work has been supported by the Canada Research Chair Program and a Canadian Foundation for Innovation (CFI) Grant.

References

1. K.B. Beckman and B.N. Ames, J. Biol. Chem. **272**, 19633 (1997); S. Loft and H.E. Poulsen, J. Mol. Med. **74**, 297 (1996); A.P. Grollman and M. Moriya, Trends in Genetics **9**, 246 (1993); C.J. Burrows and J.G. Muller, Chem. Rev. **98**, 1109 (1998).
2. E. Braun and K. Keren, Adv. Phys. **53**, 441 (2004).
3. C. Dekker and M.A. Ratner, Phys. World **14**, (8), 29 (2001).
4. J. Lagerqvist, M. Zwolak and M. Di Ventra, Nano Letters **6**, 779 (2006).
5. M. Xu, R.G. Endres and Y. Arakawa, Chapter 9, this volume.
6. M. Zwolak and M. Di Ventra, Nano Letters **5**, 421 (2005).
7. K.F. Herzfeld, J. Chem. Phys. **10**, 508 (1942).
8. J. Ladik, Acta Phys. Acad. Sci. Hung. **11**, 239 (1960).
9. D.D. Eley and D.I. Spivey, Trans. Faraday Soc. **58**, 411 (1962).
10. K.V. Mikkelsen and M. A. Ratner, Chem. Rev. **87**, 113 (1987); M.D. Newton, Chem. Rev. **91**, 767 (1991).
11. R.A. Marcus, Rev. Mod. Phys. **65**, 599 (1993).
12. D.N. Beratan, J. Betts and J.N. Onuchic, Science **252**, 1285 (1991); J. Evenson, M. Karplus, Science **262**, 1247 (1993); S. Steenken, S.V. Jovanovic, J. Am. Chem. Soc. **119**, 617 (1997); S.O. Kelley and J.K. Barton, Chem. Biol. **5**, 413 (1998); M. Ratner, Nature **397**, 480 (1999); E. Boone and G.B. Schuster, Nucleic Acids Res. **30**, 830 (2002); D.M. Adams, et al., J. Phys. Chem. B **107**, 6668 (2003).

13. R.G. Endres, D.L. Cox and R.R.P. Singh, Rev. Mod. Phys. **76**, 195 (2004).
14. G.B. Schuster (Ed.), Long-range charge transfer in DNA, Springer-Verlag, Berlin Heidlberg, (2004).
15. J. Jortner and M. Bixon, (Eds.), Electron transfer: from isolated molecules to biomolecules, Part Two, John Wiley & Sons, Inc., (1999).
16. D. DeVault, Quantum-mechanical tunnelling in biological systems, 2nd ed., Cambridge University Press, (1984).
17. G. Fischer, Vibronic Coupling, Academic Press, (1984).
18. C.R. Calladine, H.R. Drew, B.F. Luisi and A.A. Travers, Understanding DNA (Elsevier, London, 2004); J.D. Watson, et al., Molecular Biology of the Gene, (Benjamin Cummings, San Francisco, 2004), 5th edition; B. Alberts, et al., Molecular Biology of the Cell, (Garland Science, New York, 2002), 4th edition.
19. H. Sugiyama and I. Saito, J. Am. Chem. Soc. **118**, 7063 (1996).
20. H.Y. Zhang, X.Q. Li, P. Han, X.Y. Yu and Y.J. Yan, J. Chem. Phys. **117**, 4578 (2002).
21. E. Artacho, M. Machado, D. Sanchez-Portal, P. Ordejon, and J. M. Soler, Mole. Phys. **101**, 1587 (2003).
22. E. Maciá, F. Triozon and S. Roche, Phys. Rev. B **71**, 113106 (2005).
23. W. Ren, J. Wang, Z.S. Ma and H. Guo, Phys. Rev. B **72**, 035456 (2005).
24. S.D. Wetmore, R.J. Boyd and L.A. Eriksson, Chem. Phys. Lett. **322**, 129 (2000).
25. A.A. Voityuk, J. Jortner, M. Bixon and N. Rösch, J. Chem. Phys. **104**, 9740 (2000); ibid. **114**, 5614 (2001).
26. A. Troisi and G. Orlandi, Chem. Phys. Lett. **344**, 509 (2001).
27. R. Di Felice, A. Calzolari, E. Molinari and A. Garbesi, Phys. Rev. B **65**, 045104 (2002).
28. J. Rak, A.A. Voityuk, A. Marquez and N. Rösch, J. Phys. Chem. B **106**, 7919 (2002).
29. G. Cuniberti, L. Craco, D. Porath and C. Dekker, Phys. Rev. B **65**, 241314 (2002).
30. G.C. Liang, A.W. Ghosh, M. Paulsson and S. Datta, Phys. Rev. B **69**, 115302 (2004).
31. M. Zwolak and M. Di Ventra, Appl. Phys. Lett. **81**, 925 (2002).
32. X.F. Wang and T. Chakraborty, Phys. Rev. B **74**, 193103 (2006).
33. I. Saito, T. Nakamura, K. Nakatani, Y. Yoshioka, K. Yamaguchi and H. Sugiyama, J. Am. Chem. Soc. **120**, 12686 (1998).
34. V.M. Apalkov and T. Chakraborty, Phys. Rev. B **71**, 033102 (2005).
35. E. Maciá and R. Rodriguez-Oliveros, Phys. Rev. B **74**, 144202 (2006).
36. N. Sutin, Prog. Inorg. Chem. **30**, 441 (1983).
37. Y.A. Berlin, A.L. Burin and M.A. Ratner, Chemical Physics **275**, 61 (2002).
38. In the literature, both t and $-t$ have been used as the coupling parameter. In Ref. [83], the transfer matrix formalism is presented assuming the coupling parameters t and λ, and has a different form than in this chapter.
39. M.A. Ratner, J. Phys. Chem. **94**, 4877 (1990).
40. P. Carpena, P. Bernaola-galvan, P. Ch. Ivanov and H.E. Stanley, Nature, **418**, 955 (2002); ibid. **421**, 764 (2003).
41. S. Roche and E. Maciá, Mod. Phys. Lett. B **18**, 847 (2004).
42. S. Datta, Quantum Transport: atom to transistor, Cambridge University Press, (2005).

43. F.D. Lewis, X. Liu, J. Liu, S.E. Miller, R.T. Hayes and M.R. Waslelewski, Nature, **406**, 51 (2000).
44. B. Giese, J. Amaudrut, A. Köhler, M. Spormann and S. Wessely, Nature **412**, 318 (2001).
45. M. Taniguchi and T. Kawai, Physica E **33**, 1 (2006).
46. D. Porath, A. Bezryadin, S. de Vries and C. Dekker, Nature **403**, 635 (2000).
47. S. Priyadarshy, S.M. Risser and D.N. Beratan, JBIC, **3**, 196 (1998).
48. J. Jortner, M. Bixon, T. Langenbacher and M.E. Michel-Beyerle, Proc. Natl. Acad. Sci. **95**, 12759 (1998); M. Bixon, B. Giese, S. Wessely, T. Langenbacher, M.E. Michel-Beyerle and J. Jortner, Proc. Natl. Acad. Sci. **96**, 11713 (1999).
49. B. Giese, Acc. Chem. Res. **33**, 631 (2000).
50. V.D. Lakhno, V.B. Sultanov and B.M. Pettitt, Chem. Phys. Lett. **400**, 47 (2004).
51. E. Meggers, M.E. Michel-Beyerle and B. Giese, J. Am. Chem. Soc. **120**, 12950 (1998).
52. Y.A. Berlin, A.L. Burin and M.A. Ratner, J. Phys. Chem. A **104**, 443 (2000).
53. H.M. McConnell, J. Chem. Phys. **35**, 508 (1961).
54. J.R. Reimers and N.S. Hush in electron transfer in biology and the solid state, Advances in Chemistry series Vol. 226, M.K. Johnson, R.B. King, D.M. Kurtz, C. Kutal, M.L. Norton and R.A. Scott, American Chemical Society, Washington, DC (1990), page 27.
55. J. Olofsson and S. Larsson, J. Phys. Chem. B **105**, 10398 (2001).
56. J. Jortner, M. Bixon, A.A. Voityuk and N. Rösch, J. Phys. Chem. A **106**, 7599 (2002).
57. K. Senthilkumar, F.C. Grozema, C.F. Guerra, F.M. Bickelhaupt, F.D. Lewis, Y.A. Berlin, M.A. Ratner and L.D.A. Siebbeles, J. Am. Chem. Soc. **127**, 14894 (2004).
58. A.K. Felts, W.T. Pollard and R.A. Friesner, J. Phys. Chem. **99**, 2929 (1995).
59. A. Okada, V. Chernyak and S. Mukamel, J. Phys. Chem. A **102**, 1241 (1998).
60. W.B. Davis, M.R. Wasielewski, M.A. Ratner, V. Mujica and A. Nitzan, J. Phys. Chem. A **101**, 6158 (1997).
61. J. Yi, Phys. Rev. B **68**, 193103 (2003).
62. H. Yamada, Int. J. Mod. Phys. B **18**, 1697 (2004).
63. K. Iguchi, J. Phys. Soc. Jpn. **70**, 593 (2001).
64. D. Klotsa, R.A. Römer and M.S. Turner, Biophys. J. **89**, 2187 (2005).
65. H. Basch, R. Cohen and M.A. Ratner, Nano Lett. **5**, 1668 (2005); H. Basch and M.A. Ratner, J. Chem. Phys. **123**, 234704 (2005).
66. F. Zahid, M. Paulsson, E. Polizzi, A.W. Ghosh, L. Siddiqui and S. Datta, J. Chem. Phys. **123**, 064707 (2005).
67. J.L. D'Amato and H.M. Pastawski, Phys. Rev. B **41**, 7411 (1990).
68. X.Q. Li and Y.J. Yan, Appl. Phys. Lett. **79**, 2190 (2001).
69. C.J. Murphy and M.R. Arkin, Y. Jenkins, N.D. Ghatlia, S.H. Bossman, N.J. Turro and J.K. Barton, Science **262**, 1025 (1993).
70. F.D. Lewis, T.F. Wu, Y.F. Zhang, R.L. Letsinger, S.R. Greenfield and M.R. Wasielewski, Science **277**, 673 (1997).
71. S.O. Kelley and J.K. Barton, Science **283**, 375 (1999).
72. R.N. Barnett, C.L. Cleveland, A. Joy, U. Landman and G.B. Schuster, Science **294**, 567 (2001).
73. M. Bixon and J. Jortner, Chemical Physics **281**, 293 (2002).

74. J. Jortner, M. Bixon, A.A. Voityuk and N. Rösch, J. Phys. Chem. A **106**, 7599 (2002).
75. T. Renger and R.A. Marcus, J. Phys. Chem. A **107**, 8404 (2003).
76. T. Cramer, S. Krapf and T. Koslowski, J. Phys. Chem. B **108**, 11812 (2004); M. Rateitzak and T. Koslowski, Chem. Phys. Lett. **377**, 455 (2003); N. Utz and T. Koslowski, Chem. Phys. **282**, 389 (2002).
77. D.M. Basko and E.M. Conwell, Phys. Rev. Lett. **88**, 098102 (2002); Phys. Rev. B **66**, 094304 (2002).
78. T. Holstein, Ann. Phys. **8**, 325 (1959); **8** 343 (1959).
79. W.P. Su, J.R. Schrieffer and A.J. Heeger, Phys. Rev. Lett. **42**, 1698 (1979); Phys. Rev. B **22**, 2099 (1980).
80. G.B. Schuster, Acc. Chem. Res. **33**, 253 (2000).
81. P.T. Henderson, D. Jones, G. Hampikian, Y.Z. Kan and G.B. Schuster, Proc. Natl. Acad. Sci. **96**, 8353 (1999).
82. P. Maniadis, G. Kalosakas, K.Ø. Rasmussen and A.R. Bishop, Phys. Rev. B **68**, 174304 (2003); S. Komineas, G. Kalosakas and A.R. Bishop, Phys. Rev. E **65**, 061905 (2002).
83. X.F. Wang and T. Chakraborty, Phys. Rev. Lett. **97**, 106602 (2006).
84. P.F. Barbara, T.J. Meyer and M.A. Ratner, J. Phys. Chem. **100**, 13148 (1996).
85. X.F. Wang and I.C. da Cunha Lima, Phys. Rev. B **63**, 205312 (2001).
86. V. Apalkov and T. Chakraborty, Phys. Rev. B 72, 161102 (2005).
87. L.I. Glazman and R.I. Shekhter, Zh. Eksp. Teor. Fiz. **94**, 292 (1987) Sov. Phys. JETP **67**, 163 (1988).
88. A. Mitra, I. Aleiner and A.J. Millis, Phys. Rev. B **69**, 245302 (2004).
89. M. Galperin, M.A. Ratner and A. Nitzan, J. Chem. Phys. **121**, 11965 (2004).
90. J. Koch and F. von Oppen, Phys. Rev. Lett. **94**, 206804 (2005).
91. L.P. Kouwenhoven, D.G. Austing and S. Tarucha, Rep. on Prog. Phys. **64**, 701 (2001).
92. S. Komineas, G. Kalosakas and A.R. Bishop, Phys. Rev. E **65**, 061905 (2002).
93. S. Luryi, Appl. Phys. Lett. **47**, 490 (1985).
94. F. Marsiglio, Physica C **244**, 21 (1995).
95. A. La Magna and R. Pucci, Phys. Rev. B **53**, 8449 (1996).
96. A.S. Alexandrov and N.F. Mott, Rep. Prog. Phys. **57**, 1197 (1994).
97. E.B. Starikov, Phil. Mag. Lett. **83**, 699 (2003).
98. A. La Magna and R. Pucci, Phys. Rev. B **55**, 14886 (1997).
99. E.M. Conwell and S.V. Rakhmanova, Proc. Natl. Acad. Sci. USA. **97**, 4556 (2000).
100. S.V. Rakhmanova and E.M. Conwell, J. Phys. Chem. B **105**, 2056 (2001).
101. V. Apalkov and T. Chakraborty, Phys. Rev. B **73**, 113103 (2006).
102. C.A.M. Seidel, A. Schulz and H.M. Sauer, J. Phys. Chem. **100**, 5541 (1996).
103. X. Hu, Q. Wang, P. He and Y. Fang, Anal. Sci. **18**, 645 (2002).
104. E.M. Conwell and D.M. Basko, J. Am. Chem. Soc. **123**, 11441 (2001).
105. A. Omerzu, M. Licer, T. Mertelj, V.V. Kabanov and D. Mihailovic, Phys. Rev. Lett. **93**, 218101 (2004).
106. J. Koch, M.E. Raikh and F. von Oppen, Phys. Rev. Lett. **96**, 056803 (2006).

6 Vibronic Mechanisms for Charge Transport and Migration Through DNA and Single Molecules

Yoshihiro Asai[1,2] and Tomomi Shimazaki[1,2]

[1] National Institute of Advanced Industrial Science and Technology (AIST), Umezono 1-1-1, Tsukuba Central 2, Tsukuba, Ibaraki 305-8568, Japan
[2] Core Research for Evolutional Science and Technology (CREST), Japan Science and Technology Corporation (JST), Kawaguchi 332-0012, Japan
yo-asai@aist.go.jp, t-shimazaki@aist.go.jp

6.1 Introduction

Until the discoveries of the conductive polymers and the organic semiconductors, organic materials had been generally believed to be the most typical band insulators, which are electrically unimportant and uninteresting. However, thanks to the pioneering works by Shirakawa [1], Akamatsu and Inokuchi [2], our understanding of the electrical properties of the organic materials has been enriched significantly. The electrical conductivities of the organic materials have been improved so much since then. Now we have a large number of organic conductors, including even the superconductors! Discoveries of the new materials have opened up great opportunities both in fundamental science and in the industries. On the academic side, our understanding of the physics behind the electrical transport properties, magnetism and the superconductivity in organic materials has led us to a quite new field of the "physics of low dimensional electronic systems" [3]. On the application side, benefits of organic materials, such like mechanical flexibility, easy chemical processing and the light weight have been fully utilized in the industries, which has provided useful devices, such as the organic FET, batteries and so on. Discoveries of new materials which are accompanied by new properties are therefore important for a variety of reasons.

DNA, among all the biological molecules, is extremely interesting for the solid state physicists because of its structure. Base molecules are stacked along the double strands, which is reminiscent of the crystal structures of the organic conductors [3]. The possibility for a wide variety of chemical doping may give rise to the additional freedom, besides the choice and the arrangement of the base molecules to control the electronic structure and the conductance of DNA [4].

One of the difficulties in DNA molecules for solid state physics researches comes from the existence of the intrinsic water shell close to the phosphates. The water shell cannot be removed even in the ultra high vacuum environment and therefore it should be considered to be intrinsic. While its presence may be useful to sustain the dopants close to the DNA molecule, it makes accurate experiments conducted in such a way as to clarify the one-to-one correspondence between the transport property and the structure, highly demanding. Our theoretical approach may therefore have to be more or less phenomenological. While this is an un-welcome limitation, it may help us in other ways. The success of the theory of Marcus to account for the hole transfer reaction of DNA in solution is rather convincing. Because of the mathematical similarity between this theory and the small polaron model, it may be possible that the small polaron model is useful to describe the electric transport properties of the DNA molecule, although it has the parasitic water shell sticking to it.

Here, we will put an emphasis on this similarity and suppose that the small polaron model is the common and the simplest model to explain both the charge transport property and the hole transfer reaction in the DNA systems which are placed between the nano-gapped electrodes and in solution, respectively.

6.2 The Electron-Molecular-Vibration (E-MV) Coupling

We shall discuss the hole conduction through the DNA molecule in terms of the single-chain model composed of the equally spaced N base molecules in Sects. 6.2, 6.3.1 and 6.4 of this chapter. The spatial distance between the nearest neighbor molecules is given by a. Hence the physical length of the chain is $R = Na$. The double helix structure is neglected, if not mentioned otherwise. Only the highest occupied molecular orbitals (HOMO) of the N base molecules and the optical phonons are taken into account in our model. Electronic conduction can also be discussed if we replace the HOMO with the lowest unoccupied molecular orbital (LUMO) in our arguments. All these are sufficient to our discussion of DNA, if our interest is on the length scale and the energy scale of the problem but not on something that depends too much on the structure factor. Our basic model is then the following small polaron model

$$\mathcal{H} = t \sum_{\delta} \sum_{j=1}^{N} c_{j+\delta}^{\dagger} c_j + \varepsilon \sum_{j=1}^{N} c_j^{\dagger} c_j + \sum_{\mathbf{q}} \omega_{\mathbf{q}} b_{\mathbf{q}}^{\dagger} b_{\mathbf{q}} + \sum_{\mathbf{q}} \sum_{j=1}^{N} c_j^{\dagger} c_j e^{i\mathbf{q}\cdot\mathbf{R}_j} M_{\mathbf{q}} (b_{\mathbf{q}} + b_{-\mathbf{q}}^{\dagger}) ,$$

(6.1)

where c and b denote the annihilation operators for the electron on the HOMO and the optical phonon, j expresses the position of the base molecule, $\mathbf{q}$ represents the wave vector, δ represents the nearest neighbor displacement, t, ε, ω and M denote the transfer integral between the HOMOs of the nearest neighbor base molecules, the HOMO energy of the base molecule, the frequency of the optical phonon and the electron-phonon coupling constant, respectively. To be more specific, we will be taking into account the Einstein phonon, i.e., the molecular vibration in our case. The effect of the vibrational collective mode of the waters and the ions surrounding the DNA may be incorporated into the molecular vibration. The electron-phonon coupling for the Einstein phonon, known as the Holstein coupling, i.e., the electron-molecular-vibration (e-mv) coupling, has been widely studied, both for the molecules and the molecular solids [5]. The coupling has been confirmed to give reliable assignment of the vibrational sub-bands of the photoemission spectra of the molecules in the gas phase [6]. It has been successfully used to describe the superconductivity of the molecular solid such as the K_3C_{60} [7].

6.3 Ballistic and Weak Coupling Limits

In spite of the high simplicity, the exact solution of the small polaron model has been rarely obtained. We have to use the approximations whose validities are limited to some parameter regions. We will be using approximations, one of which is valid in the strong coupling region where $|M_{\mathbf{q}}/t| \gg 1$, and the other is valid in the weak coupling region, where $|M_{\mathbf{q}}/t| \ll 1$. In a special case of the weak coupling, the ballistic region appears when $|M_{\mathbf{q}}/t| = 0$. In the strong coupling region, the hole conduction cannot take place without exciting the vibrations, i.e., we have a hopping conduction mechanism there. In the weak coupling region, the vibration effect appears as a small correction to the band motion of the electrons. Although the small band width of DNA predicted by the band calculations [8] suggest that the DNA molecules are in the strong coupling region, it may be worthwhile to discuss the physics of the weak coupling and the ballistic regions to broaden our understanding.

6.3.1 Length Dependent Charge Transport

When $|M_{\mathbf{q}}/t| = 0$, there are no scattering sources for the electrons. The electron can then travel freely without losing memories of its momentum and phase. This is the ballistic transport which occurs only when the system size becomes smaller than the mean free path and the coherence length of the electron. For the standard doped semiconductors, it is about several μm. The system which has this length scale is often called the mesoscopic system. Due to the constriction structure in the artificially fabricated devices, the transverse degree of freedom of the electron is quantized to give the quantum conductance step-like structures [9].

In nano-scaled objects such as single molecules, single atomic wires and single clusters bridging two electrodes, however, the energy level structure for the longitudinal degree of freedom is discrete from the beginning, because of their finite system size. We have the competition between the coherence of the electron injected from the electrode and the finite system size effect, which is brought into our problem through the contact self-energy term between the molecule and the electrodes and the discrete energy level structure. The competition is most easily understood if we study the length dependence of the molecular conductance. The ballistic conductance of our single-chain model for the DNA composed of the N base molecules is given by

$$G(E_F) = \frac{2e^2}{h} \Gamma_1(E_F)\Gamma_N(E_F)|g_{1N}^R(E_F)|^2, \tag{6.2}$$

$$\Gamma_\alpha(E_F) = i\left(\Sigma_{C;\alpha\alpha}^R(E_F) - \Sigma_{C;\alpha\alpha}^A(E_F)\right),$$

where E_F is the Fermi level of the electrode and $\Sigma_{C;\alpha\alpha}^R(E_F)$ and $\Sigma_{C;\alpha\alpha}^A(E_F)$ ($\alpha = 1$ or N) denotes the retarded and the advanced electronic contact self energies between the molecule and the electrodes. The left and the right contacts to the electrodes are taken at the two terminals of the chains, i.e., $j = 1$ and $j = N$, respectively. The $(1,1)$ and (N, N) components of the electronic contact self energy $\Sigma_C^{R/A}$ are given by $\Sigma_{C;\alpha\alpha}^{R/A}(E_F) = \tau_\alpha^\dagger g_{electrode}^{R/A}(E_F)\tau_\alpha$, where $g_{electrode}^{R/A}$ denotes the retarded (R) and advanced (A) electronic Green's function of the electrodes and τ_α denotes the coupling between the molecule and the left or the right electrode. $g_{1N}^R(E_F)$ is the terminal, i.e., $(1, N)$ component of the retarded Green's function of the extended molecule

$$\mathbf{g}^R(E_F) = [E_F\mathbf{I} - \mathbf{H} - \Sigma_C^R(E_F)]^{-1}. \tag{6.3}$$

The matrix $\mathbf{g}^R$ is represented in terms of the basis set composed of the HOMOs of the N base molecules in DNA. In terms of the determinant theory, $g_{1N}^R(E_F)$ can be calculated analytically when the model consists only of the nearest neighbor interactions [10, 11]. The length dependent molecular conductance thus calculated has both the energy and the contact state dependencies [12].

The most important result of our DNA problem is the energy dependence. When the HOMO energy ε of the base molecule in DNA is away from the Fermi level E_F more than the twice the magnitude of the transfer integral $\mathbf{t}$, which may be the case for our DNA molecule, the conductance decays exponentially as a function of the chain length N. This is true for both the cases (a) and (b) discussed in the caption of Fig. 1, and is therefore the exponential decay in the tunnelling region, i.e., $\chi > 1$ is independent on the contact state. Actually, these have long been known among the theoretical chemists both in the problem of electron transfer reaction and the molecular conductance problem [10, 11]. To be more specific, it should be found in the Fig. 6.1 that the exponent of the length dependent decay in the tunnelling region does not depend on the contact state, but the pre-factor of the exponential decay and hence the absolute value of the conductance itself have this dependence.

In the resonant case $\chi = 0$, however, the length dependence is largely affected by the contact state. In the case (a), i.e., $|t_L^2/(t \cdot t_M)| \neq 1$, the perfect transmission is achieved only when $N = $ odd, where the energy matching condition between the molecular orbital energy of the chain (but not ε) and E_F is satisfied. The additional perfect transmission at $N = $ even appears

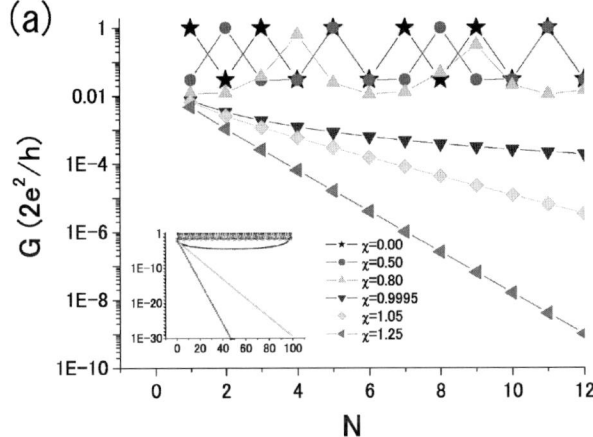

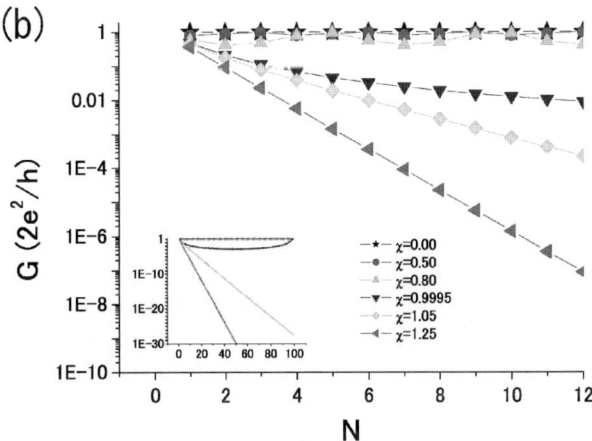

Fig. 6.1. The HOMO energy ε dependence of the length (N) dependent conductance $(G(E_F))$ [12] for the case **a** $|t_L^2/(t \cdot t_M)| \neq 1$ and **b** $|t_L^2/(t \cdot t_M)| = 1$, where t_L is a matrix element of τ, representing the transfer integral between the adjacent atoms in the molecule and the electrode. t_M is the transfer integral in the electrode. The parameter χ in the inset is given by $\chi = |(E_F - \varepsilon)/(2t)|$

only in the case (b) $|t_L^2/(t \cdot t_M)| = 1$. The latter perfect transmission is very curious because it appears irrespective of the fact that there is no state in the chain that has the same energy as the E_F. However, this state is a perfect transmission state! This curious behavior has not been noticed until quite recently [12], because of the simplifications used for calculations of the contact self energy in the previous studies.

The physics of the resonant case is interesting, but this case may have nothing to do with the DNA molecule, unless the Fermi level of the electrode

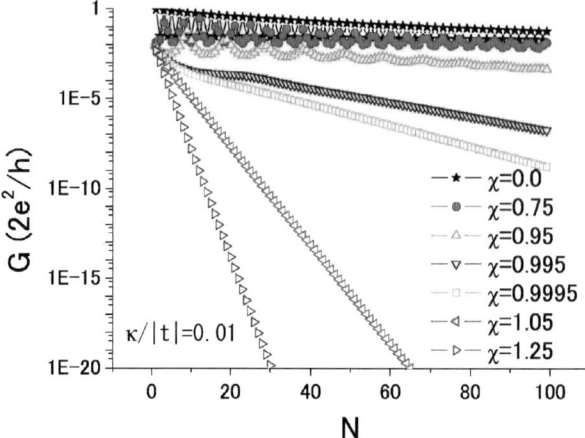

Fig. 6.2. The dephasing effect on the conductance [12]. We take $\kappa/|t| = 0.01$

and the HOMO energy level of the base molecule are engineered to be placed very close.

A weak electron-phonon coupling effect in the elastic channel may be incorporated by introducing the dephasing term into the HOMO energy of the base model, i.e., we replace ε by $\varepsilon + i\kappa$. The results are shown in Fig. 6.2. The oscillatory behavior in the $\chi < 1$ region disappears as we introduce the dephasing term. The damped oscillation behavior appears there instead. The exponent of the decay does not change up to the first order of κ in the tunnelling region, i.e., $\chi > 1$. These results indicate that the self-energy correction due to the e-mv coupling in the elastic channel is not so significant in the DNA molecule when the HOMO energy ε is away from E_F for more than twice the $|t|$.

6.3.2 Inelastic and Elastic Correction

In the weak coupling limit, i.e., $|\mathbf{M_q}/t| \ll 1$, the corrections to the band motion of the electrons due to the e-mv coupling may be rather small. It will be therefore reasonable to adopt the perturbation methods to calculate the corrections. In the transport problem, it is sometimes useful to adopt the Keldysh Green's function method combined with the steady state approximation to avoid the initial correlation problem [9]. The steady state approximation is useful unless the short-range dynamics after the incident perturbation is to be considered. In this section, the Keldysh Green's function theory and an example of the elastic and the inelastic currents for a small molecule is briefly described to give an idea of the present status of the weak coupling theory for single molecules.

In terms of the Keldysh Green's function method, the electronic current I is given by $I = I_{\text{elastic}} + I_{\text{inelastic}}$ [9],

$$I_{\text{elastic}} = \frac{e}{h} \int \text{Tr} \left[\Sigma_{\text{C}}^<(E) \mathbf{g}^>(E) - \Sigma_{\text{C}}^>(E) \mathbf{g}^<(E) \right] dE , \qquad (6.4)$$

$$I_{\text{inelastic}} = \frac{e}{h} \int \text{Tr} \left[\Sigma_{\varphi}^<(E) \mathbf{g}^>(E) - \Sigma_{\text{C}}^>(E) \mathbf{g}^<(E) \right] dE ,$$

where $\Sigma_{\text{C};\alpha\alpha}^<$ and $\Sigma_{\text{C};\alpha\alpha}^<$ are given in terms of Γ_α by $-i\Sigma_{\text{C};\alpha\alpha}^<(E) = f_\alpha(E)$ $\Gamma_\alpha(E)$ and $i\Sigma_{\text{C};\alpha\alpha}^>(E) = [1 - f_\alpha(E)] \Gamma_\alpha(E)$, respectively, $f_\alpha(E) = 1/\left[e^{\beta(E-\mu_\alpha)} + 1 \right]$, $\mu_\alpha = E_\text{F} \pm 0.5\,\text{V}$ is the Fermi distribution function in the electrode α, and Tr denotes that the trace is taken over the basis sets. V represents the bias voltage applied between the electrodes. We adopt the Born approximation to calculate the self energy due to the e-mv coupling

$$\Sigma_{\varphi}^<(E) = \frac{i}{2\pi} \sum_{\mathbf{q}} \mathbf{M}_{\mathbf{q}} \int d\omega D_{\mathbf{q}}^<(\omega) \mathbf{g}^<(E-\omega) , \qquad (6.5)$$

$$\Sigma_{\varphi}^>(E) = \frac{i}{2\pi} \sum_{\mathbf{q}} \mathbf{M}_{\mathbf{q}} \int d\omega D_{\mathbf{q}}^>(\omega) \mathbf{g}^>(E+\omega) ,$$

where $D_{\mathbf{q}}^<$ and $D_{\mathbf{q}}^>$ are the Keldysh Green's function for the vibrations, which is given by

$$D_{\mathbf{q}}^<(\omega) = -2\pi i \left[f_{\text{b}}(|\omega|)\delta(\omega - \omega_{\mathbf{q}}) + \{1 + f_{\text{b}}(|\omega|)\}\delta(\omega + \omega_{\mathbf{q}}) \right] , \qquad (6.6)$$

$$D_{\mathbf{q}}^>(\omega) = -2\pi i \left[\{1 + f_{\text{b}}(|\omega|)\}\delta(\omega - \omega_{\mathbf{q}}) + f_{\text{b}}(|\omega|)\delta(\omega + \omega_{\mathbf{q}}) \right] , \qquad (6.7)$$

where $f_{\text{b}}(|\omega|)$ is the Bose distribution function, i.e., $\left(e^{\beta|\omega|-1} \right)^{-1}$. We have neglected the damping of vibrations for simplicity, here. If we assume the steady state to neglect the initial correlations, the Keldysh Green's functions for the electron, i.e., $\mathbf{g}^>$ and $\mathbf{g}^<$ are given in terms of $\mathbf{g}^\text{R}$ and $\mathbf{g}^\text{A}$ by using the steady state relation

$$\mathbf{g}^<(E) = \mathbf{g}^\text{R}(E) \Sigma^<(E) \mathbf{g}^\text{A}(E) , \qquad (6.8)$$

$$\mathbf{g}^>(E) = \mathbf{g}^\text{R}(E) \Sigma^>(E) \mathbf{g}^\text{A}(E) , \qquad (6.9)$$

where $\Sigma^>$ and $\Sigma^<$ are the sums of the contact and the e-mv self energies, $\mathbf{g}^\text{R}$ and $\mathbf{g}^\text{A}$ are the retarded and the advanced Green's function of the extended molecule, which have already appeared in Sect. 6.3.1, but now they include the self energy due to the e-mv coupling $\Sigma_{\varphi}^{\text{R/A}}$

$$\mathbf{g}^{\text{R/A}}(E) = \left[E\mathbf{I} - \mathbf{H} - \Sigma_{\text{C}}^{\text{R/A}}(E) - \Sigma_{\varphi}^{\text{R/A}}(E) \right]^{-1} . \qquad (6.10)$$

The self energies due to the e-mv coupling are calculated by solving the self-consistent Born approximation described above. The present theory can be made more realistic by combining the theory with the electronic structure methods [13–15]. An atomistic calculation based on the extended Hückel approximation has been made for the benzene dithiol molecule placed between two Au (111) semi-infinite surfaces [13]. Our results are summarized in Fig. 6.3 as an example. In the extended Hückel calculation, E_F is taken as a parameter to fit with the experimental bulk Fermi energy of the gold and the voltage drop is supposed to occur symmetrically at the contacts. Small electric current in the elastic channel is due to the tunnelling current and behaves non-linearly as the voltage exceeds the range of the plot. The inelastic current has a typical threshold-like behavior at the vibrational energy (Fig. 6.3a), which indicates that the electric current flows in an entangled

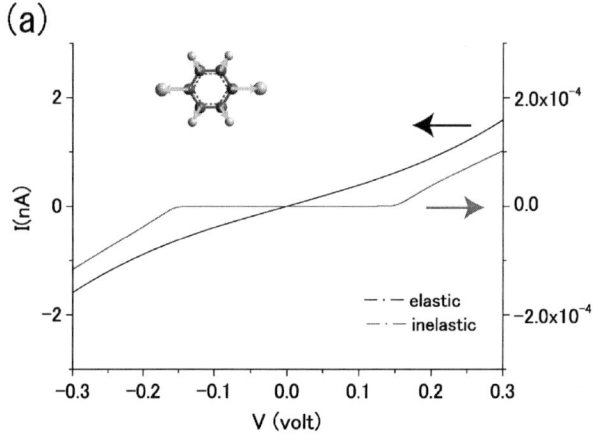

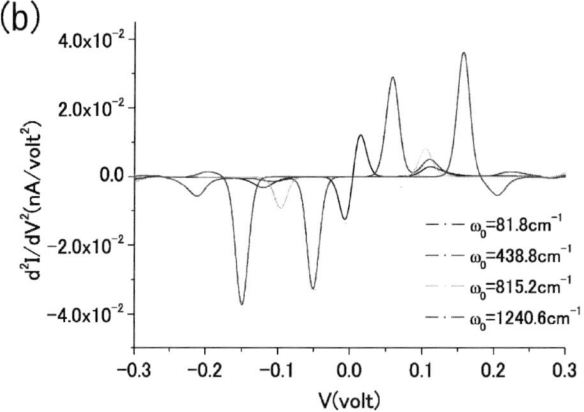

Fig. 6.3. a The current (I) versus voltage (V) plot for the elastic $I_{elastic}$ and the inelastic $I_{inelastic}$ contributions to the total electric current through a benzene dithiol molecule. The normal mode of the molecular vibration taken into account in the calculation is depicted in the inset. The frequency of the vibration mode is 1240.6 cm^{-1}, which corresponds to the threshold value of the voltage. **b** The second derivatives of the inelastic current $d^2 I_{inelastic}/dV^2$ for the totally symmetric modes. The frequencies of the modes taken into account in the calculation independently are given in the inset. The peaks appear at the voltage values corresponding to the vibrational energies

way with vibrations. This is clear in the second derivative plot shown in Fig. 6.3b.

While the inelastic electric current $I_{\text{inelastic}}$ flows only with the help of vibrational excitations, vibrations work exclusively to suppress the elastic contribution I_{elastic}. The two channels behave very differently. For example, their voltage dependence is largely different. This should not be limited to the benzene dithiol molecule but should be true in general. This observation therefore is the most important lesson from the weak coupling theory to our DNA problem.

6.4 Strong Coupling Limits

We have discussed quite extensively the weak coupling theory. This is because very reliable applications of the theory can be made in this region. We have adopted the self-consistent Born approximation there. Actually, this approximation is one of the conserving approximations [16], which should be important in the non-equilibrium calculations. In the strong coupling region, unfortunately, there are no such approximations that satisfy the conservation law exactly. We will be using some different ways to tackle the tough problem of the non-equilibrium system in the strong coupling limit.

6.4.1 Dynamical Conductivity

After the canonical transformation, our Hamiltonian is reduced to the following effective Hamiltonian which should be useful in the strong coupling limit

$$\tilde{H} = t \sum_{j\delta} c_{j+\delta}^{\dagger} c_j X_{j+\delta}^{\dagger} X_j + \varepsilon \sum_i c_j^{\dagger} c_j + \sum_{\mathbf{q}} \omega_{\mathbf{q}} b_{\mathbf{q}}^{\dagger} b_{\mathbf{q}} - \sum_j \Xi n_j , \qquad (6.11)$$

where

$$X_j = \exp\left[\sum_{\mathbf{q}} \exp(\mathrm{i}\mathbf{q} \cdot \mathbf{R}_j) \frac{M_{\mathbf{q}}}{\omega_{\mathbf{q}}} (b_{\mathbf{q}} - b_{-\mathbf{q}}^{\dagger}) \right] .$$

Here $\Xi = \sum_{\mathbf{q}} \frac{M_{\mathbf{q}}^2}{\omega_{\mathbf{q}}}$ denotes the polaron binding energy. The real part of the dynamical conductivity is calculated by applying the Kubo formula to the effective Hamiltonian

$$Re[\sigma(\omega)] = \frac{1 - \mathrm{e}^{-\beta\omega}}{2\omega} \int_{-\infty}^{\infty} \mathrm{d}t \, \mathrm{e}^{\mathrm{i}\omega t} \langle J^{\dagger}(t) J(0) \rangle ,$$

where

$$J = \mathrm{i} \frac{et}{\hbar} \sum_{j\delta} \delta c_{j+\delta}^{\dagger} c_j X_{j+\delta}^{\dagger} X_j ,$$

and β denotes the inverse of the temperature, i.e., $1/(k_B T)$. If we neglect the dynamics of electrons and suppose that electrons make just background charge for ionic motions, then the conductivity may be given in terms of correlation function of the ionic motions [5]

$$Re[\sigma(\omega)] = \frac{1 - e^{-\beta\omega}}{2\omega} t^2 \left(\frac{e}{\hbar}\right)^2$$

$$\times \sum_{\substack{j,j' \\ \delta,\delta'}} \delta\delta' \int_{-\infty}^{\infty} dt\, e^{i\omega t} \langle c_j^\dagger(t) c_{j+\delta}(t) c_{j'+\delta'}^\dagger c_{j'}\rangle \langle X_j^\dagger(t) X_{j+\delta}(t) X_{j'+\delta'}^\dagger X_{j'}\rangle$$

$$\cong t^2 \left(\frac{N}{R}\right) \left(\frac{e}{\hbar}\right)^2 c(1-c)\frac{1 - e^{-\beta\omega}}{\omega} U(\omega),$$

where

$$U(\omega) = \int_{-\infty}^{\infty} dt\, e^{i\omega t} \langle X_j^\dagger(t) X_{j+\delta}(t) X_{j+\delta}^\dagger X_j\rangle.$$

We have used the simplification, $\langle c_j^\dagger(t) c_{j+\delta}(t) c_{j'+\delta'}^\dagger c_{j'}\rangle \cong (1/\Omega)\delta_{j,j'}\delta_{\delta,\delta'}c(1-c)$. After some algebra using the coherent state, the correlation function is calculated to be

$$U(\omega) = \int_{-\infty}^{\infty} dt\, e^{i\omega t} \exp(-2S_T + \varphi(t)),$$

where,

$$\varphi(t) \equiv \sum_{\mathbf{q}} 2|u_{\mathbf{q}}|^2 [N_{\mathbf{q}}(N_{\mathbf{q}} + 1)]^{\frac{1}{2}} \cos[\omega_{\mathbf{q}}(t + i\beta/2)],$$

$$2S_T = \sum_{\mathbf{q}} |u_{\mathbf{q}}|^2 (2N_{\mathbf{q}} + 1), [\omega_{\mathbf{q}}(t + i\beta/2)],$$

$$u_{\mathbf{q}} = \frac{M_{\mathbf{q}}}{\omega_{\mathbf{q}}}(1 - e^{i\mathbf{q}\cdot\delta})$$

$$N_{\mathbf{q}} = \left(e^{\beta\omega_{\mathbf{q}}} - 1\right)^{-1}.$$

In the following, we will use the transition state approximation to the integral after making the change of the variable $Z = t + i\beta/2$ and the expansion $\varphi(Z) \cong \varphi(0) - \gamma Z^2$

$$U(\omega) = \exp(-2S_T + 1/2\beta\omega) \int dZ \exp(i\omega Z) \exp[\varphi(Z)]$$

$$\cong \exp(-2S_T + 1/2\beta\omega + \varphi(0)) \int dZ \exp(i\omega Z) \exp(-\gamma Z^2)$$

$$= \left(\frac{\pi}{\gamma}\right)^{1/2} \exp(-2S_T + \varphi(0)) e^{\frac{\beta\omega}{2}} \exp\left(-\frac{\omega^2}{4\gamma}\right)$$

$$\cong \left(\frac{\pi}{\gamma}\right)^{1/2} e^{\frac{\beta\omega}{2}} \exp(-\beta\bar{\Delta}) \exp\left(-\frac{\omega^2}{4\gamma}\right).$$

In the last equation, the following relation is used

$$2S_T - \varphi(0) = \sum_q |u_q|^2 \left([N_q + 1]^{1/2} - [N_q]^{1/2}\right)^2 \cong \sum_q |u_q|^2 \frac{\beta\omega_q}{4} = \frac{\bar{\Delta}}{k_B T} ,$$

which is valid at high temperature where $2N_q + 1 \cong 2/(\beta\omega_q) \cdot \bar{\Delta} \equiv \frac{1}{4}\sum_q |u_q|^2 \omega_q$. In the same high temperature limit

$$\gamma = \sum_q |u_q|^2 \omega_q^2 [N_q(N_q + 1)]^{1/2} \cong 4\bar{\Delta}k_B T .$$

Thus we get the following expression for the real part of the dynamical conductivity at high temperature,

$Re\,[\sigma(\omega)]$

$$\cong \frac{1}{2} z t^2 \left(\frac{M_0}{\Omega}\right) \left(\frac{e}{\hbar}\right)^2 \left(\frac{\pi}{\gamma}\right)^{\frac{1}{2}} c(1-c)\beta \frac{\sinh(\beta\omega/2)}{1/2\beta\omega} \exp\left(-\frac{\omega^2}{4\gamma}\right) \exp\left(-\frac{\bar{\Delta}}{k_B T}\right)$$

$$\sim \frac{1}{2} z t^2 \left(\frac{M_0}{\Omega}\right) \left(\frac{e}{\hbar}\right)^2 \left(\frac{\pi}{\gamma}\right)^{1/2} c(1-c)\beta \frac{\sinh(\beta\omega/2)}{1/2\beta\omega} \exp\left(-\frac{\bar{\Delta}}{k_B T}\right)$$

$$\cong \frac{1}{2} z t^2 \left(\frac{M_0}{\Omega}\right) \left(\frac{e}{\hbar}\right)^2 \left(\frac{\pi}{\gamma}\right)^{1/2} c(1-c)\beta \frac{1 - e^{-\beta\omega}}{1/2\beta\omega} \exp\left(-\frac{(\omega - \beta\gamma)^2}{4\gamma}\right) .$$

$$(6.12)$$

The frequency dependent current density $i(\omega) = \sigma(\omega)E(\omega)$ may provide a good estimate for the static current density under the bias voltage V [17]

$$i(eV) \cong Re\,[\sigma(eV)]\,E(0) \cong Re\,[\sigma(eV)]\,V/d$$

$$\cong z t^2 \left(\frac{M_0}{\Omega}\right) \left(\frac{e}{\hbar}\right)^2 \left(\frac{\pi}{\gamma}\right)^{\frac{1}{2}} \frac{c(1-c)}{ed} \sinh\left(\frac{\beta eV}{2}\right) \exp\left(-\frac{\bar{\Delta}}{k_B T}\right) ,$$

$$(6.13)$$

where V is the voltage difference between the two electrodes and d is the distance between them. The charge current thus obtained obeys the relation, $I \propto \sinh(bV)$, where the coefficient b is proportional to the inverse of the temperature. The temperature dependence of b besides the Ahrenius factor is characteristic to this mechanism. In this calculation, the e-mv coupling is treated diabatically. The calculation therefore should involve various processes including the inelastic processes. To make the counterpart clear, we will estimate the elastic current of the effective Hamiltonian in the following section by limiting our problem within the elastic subspace.

6.4.2 Elastic Conductance

If $m = \sum_q m_q$ phonons are absorbed during the hopping, the thermal hopping probability P due to the phonon part of the correlated hopping term in

(6.11), i.e., $X_{j+\delta}^{\dagger} X_j$ is given as follows [5]

$$P = \sum_{m_{\mathbf{q}}} \Pi_{\mathbf{q}} \exp\left\{-|u_{\mathbf{q}}|^2 (2N_{\mathbf{q}} + 1)\right\} \exp\left(\frac{m_{\mathbf{q}} \beta \omega_0}{2}\right)$$

$$I_{m_{\mathbf{q}}}\left(2|u_{\mathbf{q}}|^2 \left[N_{\mathbf{q}}(N_{\mathbf{q}} + 1)\right]^{1/2}\right) \times 4^{m_{\mathbf{q}}} \ .$$

We have considered here the Einstein phonon $\omega_{\mathbf{q}} = \omega_0$. When the phonon excitation energy in each $\mathbf{q}$ cancels each other and the total the phonon energy is conserved during the hopping, i.e., the conduction takes place elastically, the total number of phonons does not change, $m = \sum_{\mathbf{q}} m_{\mathbf{q}} = 0$. Owing to the additive property of the Bessel functions $\sum_m I_m(\xi_1) I_{n-m}(\xi_2) = I_n(\xi_1 + \xi_2)$, the elastic hopping probability P_{elastic} is given as follows

$$P_{\text{elastic}} = \exp(-2S_{\text{T}}) I_0(\xi) \cong \frac{1}{(2\pi\xi)^{1/2}} \exp(-2S_{\text{T}} + \xi) , \tag{6.14}$$

where $\xi = \sum_{\mathbf{q}} 2|u_{\mathbf{q}}|^2 [N_{\mathbf{q}}(N_{\mathbf{q}} + 1)]^{\frac{1}{2}}$. We have assumed that $|\xi| \succ 1$ and then $I_0(\xi) \cong \frac{1}{\sqrt{2\pi\xi}} \exp(\xi)$. Since $2N_{\mathbf{q}} + 1 = \coth(\beta\omega_{\mathbf{q}}/2)$, it is not difficult to find that $2S_{\text{T}} - \xi = \sum_{\mathbf{q}} |u_{\mathbf{q}}|^2 \tanh(\beta\omega_{\mathbf{q}}/4)$. Denoting, $\Delta = k_{\text{B}}T \sum_{\mathbf{q}} |u_{\mathbf{q}}|^2 \tanh(\beta\omega_{\mathbf{q}}/4)$, we obtain the following relation

$$P_{\text{elastic}} \propto \exp\left(-\frac{\Delta}{k_{\text{B}}T}\right) . \tag{6.15}$$

When the energy gap parameter is large, i.e., $\chi = |(E_{\text{F}} - \varepsilon)/(2t)| \gg 1$, the self energy due to the e-mv coupling may not give rise to a large change in the elastic conductance for the long chain molecules, as is the case in the dephasing model which is discussed in Sect. 6.3.1. If that happens, the most significant change due to the small polaron effect in the elastic conductance may be given in terms of the phonon-assisted transfer integral $\tilde{t} = t \exp\left(-\frac{\Delta}{2k_{\text{B}}T}\right)$. If we use the recursive Green's function method [18], the $(1, N)$ component of the retarded Green's function of the extended molecule $g_{1N}^{\text{R}}(E_{\text{F}})$ may be given in terms of the phonon assisted transfer

$$g_{1N}^{\text{R}}(E) = \tilde{t}^{N-1}$$
$$\times \left[\left(E - \varepsilon + i\eta - \Theta_{N-1}^{\text{R}}\right)\left(E - \varepsilon + i\eta - \Theta_{N-2}^{\text{R}}\right)\right.$$
$$\left. \cdots \left(E - \varepsilon + i\eta - \Theta_1^{\text{R}}\right)\left(E - \varepsilon + i\eta\right)\right]^{-1} . \tag{6.16}$$

The intramolecular self-energy term $\Theta_P^{\text{R}}(E)$ appeared in the recursive Green's function method is given by the following continued fraction

$$\Theta_P^{\text{R}}(E) = \tilde{t}^2 \cfrac{1}{E - \varepsilon + i\eta - \tilde{t}^2 \cfrac{1}{E - \varepsilon + i\eta - \tilde{t}^2 \cfrac{1}{E - \varepsilon + i\eta - \tilde{t}^2 \cfrac{1}{\ddots}}}} \tag{6.17}$$

where the continued fraction is taken up to the p-th order. In the large gap limit, i.e., $|(E - \varepsilon)/(2t)| \gg 1$, $\Theta_p^R(E)$ may be neglected. Then the terminal component of the retarded Green's function is approximated by

$$g_{1N}^R(E) = \frac{1}{\tilde{t}}\left(\frac{\tilde{t}}{E - \varepsilon}\right)^N \approx \frac{1}{\tilde{t}}\exp\left(-\bar{\beta}_E R\right)\exp\left(-\frac{E_a}{2k_BT}\right) , \qquad (6.18)$$

$$\bar{\beta}_E = \frac{1}{a}\log\left(\frac{E - \varepsilon}{t}\right) ,$$

where $E_a = k_BT\sum_{\mathbf{q}}|\tilde{u}_{\mathbf{q}}|^2\tanh(\beta\omega_{\mathbf{q}}/4)$ and $|\tilde{u}_{\mathbf{q}}|^2 = (N-1)|u_{\mathbf{q}}|^2$. While at high temperature $E_a \simeq \frac{1}{4}\sum_{\mathbf{q}}|\tilde{u}_{\mathbf{q}}|^2\omega_{\mathbf{q}}$, at low temperature $E_a \simeq k_BT\sum_{\mathbf{q}}|\tilde{u}_{\mathbf{q}}|^2$. The conductance at a finite voltage is given by

$$G(V) \cong \frac{2e^2}{h}\Gamma_L\Gamma_R\left\{\eta|g_{1N}^R(\mu_1)|^2 + (1-\eta)|g_{1N}^R(\mu_2)|^2\right\} ,$$

where $\mu_1 = E_F - \eta eV$, $\mu_2 = E_F + (1-\eta)eV$. Here we have neglected the energy dependency of Γ_L and Γ_R, η is the voltage division factor which should be taken to be 0.5 for the symmetric electrodes. For the symmetric case, the conductance of the DNA molecule $G(V)$ is given by the following equation [17]

$$G(V) \cong \frac{2e^2}{h}\left(\frac{\Gamma}{t}\right)^2\exp\left(-\bar{\beta}_{E_F}R\right)\cosh(\lambda eVR)\exp\left(-\frac{E_a}{k_BT}\right) ,$$

where

$$\lambda = \frac{1}{a}\frac{1}{(E_f - \varepsilon)} \quad\text{and}\quad \bar{\beta}_{E_F} = \frac{1}{a}\log\left(\frac{E_F - \varepsilon}{t}\right) .$$

Here, we have put $\Gamma = \Gamma_R = \Gamma_R$. We finally obtain the formula for the electric current that flows through the DNA molecule as follows

$$I \cong \frac{2e^2}{h}\frac{(\Gamma/t)^2}{\lambda eR}\exp(-\bar{\beta}_{E_F}R)\sinh(\lambda eVR)\exp\left(-\frac{E_a}{k_BT}\right) . \qquad (6.19)$$

This formula may be simplified such that $I \propto \sinh(bV)$ and $b = \lambda eR$. It should be noted that the coefficient b is independent of the temperature, which is in clear contrast to the result derived from the dynamical conductivity calculation where the inverse temperature dependence has been obtained. There are two distinct conduction mechanisms in the strong coupling region, as is the case of the weak coupling region discussed in Sect. 6.3.2. The difference in the conduction mechanisms between the elastic and the hopping ones may be most clearly seen in the temperature dependence of the voltage coefficient in the strong coupling region. Another point in this formula is the length dependence of the conductance: It decays exponentially, as it has been discussed in Sect. 6.3.1. The elastic mechanism may not be dominant for the long chain DNA molecules.

Both the inverse temperature dependent and the temperature independent voltage coefficients b have been observed in the dc electric current measurements done in the nano-gap system. The former was found in the poly(dG)-poly(dC) molecule while the latter was found in the poly(dA)-poly(dT) molecule [19]. It seems, however, that a consensus among the experiments has not yet been met. Experiments conducted on the well defined systems, especially with careful attention to the humidity control, would be very useful.

6.5 Hole Transfer Reactions Through DNA in Solution

It has long been well known that the Marcus electron transfer reaction theory can be reformulated in a similar way as discussed in Sect. 6.4.1 [20]. In fact, the formula for the real part of the dynamical conductivity $\text{Re}[\sigma(\omega)]$, i.e., (6.12) includes the exponential factor $\exp\left(-(\omega - \beta\gamma)^2/4\gamma\right)$, which may correspond to the Franck-Condon factor $\frac{1}{\sqrt{4\pi\lambda k_B T}}\exp\left(-\Delta G^*/k_B T\right)$ in the Marcus electron transfer reaction rate theory, where $\Delta G^* = \frac{(\lambda+\Delta E)^2}{4\lambda}$. The Marcus reorganization energy λ and the energy gap ΔE may correspond to $\beta\gamma = 4\bar{\Delta}$ and $-\omega$ in (6.12), respectively.

Experimental studies of the intramolecular hole transfer reaction between the singlet guanine G and the triplet guanine GGG initiated by the photo-carrier injection have been reported. The reaction rates were measured for various sequences of the DNAs. The hole transfer reaction rate depends on the distance between the singlet guanine G and the triplet guanine GGG pairs separated by adenine-thymine (A-T) bridges. Experimental studies have concluded that there are two distance dependence in the reaction rates [21]. That conclusion suggests the existence of the two distinct hole transfer reaction mechanisms. Two mechanisms have been also proposed theoretically: one is the thermally induced hopping mechanism and the other is the superexchange mechanism [22]. The latter is a tunnelling mechanism while the former is essentially a hopping mechanism. The elementary reaction processes assumed in these two mechanisms are shown in Fig. 6.4a.

The chemical reaction rate of the elementary hole transfer process can be calculated in terms of the Marcus theory

$$k = \frac{2\pi}{\hbar}V^2\frac{1}{\sqrt{4\pi\lambda k_B T}}\exp\left(-\frac{\Delta G^*}{k_B T}\right) , \qquad (6.20)$$

where V is the hole coupling term, which can be estimated in various ways.

If the sequential hole transfer between the thymine molecules is the rate-limiting process of the hole transfer reaction in the superexchange mechanism, the hole coupling terms in the elementary processes may be unified to give

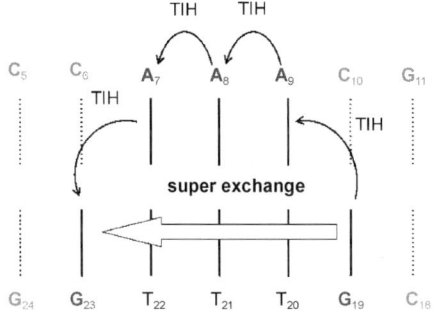

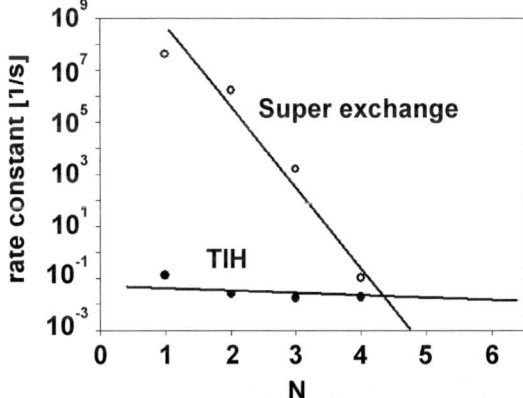

Fig. 6.4. a Elementary hole transfer reaction processes in the superexchange mechanism and the thermally induced hopping mechanism (TIH). G, C, A, T denote guanine, cytosine, adenine and thymine base molecules. The number denotes the sequence **b** Hole transfer reaction rates due to the superexchange mechanism and the thermally induced hopping mechanism [23]. N denotes the number of the A-T bridge intervening the singlet G and the triplet GGG guanine pairs

the effective hole coupling term V_{sup} which represents the hole coupling in the overall reaction

$$V_{\text{sup}} = \frac{V(G - T_1)V(T_n - G)}{\Delta(G - T_1)} \times \Pi_{j=1}^{N-1} \frac{V(T_j - T_{j+1})}{\Delta(G - T_{j+1})} \ ,$$

where $V(T_j - T_{j+1})$ is the hole coupling term between the neighboring thymine molecules, $V(T_n - G)$ and $V(G - T_1)$ are the hole coupling terms between the thymine and the guanine molecules at the two terminals of the sequence $G - T_1 - T_2 - T_3 - \ldots - T_N - G$, and $\Delta(G - T_j)$ is the difference in ionized energy between G and T. By using similar arguments as made in Sect. 6.4.2 (the arguments given to explain (6.16)–(6.18)), it is easy to show that V_{sup} gives rise to an exponential decay as a function of the distance, i.e., the tunnelling factor

$$V_{\text{sup}} \propto \exp\left(-\bar{\beta}_{\text{DA}} R\right) ,$$

where

$$\bar{\beta}_{\text{DA}} = \frac{1}{a} \log\left(\frac{\Delta(\text{G} - \text{T})}{V(\text{T} - \text{T})}\right) .$$

We have assumed that all thymines are equivalent and are equally spaced by a distance a, i.e., $V(\text{T} - \text{T}) = V(\text{T}_j - \text{T}_{J+1})$ and $\Delta(\text{G} - \text{T}) = \Delta(\text{G} - \text{T}_j)$. Because the Frank-Condon factors for hole transfers between the thymine molecules do not differ, the overall superexchange reaction rates made up of many elementary processes can be described by a single modified Marcus formula, where V is replaced by V_{sup}. Therefore, it is expected that the reaction rate due to the superexchange mechanism decays very rapidly as a function of the distance.

Elementary hole transfer reaction rates were estimated in terms of the *ab initio* molecular orbital calculations for cluster models of DNA. Those calculations show that the sequential hole transfer between the thymine molecules is the rate-limiting process in the superexchange mechanism and that the hole transfer between the guanine and the adenine is the limiting process in the thermally induced hopping process [23]. Because of these limiting processes, we found that while the reaction rate due to the superexchange mechanism decays very rapidly as a function of the distance, the reaction rate due to the thermally induced hopping has no such distance dependence. Our result is summarized in Fig. 6.4b, which is in fair agreement with the experimental result. It should be noted that the hole transfer rate due to the superexchange mechanism decays very rapidly such that the rate becomes smaller than that due to the thermally-hopping mechanism as fast as $N = 4$.

6.6 Discussions

Because the hole transfer occurs between the base molecules which are located in the hydrophobic groove rather than the hydrophilic region composed of the sugar molecules in DNA, the holes may be rather free from the water molecules in the solution. Even though the hole transfer reaction takes place in a wet environment, it may therefore be possible to expect that similar transport mechanisms work both in the nano-gap system and the hole transfer reaction system in solution. The water molecules sticking to the hydrophilic region composed of the sugar molecule in DNA, which can never be removed even in the ultra high vacuum environment, may cause additional difficulty at the contact in the nano-gap system. It might therefore be worthwhile to learn from the hole transfer reaction system which is *contact free*. It has turned out that there are two reaction mechanisms, i.e., the tunnelling mechanism and the hopping mechanism in the hole transfer reaction in DNA. Our calculation suggests that the former can be seen only when the distance between the adjacent base molecules is as small as about four base units.

In the transport problem, we have two mechanisms in DNA, i.e., the elastic conduction and the hopping conduction. The former is accompanied by the exponential decay factor as a function of the length of DNA. The exponents of the length dependence of the elastic conductance and the hole transfer reaction in terms of the superexchange mechanism differ by $\bar{\beta}_{E_F} - \bar{\beta}_{DA} = \log\left(\frac{E_F - \varepsilon}{\Delta(G-T)} \cdot \frac{V(T-T)}{t}\right)$. It may be possible that the superexchange exponent $\bar{\beta}_{DA}$ discussed in Sect. 5 could be reduced in the transport experiment if we had $|\frac{E_F - \varepsilon}{\Delta(G-T)} \cdot \frac{V(T-T)}{t}| \leq 1$. It may not be easy to observe the elastic conductance unless a very short DNA segment is synthesized and is measured, which may be difficult in the case of the nano-gap, however. Recent experimental developments in the conductance histogram measurements of single molecules and very short DNA using the STM, both in solution [24] and in the ultra high vacuum environment [25] may improve the experimental situation. Electrodes composed of the organic material may provide another possible way to access the elastic conductance in single molecules.

Some theoretical calculations on the ballistic current of DNA attached in various ways to the metallic electrode were made [26, 27]. They found small conductance in all cases. The dependence of the conductance on the contact state which they found should accompany the independence of the exponent on the contact state which was discussed in Sect. 6.3.1, however. In any case, the ballistic conductance of DNA attached to the metallic electrode is very small unless the length of DNA is short enough. Hopping conduction may therefore be the primary mechanism for the long chain DNA molecule placed between the metallic electrodes.

6.7 Summary

We have discussed the vibronic mechanisms of charge transport and migration in a single DNA molecule. Unless the length of the DNA molecule is short enough, the hopping conduction mechanism may be dominant over the elastic conduction mechanism.

Acknowledgement. One of the authors (Y.A.) is thankful to Prof. Hidetoshi Fukuyama for long standing collaborations and continuing discussions made in these years. The authors also deeply appreciate collaborations with Prof. Koichi Yamashita. Some parts of the work were made with financial supports from Japan Science and Technology Corporation (Y.A. and T.S.), Grant-in-aid from the Ministry of Education in Japan (Y.A.) and from the Designated Visiting Professor Program of the School of Engineering, University of Tokyo (Y.A), to which we would like to thank.

References

1. H. Shirakawa and S. Ikeda, Polym. J. **2**, 231 (1971).
2. H. Akamatsu and H. Inokuchi, J. Chem. Phys. **18**, 810 (1950).
3. T.Ishiguro and K. Yamaji, *Organic Superconductors* (Springer-Verlag, Tokyo, 1989).
4. H. Kino, M. Tateno, M. Boero, J.A. Torres, T. Ohno, K. Terakura and H. Fukuyama, J. Phys. Soc. Jpn. **73**, 2089 (2004).
5. G.D. Mahan, *Many-Particle Physics* (Plenum Press, New York, 1981).
6. N.O. Lipari, C.B. Duke and L. Pientronero, J. Chem. Phys. **65**, 1165 (1976).
7. Y. Asai and Y. Kawaguchi, Phys. Rev. B **46**, 1265 (1992); Y. Asai, *ibid.* **49**, 4289 (1994); **68**, 014513 (2003); W.E. Pickett, in *Solid State Physics*, vol. 49, edited by H. Ehrenreich and F. Spaepen (Academic Press, Boston, 1994); M.S. Dresselhaus, G. Dresselhaus, P.C. Eklund, *Science of Fullerenes and Carbon Nanotubes* (Academic Press, San Diego, 1996); O. Gunnarsson, *Alkali-Doped Fullerides* (World-Scientific, Singapore, 2004).
8. R.G. Endres, D.L. Cox and R.R.P. Singh, Rev. Mod. Phys. **76**, 195 (2004); M. Taniguchi and T. Kawai, Phys. Rev. E **70**, 011913 (2004); *ibid.* **72**, 061909 (2005).
9. S. Datta, *Electronic Transport in Mesoscopic Systems* (Cambridge University Press, Cambridge, 1995).
10. J.W. Evenson and M. Karplus, J. Chem. Phys. **96**, 5272 (1992).
11. V. Mujica, M. Kemp and M.A. Ratner, J. Chem. Phys. **101**, 6856 (1994).
12. Y. Asai and H. Fukuyama, Phys. Rev. B **72**, 085431 (2005).
13. Y.-C. Chen, M. Zwolak and M. Di Ventra, Nano Letters, **4**, 1709 (2004).
14. Y. Asai, Phys. Rev. Lett. **93**, 246102 (2004); **94**, 099901 (2005).
15. M. Paulsson, T. Frederiksen and M. Brandbyge, Nano Lett. **6**, 258 (2006).
16. L. Kadanoff and G. Baym, *Quantum Statistical Mechanics* (Benjamin, Massachusetts, 1962).
17. Y. Asai, J. Phys. Chem. B **107**, 4647 (2003).
18. D.K. Ferry and S.M. Goodnick, *Transport in Nanostructure* (Cambridge University Press, Cambridge, 1997).
19. K.-H. Yoo, D.H. Ha, J.-O.Lee, J.W. Park, J. Kim, J.J. Kim, H.-Y. Lee, T. Kawai and H.-Y. Choi, Phys. Rev. Lett. **87**, 198102 (2001).
20. G.C. Schatz and M.A. Ratner, *Quantum Mechanics in Chemistry* (Prentice Hall, New Jersey, 1993).
21. B. Giese, Acc. Chem. Res. **33**, 631 (2001).
22. M. Bixon and J. Jortner, J. Am. Chem. Soc. **123**, 12256 (2001).
23. T. Shimazaki, Y. Asai and K. Yamashita, J. Phys. Chem. B **109**, 1295 (2005).
24. B.X. Xu and N.J. Tao, Science **301**, 1221 (2003); B. Xu, P. Zhang, X. Li, and N. Tao, Nano Lett. **4**, 1105 (2004).
25. M. Fujihira, M. Suzuki, S. Fujii and A. Nishikawa, Phys. Chem. Chem. Phys. **8**, 3876 (2006).
26. T. Tada, M. Kondo and K. Yoshizawa, Chem. Phys. Chem. **4**, 1256 (2003).
27. O.R. Davies and J.E. Inglesfield, Phys. Rev. B **69**, 195110 (2004).

7 The Role of Charge and Spin Migration in DNA Radiation Damage

David Becker, Amitava Adhikary, and Michael D. Sevilla

Department of Chemistry, Oakland University, Rochester, MI 48309, USA
sevilla@ouchem.chem.oakland.edu

7.1 Introduction

Ionization via high energy radiation of the components of DNA initially occurs roughly in proportion to the local electron density. As a consequence, each portion of the DNA molecule and its environs such as the bases, the sugar phosphate backbone, and the waters of hydration are ionized in a near random fashion. However, owing to charge and spin transfer the chemical damage in irradiated DNA often occurs at sites other than where the original ionization takes place [1]. There are a variety of charge migration processes which occur on different time scales and which play a role in the eventual location of damage sites in irradiated DNA [2]. These involve short-range fast processes on times scales of picoseconds, which are dominated by tunneling and which include transfer of charge from the first solvation shell to DNA as well as from a base to a near neighbor base [1–12]. Longer range hole and electron transfers, especially those over a few bases, are, at room temperature, dominated by activated processes that take place over nanoseconds and longer [9–11].

When high energy radiation (γ-irradiation, X-irradiation, high energy ion beams) traverses a solvated DNA molecule, the immediate effect is the ionization and excitation of the moieties of the molecule [12–30]. Ion particle radiations and photons greatly differ in the spatial distribution of the ionizations and excitations. Whereas electromagnetic radiations such as gamma, and X-rays ionize molecules via direct absorption and the Compton effect they are considered sparsely ionizing compared to charged particles e.g., beta, alpha, proton etc. which produce a dense trail of ionizations [13, 14, 25, 28].

The rate at which the energy is transferred to the medium through which it passes, is determined by linear energy transfer (LET), i.e. the energy lost per unit length of the track [23–30]. Low LET radiation such as gamma and X-rays ($0.3 - 3$ keV/micron) produce isolated primary ionizations in "spurs" while high LET radiations (such as, alpha particles (100 keV/micron), neutrons (20 keV/micron) etc.) produce a track of closely spaced ionizations in a track "core" [25, 28].

The initial ionizations from the impinging radiation result in release of high energy electrons which then cause further ionizations and excitations producing a cascade of medium and low-energy electrons (LEE) as well as numerous excited states [31–47]. Energetic secondary electrons are responsible for causing most of the ionizations and subsequent damage to the DNA [32]. The ionizations result in DNA anion and cation radicals as well as excitations [1]. However, the LEEs can also directly damage DNA causing sugar-phosphate bond cleavage resulting in strand breaks [31–40]. Recently, it has been discovered that excited states in combination with ion-radicals also lead to DNA damage to the sugar phosphate backbone [41–47]. For all of these processes, the chemical damage in DNA initially is chiefly in the form of free radicals (cation, anion and neutral radicals), which are highly reactive but can be stabilized at low temperatures and studied using ESR spectroscopy [7–10, 12]. The specific location of the chemical damage and, consequently, the ultimate biological damage from DNA irradiation depends on the charge transfer processes that occur following ionization [1–11]. These processes and their very approximate time scales are shown in Table 7.1.

Note that the actual time scale for charge transfer depends on (i) temperature, (ii) DNA strandedness, (iii) sequence, (iv) solvation extent, and (v) complexing ligands (e.g., histones, and small molecules e.g., spermine, spermidine etc.) [1–12,48–54]. It is also noteworthy that the times for intra-duplex vs. inter-duplex transfer likely differ [8,9,51].

In this Chapter, we will present the results of Electron Spin Resonance (ESR) experiments and Density Functional Theory (DFT) calculations which have clarified the nature of charge transfer processes in DNA. Specifically, the role of hole and electron tunneling from the first solvation shell to DNA, from sugar to base, from base to base and from the excited base cation radical to sugar will be discussed. Proton transfer reactions and irreversible protonations are shown to modulate or quench electron and hole transfer processes. Mechanisms which combine excited states and charge transfer will be discussed as mechanisms for strand breaks, an important biological lesion.

Table 7.1. Charge transfer process and their relative approximate time scales

Event	Approximate time scale
Recombination in spurs	$10^{-13} - 10^{-9}$ s
Hole/electron transfer from first Solvation shell	$10^{-13} - 10^{-11}$ s
Hole transfer from sugar to base	$< 10^{-13} - 10^{-11}$ s
Long distance electron and hole transfer	10^{-12} s to long times
$C^{\bullet -} \longrightarrow T(C6)H^{\bullet}$	$< 10^{-6}$ s slow
$G^{\bullet +}$, $A^{\bullet +} \longrightarrow$ sugar radicals	ca. 10^{-12} s fast via excitation

Charged and Protonated Base Radicals*

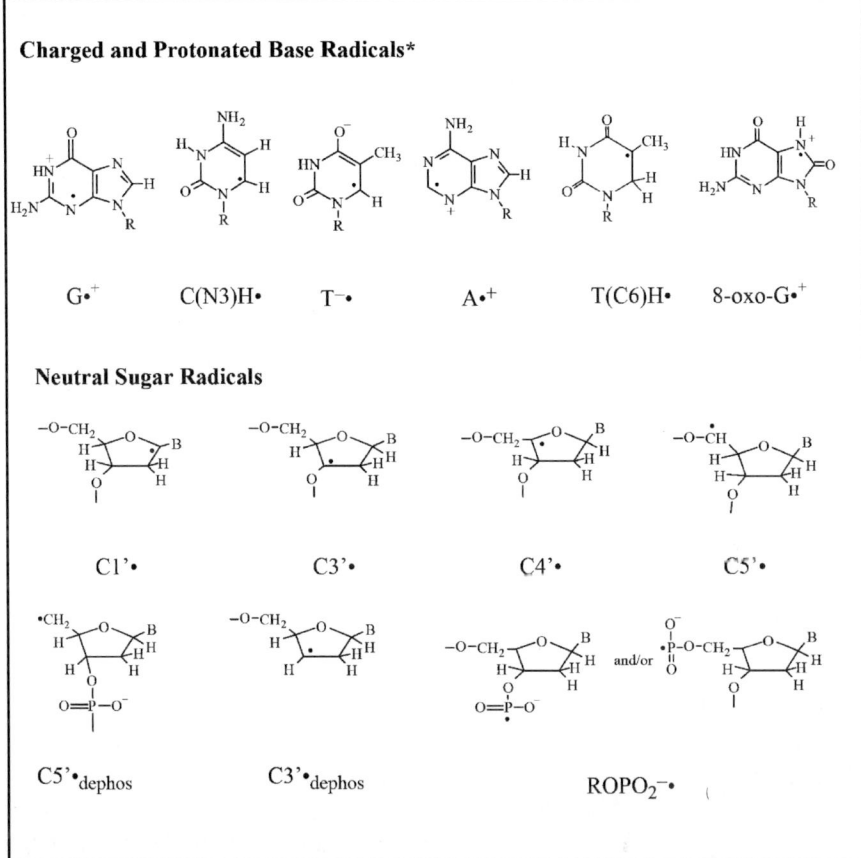

| G•⁺ | C(N3)H• | T⁻• | A•⁺ | T(C6)H• | 8-oxo-G•⁺ |

Neutral Sugar Radicals

| C1'• | C3'• | C4'• | C5'• |

| C5'•_dephos | C3'•_dephos | ROPO₂⁻• |

Scheme 7.1. Structure of Radicals Discussed. *C(N3)H• is reversibly protonated C⁻• and T(C6)H• is irreversibly protonated T⁻•

7.2 Radical Stabilization at 77 K and Higher Temperatures

From the early events following the deposition of energy by ionizing radiation in DNA, to the final biological effects of the radiation, charge transfer processes play an important role in determining the radiation induced damage that results in deleterious effects for living organisms. The rates, energetics, and nature of charge transfer, be it hole transfer, electron transfer or proton transfer are, in fact, critical determinants of the final spectrum of radiation damage and, thereby, of the possible deleterious effects to the living cell and/or organisms.

7.2.1 Fast Ion Radical Recombination

When ionizing radiation (γ-ray, X-ray, heavy ion) impinges on DNA, high energy electrons are ejected from those molecules which are ionized [13,24,28]. These electrons, largely through Compton scattering, encounter and ionize other molecules; the cascade of low to medium energy electrons which are generated by a series of sequential ionizations are responsible for most of the radiation damage that occurs to an irradiated sample [13–30]. Depending on the amount of energy deposited by a secondary electron and the proximity of the nearby ionizations that occur (if there is more than one proximate ionization), an entity called a spur ($6-100\,eV$), blob ($100-500\,eV$) or short track ($500-5000\,eV$) is formed [14, 15, 27, 28]. Each of these is a volume of space in which one or more nearby ionizations occur. For low linear energy transfer radiation (γ-ray, X-ray), the energy deposition pattern is envisioned as a series of widely separated spurs, blobs and tracks [13–30]. This pattern of energy deposition results in an interesting effect, *viz.*, that the dominant radical process that occurs when ionizing radiation impinges on DNA is ion radical recombination [7–10, 12].

Theoretical studies combined with experimental results have borne out this conclusion [13–28]. Based on ionization cross sections for interactions with electrons, the total initial yield of ionizations in H_2O has been determined to be $1.18\,\mu mol/J$ [25, 27, 28]. A similar value is expected for DNA. For hydrated ($\Gamma = 14$ H_2O/nucleotide) salmon sperm DNA γ-irradiated at $77\,K$, a yield of $0.24\,\mu mol/J$ has been reported by our laboratory [41, 55–57]. Thus, about 75% of the initial ionizations do not end up as stabilized radicals at $77\,K$ [7–10, 12]. Geminate recombination, in which the electron ejected in an ionization event does not escape from its positive ion partner, largely accounts for the 75% of ionizations which do not result in radical formation and stabilization [7–10, 12].

Thus, recombinations in a spur, blob, or track on a very fast time scale constitute the first instance in which charge transfer, through electron transport, affects the eventual outcome of DNA damage in irradiated samples [7–10, 12–30].

7.2.2 Charge Transfer from the DNA Hydration Shell

The DNA Hydration Shell. Fully hydrated DNA has ca. 12 waters of hydration per nucleotide in its primary hydration shell ($\Gamma = 12$) and an additional ca. 10 waters per nucleotide in its secondary hydration shell. Water molecules above $\Gamma = ca.\ 22$ behave as (crystalline) bulk waters with a near normal water structure. Experimentally, the number of hydration water molecules present in a sample of DNA can be controlled by equilibrating the sample, at room temperature, with water vapor at a specific relative humidity [56, 58, 59]. However, for DNA samples hydrated at room temperature and subsequently cooled to $77\,K$, it has been found that the structure, at $77\,K$, of the frozen water surrounding a DNA molecule changes as water molecules are added. For example, for DNA hydrated to Γ = ca. 22, all the water molecules are in a glassy phase. As more water is added to the hydration shell the additional waters form a crystalline ice phase. The formation of this ice phase causes ca. 8 of those molecules that were previously in a glassy state join to the crystalline ice state, with the result that the glassy phase occurs only up to Γ = ca. 14. This is pictorially illustrated in Fig. 7.1, in which it is noted that the hydration levels reported have an uncertainty of 1 to 2 waters/nucleotide [56, 58, 59].

One of the principal radical products formed in bulk water by ionizing radiation is the hydroxyl radical, •OH; other radical species formed are electrons, and hydrogen atoms. At 77 K, in γ-irradiated water, the hydroxyl radical is the only one of these species stabilized and trapped. It also gives rise to a well-known ESR spectrum that depends on the nature of the frozen water, i.e. whether it is in a crystalline ice or in a glassy state [56, 58, 59]. An important finding regarding the radiation chemistry of hydrated DNA is that, at 77 K, the ESR spectrum from •OH is not observed in samples with $\Gamma \leq$ ca. 8 H_2O/nucleotide, despite the significant number of water molecules present. For samples with more than ca. 8 water molecules per nucleotide and less than ca 22 waters, i.e., for the secondary hydration layer, an •OH ESR spectrum is observed [58, 59]. This spectrum corresponds to that expected for an •OH in a glassy water phase. As more water is added, the spectra from both •OH in polycrystalline bulk water ($\Gamma > 14$) and •OH in a glassy water phase ($8 < \Gamma < 14$) are observed, since for extensively hydrated DNA all waters above Γ = ca. 14 are in the normal polycrystalline ice phase (Fig. 7.1) [56, 58, 59].

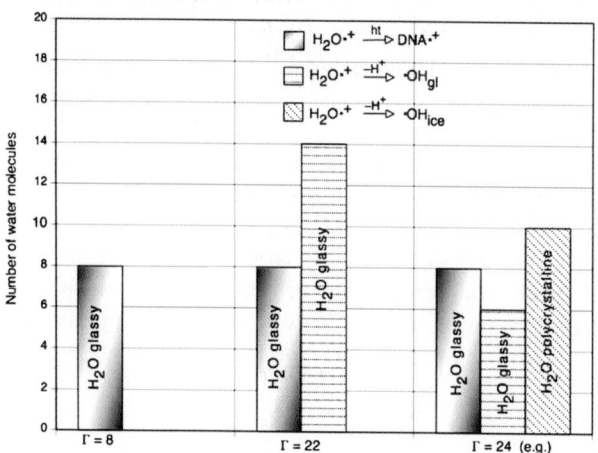

Fig. 7.1. The fate of the radiation-induced holes formed in the water hydration layer as a function of the nature of the DNA hydration layer and number of waters of hydration/nucleotide (Γ)

The lack of an observable •OH ESR spectrum in the first ca. 8 water molecules per nucleotide suggests that an important hole transfer process was occurring in these strongly bound waters, from the water layer to the DNA. As shown in Scheme 7.2, the situation is more complex than a straightforward hole transfer to the DNA. A competitive process occurs in the first hydration shell in which hole transfer from $H_2O^{\bullet+}$ to the DNA competes with deprotonation of $H_2O^{\bullet+}$ to form •OH [58,59].

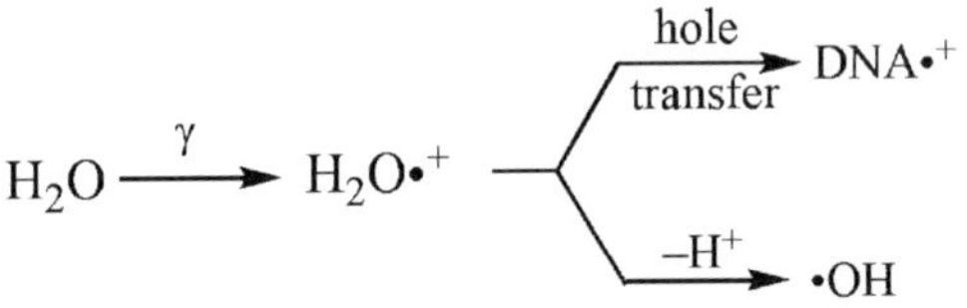

$$H_2O \xrightarrow{\gamma} H_2O^{\bullet+} \begin{cases} \xrightarrow[\text{transfer}]{\text{hole}} DNA^{\bullet+} \\ \xrightarrow{-H^+} \bullet OH \end{cases}$$

Scheme 7.2.

For tightly bound waters, hole transfer must be fast relative to the rate of $H_2O^{\bullet+}$ deprotonation which, itself, occurs on a sub-picosecond time scale. For loosely bound water molecules, which are further away from the DNA than the first solvation shell, deprotonation is faster than hole transfer and hydroxyl radical (•OH) forms [58,59].

Since many strand breaks are thought to originate with the electron loss path in hydrated DNA, the "excess" holes that form on the DNA molecule as a result of hole transfer from the first 8 waters of hydration will add to the radiation damage done to cellular DNA from ionizing radiation. This additional damage may be significant. It is generally assumed that a chemical moiety is ionized by ionizing radiation in proportion to its number of valence electrons. Eight water molecules have 64 valence electrons; on average a DNA nucleotide has 110 valence electrons. Thus, hole transfer from bound water has the potential to increase the radiation damage to the DNA by ca. 50% [7–12,56].

Hole transfer does not have to result in a frank strand break to be potentially damaging. If transfer were to occur to the base stack and result in a chemically damaged base near (within ca. 10 base pairs) of a strand break or strand break precursor, a multiply damaged site (MDS) will result [60,61]. It is known that sites with such complex damage are often not repairable by cells. In summary, hole transfer from bound waters to DNA likely contributes significantly to the chemical processes that lead to radiation damage on the irradiation of hydrated DNA.

7.2.3 Trapped Radicals at 77 K

At 77 K in hydrated DNA, the composition of the radical cohort stabilized is a result of charge transfer processes that occur after irradiation. In order to determine the cohort of radicals present at 77 K after γ-irradiation at 77 K, the ESR spectra of gamma irradiated hydrated salmon sperm DNA [62–64] (Fig. 7.2), is analyzed for individual radicals using carefully developed individual benchmark spectra [64].

The choice of benchmark spectra reflects the radicals found at 77 K, $G\bullet^+$, $T\bullet^-$ and $C(N3)H\bullet$ from the DNA bases and a mixture of neutral sugar radicals labeled $\Sigma N_i\bullet$ [57,62–64]. The specific quantitative make-up of the mixture of neutral radicals is not fully known at this time, but there is good evidence for $C1'\bullet$, $C3'\bullet$, $C5'\bullet$ and $C3'\bullet_{dephos}$ (Scheme 7.1) [41–46]. Using the benchmark spectra shown in Fig. 7.3, the low dose composition of the radicals are determined. For salmon sperm DNA (77 K, $\Gamma = 14$), the best estimates from previous [41,55,64] and recent unpublished work for the percentage of each radical found at 77 K are: $G\bullet^+(35 \pm 5\%)$; $T\bullet^-(25 \pm 5\%)$; $C(N3)H\bullet(25 \pm 5\%)$; and $\Sigma N_i\bullet(15 \pm 5\%)$. After annealing to 130 K [57] the percentages change to: $G\bullet^+(40 \pm 5\%)$, $T\bullet^-(10 \pm 5\%)$, $C(N3)H\bullet(38 \pm 5\%)$ and $\Sigma N_i\bullet(12 \pm 5\%)$. These results differ only on the amounts of $T\bullet^-$ and $C(N3)H\bullet$ as it is well established on annealing from 77 K to 130 K that electron transfer occurs from $T\bullet^-$ to C [41,56] (*vide infra*).

The yields just cited force the conclusion that hole and electron transfer must occur in the radical stabilization dynamics at 77 K. As stated previously, for low LET radiation, the extent of ionization at a specific chemical moiety, caused by the cascade of low to medium energy electrons that result from the original γ- or X-ray photon, is approximately proportional to the number of valence electrons on the moiety. Thus in DNA, ca. 50% of the ionizations occurs on the sugar-phosphate backbone and ca. 50% on the four bases (Scheme 7.3).

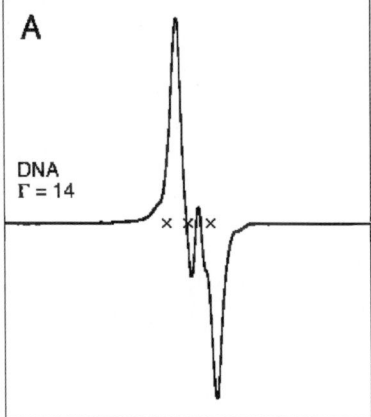

Fig. 7.2. Experimental ESR spectrum for hydrated DNA at 77 K after irradiation at 77 K. The three *crosses* are the positions of the Fremy Salt resonances, with $g = 2.0056$ and $A_N = 13.09$ G

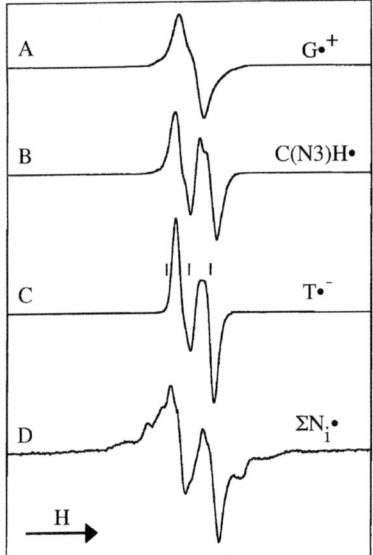

Fig. 7.3. Benchmark ESR spectra for individual DNA radicals. **A** Guanine cation radical, **B** One electron reduced cytosine, (i.e., protonated anion radical). **C** One electron reduced thymine (i.e.,anion radical), **D** Mixture of neutral sugar back-bone radicals. Reprinted with permission from [64], Radiation Research, Copyright (2005), Radiation Research Society

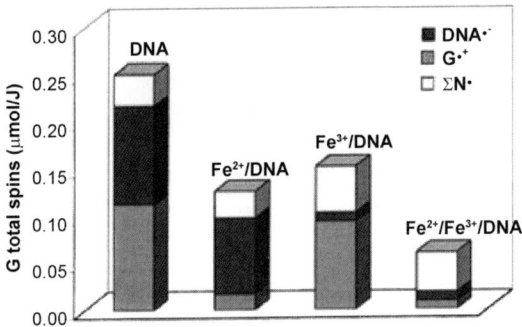

Fig. 7.4. This bar graph gives the relative yields of DNA anion radicals ($T^{\bullet -}$ and $C(N3)H^{\bullet}$, labelled collectively $DNA^{\bullet -}$), guanine cation radical, $G^{\bullet +}$, and sugar radicals for DNA and DNA with specific scavengers for holes, Fe^{+2}, electrons, Fe^{+3}, and both holes and electron, Fe^{+2}/Fe^{+3}. From these studies the relative distributions of the various radical components of DNA could be determined. Reprinted with permission from [64], Radiation Research, Copyright (2005), Radiation Research Society

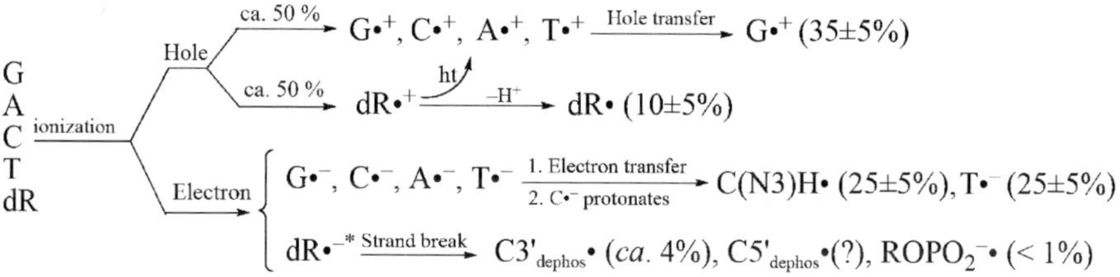

Scheme 7.3. Abundances of trapped radicals after ionization of DNA

As shown in Scheme 7.3, ionizing radiation creates holes on each of the DNA moieties; energetic electrons are ejected as a result. As the electrons thermalize through scattering processes, they eventually attach to the DNA moieties with the highest electron affinities. In order to arrive at the observed cohort of trapped radicals, a scrambling of holes and electrons must occur at this stage. Within the limits of experimental observation, most of the trapped holes end up transferring to guanine. This is expected because guanine has the lowest ionization energy of the four bases, and lower ionization energy than the sugar/phosphate backbone [1–12, 65–76]. However, roughly 20% of the trapped holes are trapped as neutral sugar radicals via the deprotonation of sugar cation radicals [64]. In a similar vein, the electrons move to and are largely trapped on the pyrimidine bases, thymine and cytosine, which have higher electron affinities than the purines and sugar/phosphate backbone [65–76]. The cytosine undergoes reversible protonation at N3 from its guanine partner to form C(N3)H• [1] (Scheme 7.4 below).

A very small percentage of the electrons initiate dissociative electron attachment at the sugar phosphate backbone, resulting in frank strand breaks which amount to ca. 4% of all stabilized radical species [42].

A statistical analysis comparing the stabilized radical composition in salmon sperm DNA to the initial sites of ionization results in the conclusion that hole and electron transport over only a few base pairs (ca. 2 or 3) is sufficient to result in the composition of radicals found at 77 K. As will be shown (Sect. 7.3), electron and hole transport over 8–10 base pairs on the time scale in question at 77 K is facile so this is well within the transfer distances allowed [9, 48–54]. In summary, hole and electron transfer play

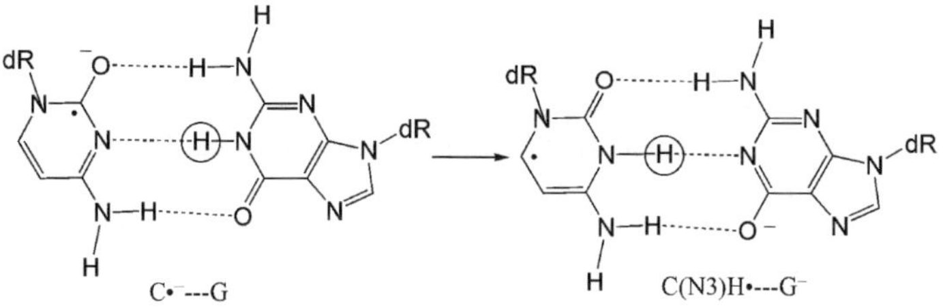

Scheme 7.4.

$$S(H) \xrightarrow{\text{ionization}} S(H) \bullet^+ \xrightarrow{-H^+} S \bullet \xrightarrow[\text{break}]{\text{strand}} \text{base release}$$

Scheme 7.6.

as shown in Scheme 7.6, in which S(H) represents the sugar, with a single hydrogen atom shown (in principle, the hydrogen explicitly shown could be bonded to any of the sugar carbon atoms).

Once the percentage of initial ionizations, the percentage of final products, and the reaction paths from ionizations to products known or assumed, the percentages for each process shown in Fig. 7.5 is uniquely determined [79]. As can be seen, the paths to base damage products and base release (strand breaks) is heavily influenced by charge transfer processes. In this model, approximately 56% of the initial ionizations occur on the sugars. However, in order to match the base release observed, 20% of the positive charges on the sugar cation radicals initially formed must transfer to the sugar's companion base. In the next step for charge transfer, considerable positive charge transfers from cationic base radicals to guanine to form $G \bullet^+$. In fact, with this model, 39% of the total positive charge from adenine, thymine and cytosine cation radicals undergoes this charge transfer process. This is required to rationalize the large amount of 8-oxo-guanine and Fapy-Guanine found in the observed products. (Reference 79 should be consulted for the structures of each of the base damage products.) In summary, this model and analysis indicates that charge transfer processes play a significant role in determining the amount of each of the products found in irradiated DNA.

It should be noted that at the time these experiments were done, the problem of forming 8-oxo-guanine as an artifact in the work-up of base damage products was well understood and scrupulously avoided [79]. The percentage of initial ionizations takes into account the GC:AT ratio found in salmon sperm DNA.

7.2.6 Is DNA a Conductor?

Simply put, the fact that trapped radicals are observed in irradiated DNA at 77 K is definitive proof that DNA cannot be an electrical conductor. The initial ionization events from ionizing radiation result in a cascade of free electrons and localized holes [1]. If DNA were a conductor, these electrons would easily and rapidly migrate to the holes (which themselves can transfer to other than their original locations), resulting in highly efficient radical recombination. Thus, few or no trapped radicals would be observed. Since classical conductivity increases with a decrease in temperature, the lack of conductivity at 77 K would imply an even lower conductivity at room temperature. It is quite clear from low temperature Electron Spin Resonance studies that DNA is not a conductor [1–12, 80–83].

It is also quite clear that hole transfer and electron transfer do occur in irradiated DNA, but the mechanism for these transfers changes from tunneling at low temperatures to activated hopping at higher temperatures (*vide infra*). Comparisons of low versus high temperature charge transfer behavior must take into account the potential differences in the mechanisms involved [1–12, 77, 80–90].

7.3 Studies of Excess Electron and Hole Transfer in DNA at Low Temperatures

7.3.1 Introduction

The process of electron and hole migration within DNA has been a topic of intense experimental [1–12, 48–54, 77, 80–90] and theoretical [90–99] interest and dispute. The overall picture is now becoming increasingly understood as the interplay of both tunneling and multistep activated hopping processes. In this section we review recent electron spin resonance studies that follow the time-dependent transport of electrons and holes from DNA base trap sites to acceptors such as intercalators or modified bases at low temperatures under conditions for which only the tunneling mechanism is operative [8, 9]. In these conditions electron transfer through the DNA stacked bases and between duplexes has been elucidated, as have the effects of DNA hydration, complexing agents, base sequence, and H/D isotope exchange on electron-transfer distances and rates [9, 48–54]. Studies which vary the temperature have shown that activated mechanisms such as hopping dominate over tunneling at temperatures of 200 K and above [9, 48–52, 77].

7.3.2 Electron and Hole Transfer from Trapped Ion Radical Species of DNA to Intercalators or Modified Bases

In order to investigate hole and electron transfer we have employed a number of techniques to produce holes and electron adducts within DNA [9, 48–54]. We found that frozen 7 M LiBr aqueous solutions containing DNA were useful systems for investigation of electron transfer that is largely free from the effect of holes. These frozen solutions are glassy in nature and γ-irradiation at low doses (700 Gy) at 77 K results in predominantly electrons and $Br_2 \bullet^-$ [48]. The excess electrons generated by the irradiation attach to DNA forming $DNA \bullet^-$ [a mixture of $C(N3)D \bullet$ and $T \bullet^-$], whereas the holes remain trapped in the solution as $Br_2 \bullet^-$. $Br_2 \bullet^-$ shows a broad ESR signal that does not interfere in the region of interest [48, 53, 54]. A number of intercalators were employed, such as mitoxantrone (MX), Ethidium bromide (EtBr), and 5-nitro-1,10-phenanthroline (NPa) [48]. Best results were found for MX because of its high electron affinity and large binding constant with DNA. Our use of intercalators applies standard techniques first employed by Peschak et al. for mitoxantrone (MX) in DNA [100]. MX in 7 M LiBr aqueous solutions was found to bind well to salmon testes DNA and to a number of oligos including, polydAdT-polydAdT, polydIdC-polydIdC, polyA-polyU, and polyC-polyG [52]. MX was found to bind poorly to polydGdC-polydGdC and polydG-polydC [52]. We find that intercalators appear to randomly insert within DNA if added at low ratios of intercalator to DNA base pairs of ca. 1/20 or less [48]. Over time, electron transfer from DNA anion radicals to the intercalator occurs. The ESR spectra of MX-DNA systems taken immediately after irradiation (20 min) and at time intervals of increasing length (up to weeks) leads to the direct observation of electron and hole transfer in the DNA [8, 9, 48–54]. In Fig. 7.6 the first-derivative ESR "benchmark" spectra of one-electron reduced DNA (a mixture of C(N3)D• and T•⁻) and one-electron reduced MX radical are shown. The clear distinction between the two spectra allows the direct observation of electron and hole transfer in DNA by following the spectra of intercalated DNA immediately after irradiation and at time intervals of increasing length (up to weeks) (see Fig. 7.7) [48].

Under the assumption of random intercalation and low loading (where the mole ratio of intercalator or modified base to DNA base pairs, ν, much smaller than 1) the probability that at least one intercalator is within distance, D, in base pairs (bp), from the site of a trapped electron is given by [9, 48]

$$F(t) = 1 - (1 - \nu)^{2D(t)} . \tag{7.1}$$

$F(t)$ represents the fraction of all electrons or holes captured by an intercalator at time t relative to all electrons originally captured by DNA. D is time dependent and increases with time as given by the relationship [9, 48]

$$D(t) = (1/\beta) \ln(k_0 t) , \tag{7.2}$$

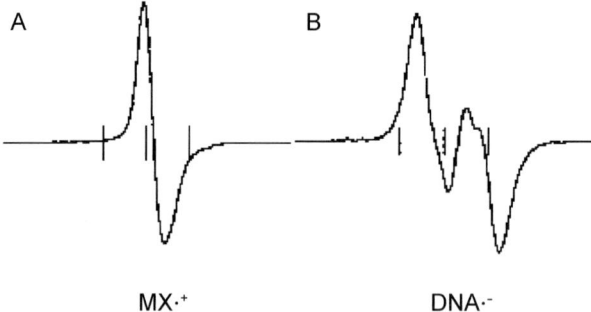

Fig. 7.6. First-derivative electron spin resonance "benchmark" spectra of A one-electron reduced $MX^{(2+)}$ ($MX^{\bullet+}$) in frozen 7 M LiBr aqueous solution and B DNA anion radical in frozen 7 M LiBr aqueous solution. The three markers are each separated by 13.09 G. The central marker is at $g = 2.0056$ [48]. Reprinted with permission from [48], J. Phys. Chem. Copyright (2000) American Chemical Society

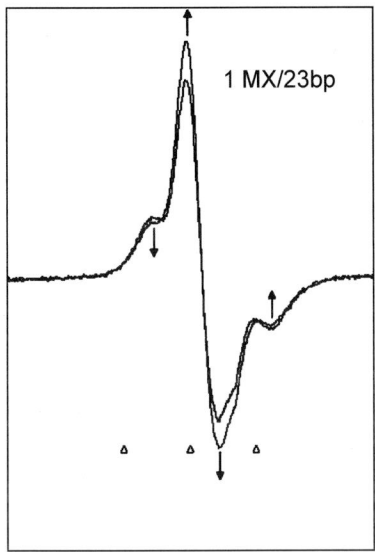

Fig. 7.7. This figure shows results for a sample with 1 MX/23 bp loading in 7 M LiBr immediately after irradiation and 15 days later [54]. An absolute increase in the intensity of MX anion radical (the *central peaks*) with a concomitant decrease in the DNA anion radical (the *outer peaks*) is found with time. Reprinted with permission from [54], J. Phys. Chem. Copyright (2006) American Chemical Society

where k_0 is the pre-exponential constant in the relationship, $k = k_0 e^{-\beta t}$ and β (the tunneling constant) is in bp^{-1}. For a single step tunneling process, plots of D vs $\ln(t)$ are expected to be linear with the slope equal to $1/\beta$ [48].

As expected from (7.1), the fraction of electrons captured by the intercalator relative to all electrons originally trapped on DNA bases was found to increase with increased loading of the intercalator. Figure 8 shows an example of first-derivative ESR spectra observed immediately after γ-irradiation of $20\,\mathrm{mg\,mL^{-1}}$ DNA in 7 M LiBr with various loadings of MX. At the lowest loading of MX (1 MX/228 bp) 8.7% of the electrons are found on MX, whereas at the highest loading (1 MX/23 bp) 59% are captured by MX with the remainder on DNA.

At loadings lower than 1 MX per 20 DNA bps, the fraction of the electrons captured by the intercalator was found to follow (7.1). Electron transfer distances D are found to be ca. 9 to 10 bps at 1 min and increase with $\ln(t)$ as expected for a single-step tunneling process. Analyses based on the time dependence of the yield of MX• for salmon sperm DNA and a variety of oligos gave tunneling constants β in the range $0.7-1.2\,\text{Å}^{-1}$ [9, 48–52]. These results do not suggest that tunneling through the DNA base stack provides a particularly facile route for transfer of excess electrons through DNA – at least at low temperatures. The transfer distances were found to increase with increasing electron affinity of intercalators [48].

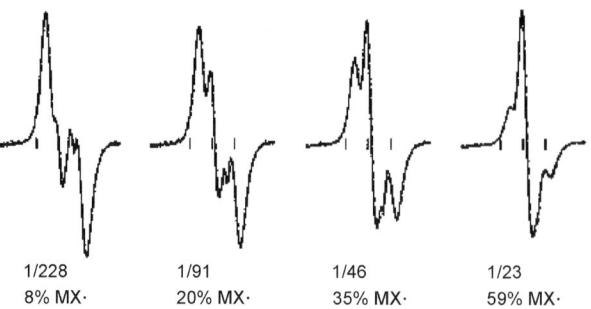

1/228   1/91   1/46   1/23
8% MX·   20% MX·   35% MX·   59% MX·

Fig. 7.8. First-derivative electron spin resonance spectra found immediately after γ-irradiation of samples of 20 mg/mL DNA in 7 M LiBr with various loadings of MX. The *dashed* spectra are simulations made by linear least-squares fits of the benchmark functions (Fig. 7.6a and b) to experimental spectra. The spectra clearly show that MX• increases in relative amount to the DNA anion radical with increased loading of MX. The fraction of electrons captured by MX increases with time. Reprinted with permission from [48], J. Phys. Chem. Copyright (2000) American Chemical Society

7.3.3 Electron Transfer Between DNA Duplexes

In a frozen 7 M LiBr aqueous solution only one glassy phase is formed, in which DNA double strands are homogeneously distributed and separated; the average separation between duplex DNA molecules is about 200 Å at 10 mg DNA/mL [49]. Hydrated DNA can take up to 21 waters/nucleotide without a crystalline ice phase forming (see Fig. 7.1). However in frozen aqueous solutions an ice phase forms with separation of the DNA into a phase with about 14 waters/nucleotide (see Fig. 7.1). Figure 7.9 is the schematic diagram depicting the spatial arrangement of DNA in a glass (frozen 7 M aqueous solution), an ice (frozen aqueous solution), and a hydrated solid (21 D_2O/nucleotide). Cai et al. studied the effect of increasing DNA concentration on the electron transfer from one-electron reduced DNA bases to MX in 7 M LiBr glass and observed a clear dependence of electron transfer (ET) on DNA concentration [49]. As the concentration of DNA increases, the average distance between DNA double strands (D_{ds}) decreases, the apparent ET distance (D_a) increases, and the apparent tunneling constant decreases. Investigations of electron transfer in γ-irradiated frozen aqueous solutions (D_2O ices) containing different concentrations of DNA were also performed. In these frozen solutions both electron and hole transfer mechanisms are at work. Significantly, the apparent electron transfer distances and tunneling constants were found to be *independent* of the concentration of DNA when DNA water solutions were cooled to form frozen ice samples. These samples produced results identical to pure DNA which was hydrated to 14 waters/nucleotide ($\Gamma = 14$). The apparent electron transfer distances (D_a) in icy and solid samples are far larger in these systems than in glassy samples owing to interduplex transfer.

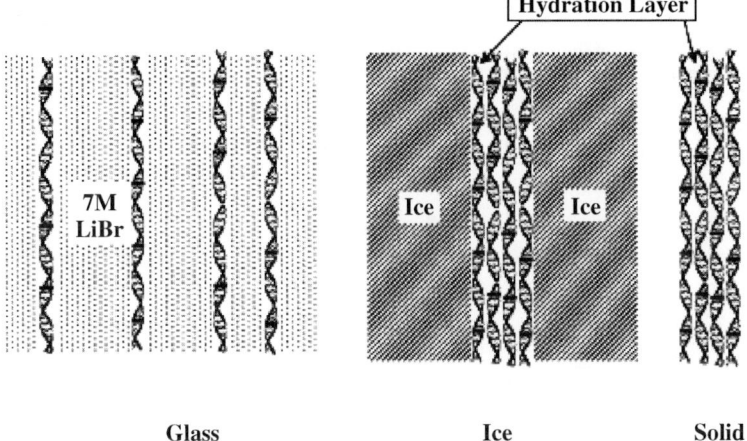

Fig. 7.9. Schematic diagram depicting the spatial separations of DNA duplexes in a glass (frozen 7 M aqueous solution), an ice (frozen aqueous solution), and a hydrated solid (14 D_2O/nucleotide) [49]. Reprinted with permission from [49], J. Phys. Chem. Copyright (2000) American Chemical Society

The experimentally found apparent electron transfer distances (D_a) are actually a count of all base pairs within the tunneling range of the inter-calator. In order to treat solid DNA systems, Cai et al. proposed a three-dimensional tunneling model that assumes electron transfer both along a primary DNA duplex and between neighboring duplexes and that D_a reflects the sum of these two processes [49]. Even though this model did not account for some of the details of the transfer, e.g., the individual strands of the double strand are not distinguished but only the average of distances is employed, the model was found to fit experimental results for the DNA concentration dependence of the apparent ET distances reasonably well. The results, again, yield about a 10 bp electron transfer distance at short times and a β value near 1 bp^{-1}, even in these icy systems. Furthermore, the fact that the apparent transfer distance was very sensitive to inter-duplex distance resulted in a sensitive measure for duplex to duplex distances resulting from changes in hydration or changes in DNA counterions (see below).

7.3.4 The Effect of DNA Hydration and Complexing Agents

Cai et al. [51] studied the effect of the level of DNA hydration on electron and hole transfer in the MX-DNA system. ESR spectra show that MX radicals decrease relative to the DNA radicals with increasing hydration levels from 2.5 to 22 waters/nucleotide and therefore show that, as the hydration level increases, the apparent transfer distance for electron and hole transfer substantially decreases. Figure 7.10 shows plots of the transfer rates of electrons, holes, and overall DNA radicals at 77 K one min after irradiation vs hydration

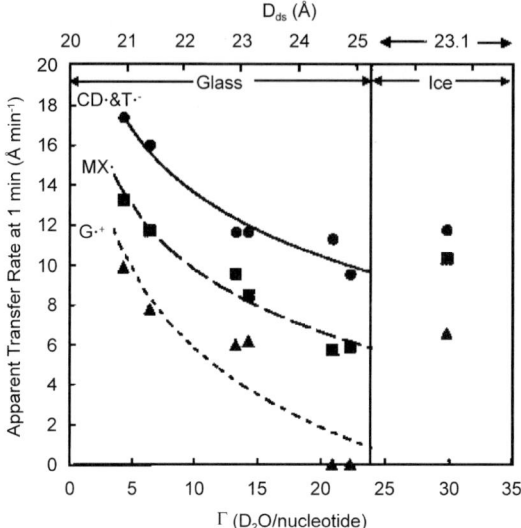

Fig. 7.10. Plots of the transfer rates of electrons and holes at 77 K vs hydration levels (*lower axis*) as vs the distance between DNA ds's (*upper axis*). Values of D_{ds} are estimated from the work of Lee et al. [51]. The results show that as amorphous (glassy) hydration increases up to $\Gamma = 22$ D_2O.nucleotide, D_{ds} increases and transfer rate decreases. Above $\Gamma = 22$ D_2O/nucleotide, an ice phase is formed, and leaves the actual amorphous hydration level at around 14 D_2O/nucleotide with the remainder in the ice phase [51]. Reprinted with permission from [51], J. Phys. Chem. Copyright (2001) American Chemical Society

levels (lower axis) as well as vs the distance between DNA duplexes (upper axis). Please note that at hydration levels higher than 22 D_2O/nucleotide, a separate ice phase is formed, which leaves only ca. 14 D_2O/nucleotide in the amorphous water layer around DNA [8, 58, 59]. Thus near-equivalent transfer rates for the hydration levels of 14 and 30 D_2O/nucleotide are observed.

Cai et al. [51] also replaced the sodium counterion of MX-DNA with various aliphatic amine cations, e.g., spermine tetrahydrochloride (SP), dodecyltrimethylammonium bromide (DOD), and octadecyltrimethylammonium bromide (OCT) and polymeric amine cations, e.g., poly-L-lysine hydrobromide (PLL) and polyethylenimine hydrochloride (PEI) to vary the separation between DNA duplexes. The radiation-produced electrons from the complexing agents readily transfer to the more electron affinic DNA. They also found that the addition of a second layer of aliphatic amine cations further suppressed the transfer of DNA holes and electrons to near that found for isolated DNA duplexes. These results support a dependence of the apparent transfer distance on the separation distance between the DNA duplexes (D_{ds}), as suggested in the 3D tunneling model. In fact, since tunneling falls off exponentially with distance, tunneling is a very sensitive measure of distances up to about 35 Å between duplexes, beyond which the rates are too small to measure. Using the 3D ET model they estimated the separation distances between DNA duplexes (D_{ds}) with complexing agents from apparent ET distances (D_a). This technique was extended to chromatin in which the DNA is wrapped around the histone proteins and an estimate of the distance between the strands was obtained. This work assumed that electrons formed on the histone proteins largely transferred to DNA as has been reported earlier [101, 102].

Cai et al. [52] investigated electron and hole transfer in various polynucleotide duplexes and compared them with previous results found for salmon sperm DNA, to examine the effect of base sequence on excess electron and hole transfer along the DNA "π-way" at low temperature. Electron and hole transfer in DNA was found to be clearly base sequence dependent. In glassy aqueous systems (7 M LiBr glasses at 77 K), excess electron-transfer rates increases in the order: polydIdC-polydIdC < salmon testes DNA < polydAdT-polydAdT. Analogous results are found in frozen ices at 77 K where excess electron and hole transfer rates increase in the order polyC-polyG < salmon

Table 7.2. Transfer distance and distance decay constants for electron and hole transfer to MX in polynucleotides-MX and DNA-MX at 77 K [9, 52]

Polynucleotide	Ref.	Medium	Donor site ET	HT	$D_a(1')$ (bp)	$D_1(1')$ (bp)	$\beta(\text{Å}^{-1})$
(PolydAdT)$_2$	[52]	D$_2$O glass[a]	T•$^-$			9.4 ± 0.5	0.75 ± 0.1
(polydIdC)$_2$	[52]	D$_2$O glass	CD•			5.9 ± 0.5	1.4 ± 0.1
DNA	[52]	H$_2$O glass	T•$^-$+CH•			8.5 ± 1.0	0.8 ± 0.1
DNA	[57]	D$_2$O glass	T•$^-$+CD•			9.5 ± 0.5	0.9 ± 0.1
PolyA-polyU	[52]	D$_2$O ice[b]	U•$^-$	A•$^+$	39 ± 10 53 ± 10	13^c 14^c	0.7^d 0.6^d
PolyC-polyG	[52]	D$_2$O ice	CD•	G•$^+$	15 ± 5 35 ± 5	8^c 11^c	1.1^d 0.8^d
DNA	[55] [103–105]	D$_2$O ice	CD•+T•$^-$	G•$^+$	17 ± 5 42 ± 5	8^c 12^c	1.1^d 0.7^d

[a] Glass indicates frozen 7 M LiBr aqueous solutions

[b] Ice refers to frozen aqueous solutions which form an crystalline ice phase and regions of DNA with ca. 14 hydration waters

[c] The apparent transfer distance, $D_a(1')$, includes transfer between duplexes as well as within the duplex. The transfer distance along one duplex is given the designation, $D_1(1')$

[d] The values of β for the ice samples have been estimated from their $D_1(1')$ values assuming $k_0 = 1 \times 10^{11}$ s^{-1} from the DNA in D$_2$O glass results

testes DNA < polyA-polyU. Transfer distances at one minute and distance decay constants for electron and hole transfer from base radicals to MX in polynucleotides-MX and DNA-MX at 77 K compiled in Table 7.2. This table shows that the electron-transfer rate from donor sites is greatest from $U\bullet^-\sim T\bullet^-$ and less from C(N3)D• owing to its protonated state. The hole transfer rate from donor sites decreases in the order $A\bullet^+ > G\bullet^+$. It is likely that proton transfer from G to $C\bullet^-$ within the $GC\bullet^-$ base pair, forming C(C6)H•, and the proton transfer from G^+ to C within the $GC\bullet^+$ base pair resulting in contributions of G(-H)•C(+H)$^+$ accounts in part for the lower transfer distances found in polyC-polyG (with analogous processes at work in polydIdC-polydIdC).

7.3.5 The Effect of Temperature

Cai et al. [52] employed MX-intercalated DNA and ESR to investigate the time dependence of free radical fractions as a function of temperature from 4 to 195 K in hydrated DNA and frozen glassy aqueous solutions containing DNA. By monitoring the ESR signals of MX and DNA radicals including $G\bullet^+$, C(C6)D•, $T\bullet^-$, and T(C6)D• with time, this work elucidated the ranges of temperature at which tunneling, protonation, hopping or recombination are dominant (Table 7.3).

The rates of electron transfer in MX-DNA were found to be nearly identical from 4 to 130 K. The lack of an observable temperature effect suggests that tunneling of electrons from DNA radicals to MX is the dominant process at low temperatures.

Table 7.3. Charge transfer process and their relative approximate time scales

Dominant processes in excess electron transfer	Temperature
Excess electron tunneling	< 190 K
Activated hopping: results in recombination of $C\bullet^-$ and $G\bullet^+$	> 190 K
Protonation processes which hinder or stop electron transfer	
Reversible protonation of $C\bullet^-$ forming C(N3)H•	4 K and up
Irreversible protonation of $T\bullet^-$ forming T(C6)H•	> 130 K
Irreversible protonation of $C\bullet^-$ forming C(N6)H•	> 200 K

7.3.6 H/D Isotope Effect

Cai et al. compared electron transfer from one-electron reduced DNA base radicals to MX in D_2O glasses with that of H_2O glasses at 77 K [50]. A slightly smaller value of β for electron transfer in DNA in H_2O relative to D_2O was observed. The greater stability of C(N3)D• over C(N3)H• should provide a slightly weaker driving force for electron transfer from C(N3)D• to MX than from C(N3)H• to MX. The small size of the effect may be a result of the fact that the dominant electron transfer process at low temperatures is from thymine anion radical which is not protonated in DNA. Shafirovich et al. [106] also reported a kinetic deuterium isotope effect on proton-coupled electron-transfer reactions at a distance in DNA duplexes.

7.3.7 Competitive Electron Scavenging in Bromine-Doped DNA

Electron transfer within DNA was investigated by employing DNA doped with randomly spaced electron traps introduced by careful bromination of DNA. The procedure is shown by NMR and GC/MS techniques to modify thymine, cytosine, and guanine bases, transforming them into 5-bromo-6-hydroxy-5,6-dihydrothymine, T(OH)Br, 5-bromocytosine, CBr, and 8-bromoguanine, GBr, derivatives [107]. The bromination products formed in molar ratio close to T(OH)Br/CBr/GBr = 0.2:1:0.23 and serve as internal electron scavengers on gamma-irradiation. Paramagnetic products that result from electron scavenging in DNA by T(OH)Br and CBr units at 77 K have been identified by ESR as the 6-hydroxy-5,6-dihydrothymin-5-yl (TOH) radical and the 5-bromocytosine σ^* anion radical, CBr•$^-$ [107]. Quantitative estimates show that electron scavenging by T(OH)Br in bromine-doped DNA is over an order of magnitude more efficient than the five fold more abundant CBr traps whereas in solution the CBr was an equally effective scavenger to the electrons as T(OH)Br. This indicates that there is a high probability that the electron survives encounters with the planar CBr traps through either transmission or reflection in the stacked DNA duplex. However, the nonplanarity of the T(OH)Br is proposed to induce a defect within the strand which produces an effective trap site [107].

7.3.8 Evidence for Multiple One-Electron "Hole" Transfers in Irradiated DNA

Recent work on γ-irradiated DNA [108] has found strong evidence that multiple oxidative steps occur at a single guanine in DNA via sequential hole transfers. The authors find that G•$^+$ undergoes hydration to form •GOH on annealing. This species is easily oxidized to 8-oxo-G via one-electron oxidation by hole transfer and this is subsequently oxidized to 8-oxoG$^+$• by another one-electron oxidation (Scheme 7.7). This multistep process (shown in Scheme 7.7) is considered evidence for long range hole transfer by thermally activated hopping.

These authors have used ^{17}O -incorporated water (i.e., H$_2^{17}$O) to elucidate the detailed mechanisms of the hydration reaction of G•$^+$ within DNA. Careful annealing studies were performed with DNA samples in H$_2^{17}$O containing Thallium (III) as an electron scavenger and these were compared with

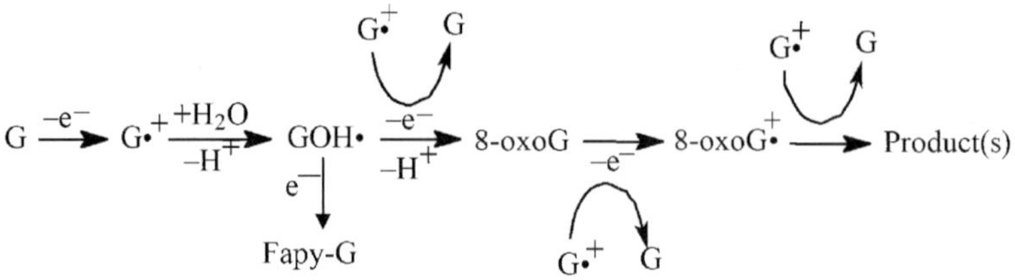

Scheme 7.7. Reprinted with permission from [108], Nucleic Acids Research, Copyright (2004) Oxford University Press

matched samples in normal water i.e, $H_2^{16}O$. Remarkably it was found that the subtraction of the two sets of ESR spectra ($H_2^{16}O$–$H_2^{17}O$) at each temperature selectively exposes the ESR spectra of radicals formed by the addition of water to the $G\bullet^+$ within DNA. Through this technique these authors have established unambiguous detection in DNA of the •GOH and 8-oxo-G^+• as well as strong evidence for their sequential production (Scheme 7.8).

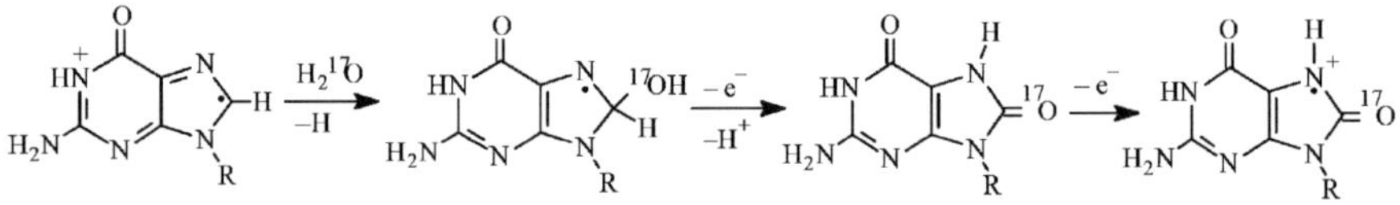

Scheme 7.8. Reprinted with permission from [108], Nucleic Acids Research, Copyright (2004) Oxford University Press

7.3.9 Overview

Low-temperature ESR studies of irradiated DNA systems have provided a clear understanding of the role of tunneling in electron transfer through DNA. Single-step tunneling is limited to transfer distances under 35 Å at a timescale of minutes. As DNA duplexes approach within this distance in solution or in the solid state, transfer between duplexes becomes competitive with transfer along the DNA duplex. The DNA hydration layer or other molecular species can serve to separate the duplexes and retard interduplex transfer [9,51]. Excess electron and hole tunneling through DNA is found to be clearly dependent on base sequence and the nature and energetics of the donor (DNA ion-radical sites) and acceptor (trap sites). Studies carried out using γ-irradiated randomly brominated DNA show that in bromine doped DNA, T(OH)Br is a more efficient electron scavenger than CBr [107]. Proton transfer between base pairs plays an important role in the energetics of electron and hole transfer. Thermal studies show that the depths of the electron and hole traps are overcome at temperatures near 200 K and hopping then dominates tunneling (see Table 7.3). However, at these temperatures irreversible

protonations at cytosine and thymine are also activated and severely limit
the range of excess electron transfer [8,9,77]. Thus, at near room temperature
conditions, activated hopping becomes the dominant charge transfer process
as shown by the sequential one-electron oxidation [108] (see Schemes 7.7
and 7.8).

7.4 Hole Transfer and Sugar Radical Formation from Excited States

7.4.1 Role of Charge and Spin Transfer in Excited State Sugar Radical Formation

In our initial work with γ-irradiated DNA we found evidence suggestive of
conversion of G•$^+$ to sugar radicals at high doses [41]. This led us to suggest
a role for excited states in sugar radical formation [41]. Our observation
of relatively high yields of neutral sugar radicals in DNA irradiated with
high-energy argon-ion beams, relative to that found in γ-irradiated samples,
also led us to hypothesize that excited states in the densely ionized ion beam
track core may lead to sugar radicals [42]. One of the possible mechanisms for
excited state processes involves charge and spin transfer from an excited state
base electron-loss radical to the DNA sugar moiety, followed by deprotonation
of the resulting sugar cationic radical to form a neutral sugar radical. This
is illustrated for the guanine to C5$'$• transformation shown in Scheme 7.9.

The possible resulting sugar radicals (Scheme 7.1) are known to be precursors of strand breaks [12] and are, thus, important lesions in DNA radiation
chemistry. The C2$'$• radical is not included in Scheme 7.1 because the relatively high C–H bond energy for the C2$'$–H bonds makes it unlikely to form
and we have not observed it in any of our investigations [43–46].

In an experiment devised to test the mechanism proposed in Scheme 7.9,
a small amount of a nucleotide is dissolved in a ca. 7 M LiCl solution that
also contains the electron scavenger potassium peroxydisulfate ($K_2S_2O_8$),
and the solution cooled to 77 K in order to form a glass. Gamma irradiation (ca. 2.5 kGy) of the sample at 77 K leads to formation of the oxidative
SO_4•$^-$ and Cl_2•$^-$ radicals [44–46]. The presence of potassium peroxydisul-
phate completely suppresses the electron gain path [44–46]. Annealing of the
sample to ca. 150 – 155 K leads to formation of essentially pure base electron-
loss radicals, because, Cl_2•$^-$ does not directly oxidize the sugar phosphate

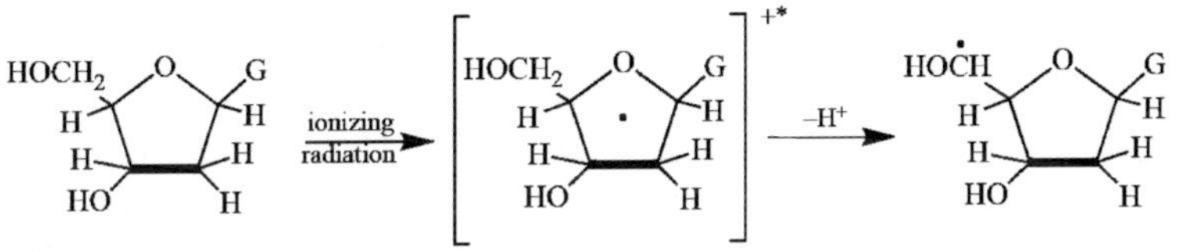

Scheme 7.9.

backbone and as a consequence only $G\bullet^+$ is produced for guanine containing systems [44]. Note that at pH 5, the pH of 7 M LiCl in D_2O, the one-electron-oxidized base radical is largely protonated in these systems because of its higher pK_a [109]. UV-visible illumination of the base electron-loss radical ($G\bullet^+$ in this case) results in excited states from which the formation of sugar radicals can be observed, as depicted in Scheme 7.10, for the formation of C5'•.

The ESR spectra that correspond to the excited state charge and spin transfer process from $G\bullet^+$ to the sugar moiety, resulting in neutral sugar radicals, are shown in Fig. 7.11 for 5'-dGMP (2'deoxyguanosine-5'-monophosphate). The spectrum for $G\bullet^+$ (same as spectrum 3A) is shown in spectrum 11A. Because the natural pH of the LiCl glass used is ca. 5, $G\bullet^+$ is not deprotonated [109]. After illumination (200 min, 77 K, visible floodlight), of $G\bullet^+$, four new weak line components from the sugar radical C3'• are vis-

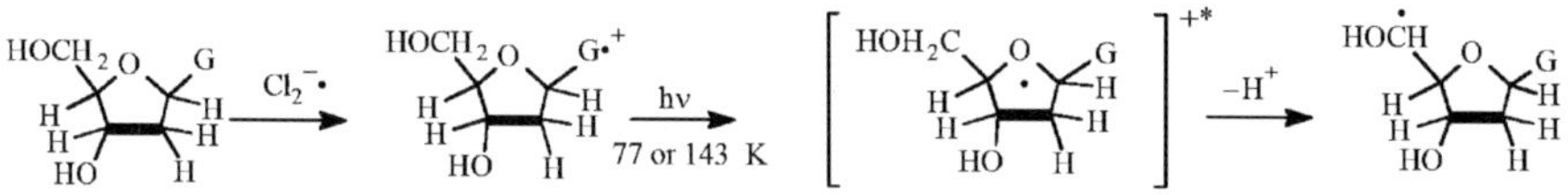

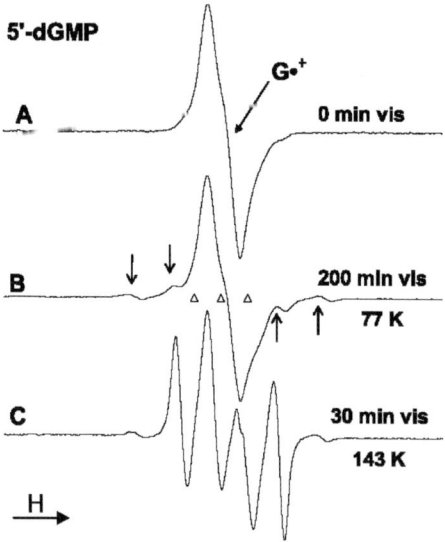

Scheme 7.10. Reprinted with permission from the [44], Nucleic Acids Research, Copyright (2005) Oxford University Press

5'-dGMP

A $G\bullet^+$ 0 min vis

B   200 min vis   77 K

C   30 min vis   143 K

H

Fig. 7.11. A ESR spectrum of $G\bullet^+$ at 77 K in 7 M LiCl formed by $Cl_2\bullet^-$ attack on 5'-dGMP. **B** Spectrum found on visible photolysis of $G\bullet^+$ at 77 K. **C** Spectrum found on visible photolysis of $G\bullet^+$ at 143 K showing predominantly the C1'• sugar radical. Reprinted with permission from [44] Nucleic Acid Research, Copyright (2005) Oxford University Press

Table 7.4. Sugar radicals formed on photo-excitation of G•+ [a,b]. Reprinted with permission from [44], Nucleic Acids Research, Copyright (2005) Oxford University Press

Compound	Temperature (K)[c]	Percent converted[d]	C1′•[e]	C3′•[e]	C5′•[e]
dGuo	143	90	10%	35%	55%
	77	30	10	40	50
5′-dGMP	143	95	95	5	–
	77	30	15	30	55
3′-dGMP	143	85	40	–	60
	77	15	40	–	60
dsDNA(ice)	143	ca. 50	100	–	–
	77	ca. 50	100	–	–

[a] Percentage expressed to ±5% relative error
[b] All glassy samples are at the native pH of 7 M LiCl (ca. 5). For DNA the pH of the aqueous solution before freezing was 7
[c] Temperature at which G•+ was illuminated
[d] Percentage of G•+ that converts to sugar radicals. The total spectral intensities before and after illumination were the same, within experimental uncertainties
[e] Each calculated as percentage of total sugar radical concentration; these sum to 100%

ible in the wings of the spectrum (arrows) [44]. Results of an experiment that used illumination at 143 K (30 min) of an identically prepared sample is shown in Fig. 7.11C [44]. Computer analysis of this spectrum using the benchmark spectra for C1′•, C5′• and C3′• respectively indicates it is a composite from C1′• (ca. 95%) and C3′• (ca. 5%) (Table 7.4). It is clear that there has been a photo-induced conversion from G•+ to the sugar radicals. Similar experiments using dGuo and 3′-dGMP resulted in a similar conversion of G•+ to sugar radicals, but with different relative amounts of radicals formed (Table 7.4) [44].

It is curious that the C4′• radical is not observed in these experiments, even though calculations show its energy is close to that of C1′• and C3′• and is, therefore, as stable as these other radicals [44]. It is possible that C4′• might be present after photo-excitation of G•+, and C4′• could have an underlying, broad ESR spectrum which is not easily observable [44]. However, photo-excitation of G•+ formed in dGuo having selective deuteration at C3′ in the sugar moiety suggests that it is clearly not a substantial contributor to the ESR spectrum [44]. Reasons for this are not clear.

7.4.2 Wavelength Dependence of Sugar Radical Formation

Experiments were performed to determine the dependence of sugar radical formation from G•+ (in dGuo) and DNA on the wavelength of light used to

Table 7.5. Wavelength dependence of radical formation from $G^{\bullet+}$ in dGuo and DNA. Reprinted with permission from [44], Nucleic Acids Research, Copyright (2005) Oxford University Press

	Wavelength of illumination at 143 K	$C1'\bullet$	$C3'\bullet$	$C5'\bullet$
dGuo	380–480 nm	7%	35%	58%
	> 540 nm	3%	40%	57%
DNA	380–480 nm	100%	0	0
	> 540 nm	0	0	0

illuminate the cation radical. No significant dependence on the wavelength of light was found for dGuo but DNA showed no production of sugar radicals above 540 nm (Table 7.5) [44].

7.4.3 pH Dependence of Sugar Radical Formation

The effect of pH on sugar radical formation from visible light illumination (143 K) of $G^{\bullet+}$ (in dGuo), was also investigated [44], and the photo-conversion of $G^{\bullet+}$ to sugar radicals is pH sensitive. In glassy aqueous(D_2O) 7 M LiCl at low temperature, the N1 hydrogen in $G^{\bullet+}$ has pK_a ca. 5 [109], so at pH ≥ 7 the N1 proton is largely dissociated (reaction 1). We have observed that in aqueous (D_2O) 7 M LiCl glassy systems, sugar radical formation via photo-excitation of $G^{\bullet+}$ occurs from pH = 2 to 6 and is suppressed for pH ca. 7 to 11 [109]. Thus, at pHs for which G(-H)$\bullet$ and G(-2H)$\bullet^-$ exist, no significant photo-conversion is found.

$$pK_a \text{ ca. } 5 \qquad 7.3$$

$$pK_a > 10 \qquad 7.4$$

In dsDNA, at pH 7, $G^{\bullet+}$ is only partially deprotonated to its base pair cytosine [43, 44]. Thus, we observe a substantial photo-conversion of $G^{\bullet+}$ to C1'$\bullet$ in DNA.

7.4.4 TD-DFT Calculations

The mechanism that we have proposed for the photo-conversion of one-electron-oxidized guanine or adenine to sugar radicals is a charge and spin transfer which entails charge and spin delocalization into the sugar moiety of the nucleoside/tide on photo-excitation followed by subsequent fast deprotonation from the transitory excited state cationic sugar radical to form a neutral sugar radical [43–45]. In order to reach a fuller understanding of this propose mechanism, we have performed time-dependent density functional theory (TD-DFT) calculations of the excited states and transition energies with B3LYP functionals and a 6–31G(d) basis set [44] for $G\bullet^+$ in dGuo. Using a geometry optimized (B3LYP, 6–31G(d)) structure for $G\bullet^+$, the first 12 states predicted throughout the UVA-visible region were calculated. Only those with substantial oscillator strengths are shown in Table 7.6. All 12 transitions are between core MOs 58–69 and the singly occupied molecular orbital (SOMO) 70 in $G\bullet^+$ [44]. Similar calculations were also performed using the much larger 6–311++G(d,p) basis set; the results obtained were very similar [44].

A key finding in the calculation is that the inner filled molecular orbitals, often have substantial contributions from atomic orbitals on the sugar ring. The fact that transitions with significant delocalization on the sugar ring are found throughout the UVA-visible region is in good agreement with our results with model compounds, in which no significant wavelength dependence for sugar radical formation is found (see Table 7.5) [44].

Table 7.6. TD-DFT b3lyp 6-321G* calculated electronic transitions for $G\bullet^+$ in dGuo. Reprinted with permission from [44], Nucleic Acids Research, Copyright (2005) Oxford University Press

ΔE (eV)	λ (nm)	f[a]	Transition(density)	Delocalozation into sugar [b]
2.51	494	0.0067	$63\beta \rightarrow 70\beta(0.79)$	2
2.80	444	0.0122	$62\beta \rightarrow 70\beta(0.88)$	0
3.36	368	0.0137	$58\beta \rightarrow 70\beta(0.69)$	2
			$61\beta \rightarrow 70\beta(-0.59)$	3
3.41	363	0.0664	$59\beta \rightarrow 70\beta(0.66)$	1
			$60\beta \rightarrow 70\beta(-0.58)$	3
3.61	343	0.0021	$51\beta \rightarrow 70\beta(0.72)$	3
3.67	337	0.0015	$60\beta \rightarrow 70\beta(0.72)$	3

[a] Oscillator strength; only those transition of 0.002 or above are shown
[b] Estimate of degree of hole delocalozation from the base onto the sugar suggested by the initial MO: 0. Nearly all remains on the Guanine base, 1. Shared between base and sugar ring favoring base, 2. Equally shared between base and sugar ring, 3. Shared between base and sugar ring favoring sugar, 4. Nearly all transferred to the sugar ring

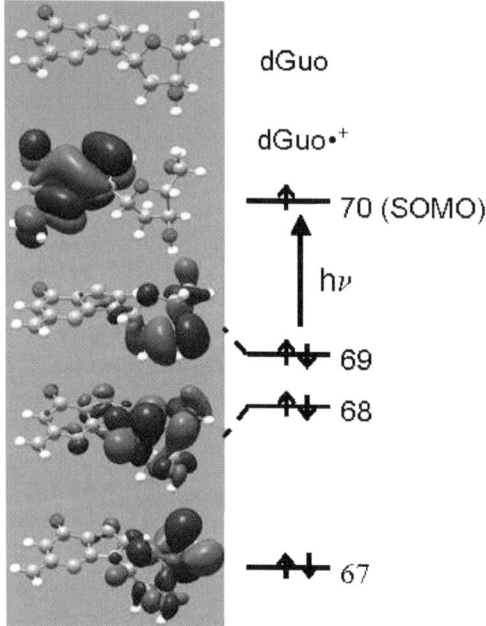

Fig. 7.12. MOs for G•+ in dGuo. Showing from *top* to *bottom*. The structure of dGuo. The MO of the SOMO in dGuo•+. The MO of the next lower three MOs. These all clearly show that a transition to the SOMO from these MOs would entail spin and charge transfer from the quanine ring to the sugar ring. Adapted with permission from [44], Nucleic Acid Research, Copyright (2005) Oxford University Press

In Fig. 7.12 we show four molecular orbitals for G•+ in dGuo computed by TD-DFT (6-31G*, B3LYP) and visualized via Gaussview. In the ground state G•+, the hole is localized on the guanine base in the SOMO. Many of the filled inner core molecular orbitals are localized on the sugar ring. Because of this, production of a core excited state results in transfer of the hole from the guanine ring to the sugar ring. It is notable that MO 61β has a large contribution from atomic orbitals at C5′. Because of this, spin and charge transfer to MO 61β would result in significant positive charge localization at C5′, which would, in turn, promote deprotonation from this site. The large concentrations of C5′• found in several cases in Tables 7.4 and 7.5 may be partially explained by this phenomenon. It is also notable that the transition oscillator strengths are largest for transition from those MOs which are located on both the base and the sugar ($58\beta \rightarrow 70\beta$ for example).

7.4.5 Effect of Phosphate Groups on Sugar Radical Formation

A phosphate group at the C3′ or C5′ atom of the sugar group appears to deactivate the site to radical formation [44]. In fact, earlier theoretical calculations indicated that replacing an OH group at C3′ or C5′ with a phosphate group increased the energy of the resulting C3′• and C5′• radicals, respectively, making the phosphate containing radical less stable than the same radical with an OH group in place of the phosphate [110,111]. However, experimental results with dGuo, 3′-dGMP and 5′-dGMP indicate that the effect of a phosphate group on radical formation at the specific carbon atom in the sugar moiety is determined by more than radical stability. In dGuo, either at 77 K or at 143 K, a phosphate group at C3′ or C5′ does discourage radical formation at C3′ or C5′, respectively, on photo-excitation of G•+ (Table 7.4) [44], likely reflecting the higher energy of the radicals when a phosphate is present. On the other hand, at 77 K, phosphate substitution at C5′ does not reduce the percentage of C5′• found. The large charge and spin transfer (*vide supra*) to the C5′ of the sugar group on photo-excitation (see Fig. 7.12) seems to account for the C5′• found, despite the fact that it is not the most stable radical site [44].

7.4.6 Photo-Induced Hole Transfer in Dinucleoside Phosphates

The experimental work involving photo-excitation of one-electron oxidized base in monomers (*vide supra*) [44,45] and the TD-DFT/6-31G(d) calculations in monomers and in oligomers [46,47] established that the formation of sugar radicals is due to the transition between core MOs to the SOMO; we have proposed that these transitions are also applicable to DNA as well [46]. Therefore, we have extended these experimental and theoretical studies to dinucleoside phosphates [46,47]. Photo-excitation of G•+ formed in TpdG at 143 K shows ca. 85% conversion to sugar radicals (75% C1′• and 10% C3′•) (Fig. 7.13). The hyperfine couplings and the *g* value of the C1′• formed in TpdG is found to be identical with those for the same radical in 5′-dGMP. As indicated by the transition energies, we expect base to base radical transfer from G•+ at higher wavelength and the sugar radical formation from G•+ at UVA-vis wavelengths. However, we have not observed any line components of thymine radicals via photo-excitation of G•+ in TpdG (Fig. 7.13).

The transition energies calculated using TD-DFT/6–31G(d) suggests that at higher wavelengths, base to base hole transfer is likely [46]. For double stranded oligos lacking Gs, formation of the allylic thymyl i.e. UCH$_2$• radical has been observed by Schuster and his co-workers [112]. This finding adequately supports these TD-DFT calculations [46,47].

We have already mentioned in the photo-excitation studies carried out in monomers [44,45] and in dinucleoside phosphates so far [46], that apart from the formation of C1′•, C3′• and C5′• sugar radicals, production of C4′• is not observed. The product analyses for X-ray irradiated crystalline CG-CACG:CGTGCG and CACGCG:CGCGTG do not show any C4′• intermediate as well [113]. This commonality has also led us to propose that γ-ray induced hole formation and photo-excitation of the cation-radical are equivalent as both of which result in holes in inner-shell MOs which are localized on the sugar phosphate backbone [46].

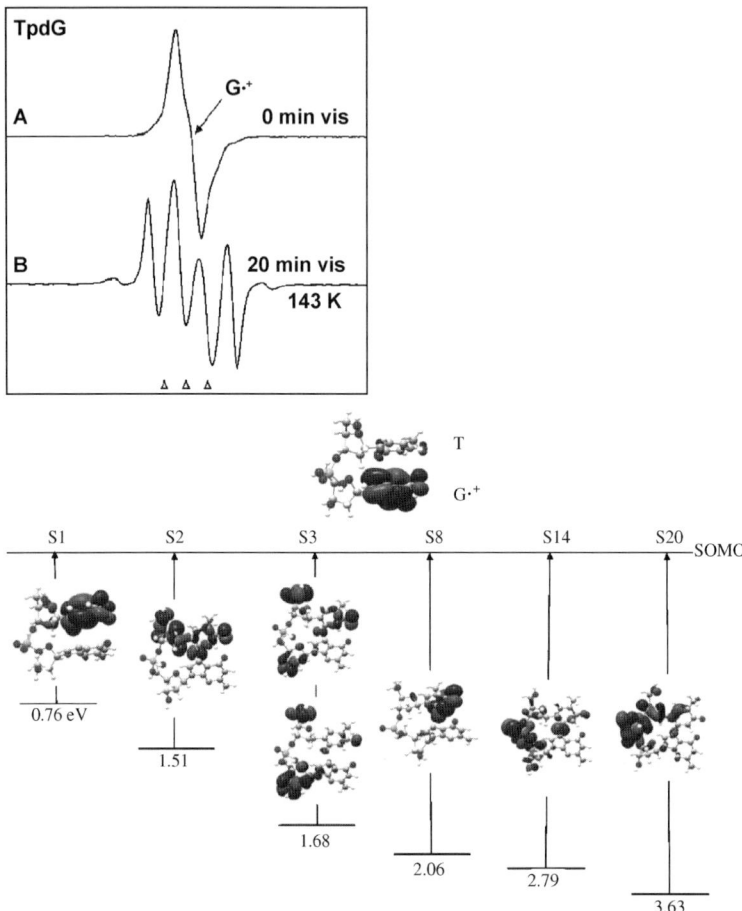

Fig. 7.13. (*Top*) ESR spectra resulting from photo-excitation of G•⁺ (obtained via one-electron oxidation of TpdG) to Cl′• (prominent quartet in the *middle*) and C3′• (*line components at the wings*). (*Bottom*) Schematic representation of the MOs and their energies from TD-DFT/6–31G(d) calculations. The transition energies of inner-shell (core) MOs to SOMO suggest sugar radical formation at lower (UVA-vis) wavelengths, and base-to-base transitions at longer wavelengths. Reprinted with permission from [46], Radiation Research, Copyright (2006), Radiation Research Society

7.4.7 Sugar Radical Formation in DNA

Sugar radicals in DNA are also induced by visible photo-excitation of G•⁺ [43, 44]. Figure 7.14 shows the ESR spectra obtained for this transformation. In this case, as shown in Fig. 7.14A, γ-irradiation of hydrated DNA ($\Gamma = 14$ D₂O/nucleotide 77 K) results in a mixture of G•⁺, C(N3)H•, T•⁻, and a small percentage of sugar radicals (*vide supra*, Scheme 7.3). Illumina-

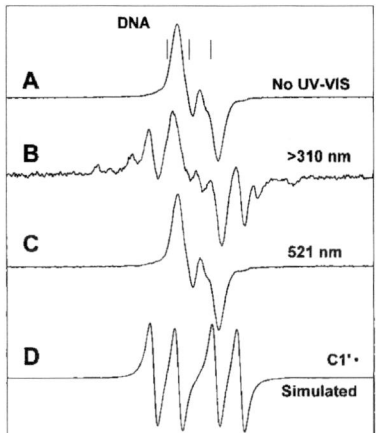

Fig. 7.14. A ESR spectrum of γ-irradiated hydrated DNA. **B** The sample in A after UVA-vis exposure showing C1′• and T(C-6)D• radicals. **C** Identical sample after exposure to 521 nm light. **D** The C1′• radical simulated ESR spectrum. Reprinted with permission from [44], Nucleic Acids Research Copyright (2005) Oxford University Press

tion with light of wavelength between 310 nm and 480 nm results in the spectrum shown in Fig. 7.14B. This spectrum displays considerable intensity from line components from the C1′• sugar radical, for which a simulated spectrum [44] is shown in Fig. 7.14D. In the companion Fig. 7.15, the transformation from $G•^+$ to C1′• in double stranded DNA is shown as a function of time of illumination. In this case, Tl^{3+} was added to the DNA as an electron scavenger in order to highlight the electron loss path more clearly. As indicated in Table 7.4, there is approximately a 50% conversion of $G•^+$ to C1′•. In dsDNA, only C1′• is observed via photoexcitation of $G•^+$. In DNA, both the C3′ site and C5′ site have a phosphate group and hence in DNA the formation of C1′• via photo-excitation of $G•^+$ only is consistent with the destabilization of the C3′• and C5′• radicals by phosphate moieties [110, 111]. In addition, in dsDNA, sugar radical formation does depend on the wavelength of the exciting light (see Table 7.5). As illustrated in Fig. 7.14C, illumination with light having wavelength ca. 520 nm and above does not cause a change in the ESR spectrum i.e., no observable sugar radicals are formed [44]. Since base stacking is present in DNA, an additional well-known process, photo-induced charge and spin transfer to nearby bases may compete with hole transfer to the adjoining deoxyribose moiety, thereby validating the TD-DFT calculations proposing base to base hole transfer (see Sect. 7.4.6) at higher wavelengths [46, 47].

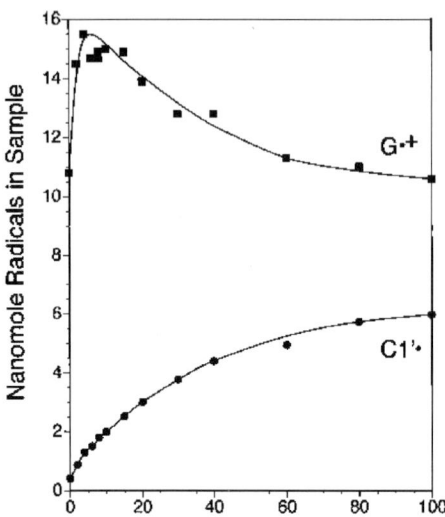

Fig. 7.15. Results for γ-irradiated hydrated DNA with $T1^{+3}$ as an electron scavenger. The time dependence of the ESR signals of $G\bullet^+$ and $C1'\bullet$ with time of exposure which clearly shows the conversion of $G\bullet^+$ to $C1'\bullet$. The initial increase in $G\bullet^+$ is a result of rapid photo-conversion of $T1^{+2}$ to $G\bullet^+$. Reprinted with permission from [43], Radiation Research Copyright (2004) Radiation Research Society

7.4.8 Charge and Spin Transfer in DNA Sugar Radical Formation and Strand Breaks

The mechanism proposed for the formation of sugar radicals via photo-excitation of DNA base cation radicals is that photo-excitation causes positive charge and unpaired spin density to delocalize from the DNA base into the sugar phosphate moiety. This is an example of the manner in which charge transfer processes may direct radical formation in DNA and its components. The molecular orbital structure of the excited cationic radical concentrates charge on a sugar carbon site that is susceptible to deprotonation, which, itself, is another charge transfer process. Deprotonation separates spin from charge, a process which renders the resulting neutral sugar radical less susceptible to recombination than the charged radical. In DNA the sugar radicals are precursors to strand breaks, which are substantial biological damage sites that must be repaired for proper cell function.

In evaluating the effect of charge transfer on sugar radical formation, the sugar sites with the highest positive charge density will deprotonate most rapidly. These sites are not necessarily correlated with the lowest energy for the resulting radical. Thus, radical formation is largely under kinetic control from the transistion state spin and charge distribution rather than thermodynamic control.

7.5 Conclusion

In this review, we presented a number of recent findings that show that the initial ion radicals and excited states formed by radiation within DNA migrate and that this migration ultimately determines the location of the chemical and, later, biological damage. Although this work and the others reported in this book suggest that much about these processes is understood, a full understanding of the time scale and nature of these processes will be necessary to fully comprehend the full spectrum of effects, both beneficial and hazardous, that radiation creates in living organisms.

Acknowledgement. We thank the National Cancer Institute of the National Institutes of Health (Grant RO1CA45424), and the Oakland University Research Excellence Program in Biotechnology for support of this work. A.A. is grateful to the authorities of the Rajdhani College and the University of Delhi for leave to work on this research program.

References

1. D. Becker and M.D. Sevilla, Adv. Radiat. Biol. **17**, 121 (1993).
2. (a) K. Senthilkumar, F.C. Grozema, C.F. Guerra, F.M. Bickelhaupt, F.D. Lewis, Y.A. Berlin, M.A. Ratner and L.D. Siebbeles, J. Am. Chem. Soc. **127**, 14894 (2005). (b) F.D. Lewis, X. Liu, J. Liu, S.E. Miller, R.T. Hayes and M.R. Wasielewski, Nature (London), **406**, 51 (2000).
3. F.D. Lewis, R.L. Letsinger and M.R. Wasielewski, Acc. Chem. Res. **34**, 159 (2001).
4. B. Giese, Ann. Rev. Biochem. **71**, 51 (2002).
5. S. Delaney and J.K. Barton, J. Org. Chem. **68**, 6475 (2003).
6. G.B. Schuster, Acc. Chem. Res. **33**, 253 (2000).
7. D. Becker and M.D. Sevilla, in *Royal Society of Chemistry Specialist Periodical Report* edited by B.C. Gilbert, M.J. Davies, D.M. Murphy; Electron Spin Resonance. **16**, 79 (1998).
8. M.D. Sevilla and D. Becker, in *Royal Society of Chemistry Specialist Periodical Report* edited by B.C. Gilbert, M.J. Davies, D.M. Murphy; Electron Spin Resonance. **19**, 243 (2004).
9. Z. Cai and M.D. Sevilla, in *Long Range Charge Transfer in DNA II* edited by G.B. Schuster; Topics In Current Chemistry; Springer-Verlag, Berlin, Heidelberg; **237**, 103 (2004).
10. W.A. Bernhard and D.M. Close, in *Charged Particle and Photon Interactions with Matter Chemical, Physicochemical and Biological Consequences with Applications* edited by A. Mozumdar, Y. Hatano (Marcel Dekkar, Inc., New York, Basel, 2004).
11. H.-A. Wagenknecht, in *Charge Transfer in DNA: From Mechanism to Application* edited by H.-A. Wagenknecht (Willey-VCH Verlag GmbH & Co. KGaA, Weiheim, 2005).
12. C. von Sonntag, in *Free-radical-induced DNA Damage and Its Repair* (Springer-Verlag, Berlin, Heidelberg, 2006).

13. A. Mozumder, in *Advances in Radiation Chemistry* edited by M. Burton, and J.L. Magee (Wiley-Interscience, New York, 1969).
14. A. Chatterjee and J.L. Magee, Radiat. Prot. Dosim. **13**, 137 (1985).
15. A. Chatterjee and W.R. Holley, Int. J. Quantum. Chem. **13**, 709 (1991).
16. H. Nikjoo, D.T. Goodhead, D.E. Charlton and H.G. Paretzke, Int. J. Radiat. Biol. **60**, 739 (1991).
17. V. Michalik, Int. J. Radiat. Biol. **62**, 9 (1992).
18. M. Pinak and A. Ito, J. Radiat. Res. **34**, 221 (1993).
19. M. Kramer and G. Kraft, Adv. Space. Res. **14**, 151 (1994).
20. H. Nikjoo, S. Uehara, W.E. Wilson, M. Hoshi and D.T. Goodhead, Int. J. Radiat. Biol. **73**, 355 (1998).
21. F. Ballarini, M. Biaggi, L. De Biaggi, A. Ferrari, A. Ottolenghi, A. Panzarasa, H.G. Paretzke, M. Pelliccioni, P. Sala, D. Scannicchio and M. Zankl, Adv. Space. Res. **34**, 1338 (2004).
22. M.K. Bowman, D. Becker, M.D. Sevilla and J.D. Zimbrick, Radiat. Res. **163**, 447 (2005).
23. A. Hummel, in *Radiation Chemistry: Principles and Applications* edited by Farhataziz and A.J.M. Rodgers (Verlag Chemie, Weinheim, 1987).
24. J.F. Ward, in *Physical and Chemical Mechanisms in Molecular Radiation Biology* edited by W.A. Glass, and M.N. Verma, **403–415** (Plenum Press, New York, 1991).
25. A. Chatterjee, and W. R. Holley, Adv. Radiat. Biol. **17**, 181 (1993).
26. D.T. Goodhead, Int. J. Radiat. Biol. **65**, 7 (1994).
27. J. Kiefer, in *Biological radiation effects* (Springer-Verlag, Berlin, Heidelberg, New York, 1990).
28. A. Mozumder, *Fundamentals of Radiation Chemistry* (Academic Press, San Diego, 1999).
29. C. von Sonntag, in *The Chemical Basis of Radiation Biology* (Taylor and Francis, London, 1987).
30. W.R. Holley, A. Chatterjee and J.L. Magee, Radiat. Res. **121**, 161 (1990).
31. B. Boudaffa, P. Cloutier, D. Hunting, M. Huels and L. Sanche, Science, **287**, 1658 (2000).
32. L. Sanche, Mass Spectrometry Reviews, **21**, 349 (2002).
33. X. Li, M.D. Sevilla and L. Sanche, J. Am. Chem. Soc. **125**, 13668 (2003).
34. J. Berdys, I. Anusiewicz, P. Skurski and J. Simons, J. Am. Chem. Soc. **126**, 6441 (2004).
35. Y. Zheng, P. Cloutier, D.J. Hunting, L. Sanche and J.R. Wagner, J. Am. Chem. Soc. **127**, 16592 (2005).
36. I. Bald, J. Kopyra and E. Illenberger, Angew. Chem. Int. Ed. **45**, 4851 (2006).
37. C. König, J. Kopyra, I. Bald and and E. Illenberger, Phys. Rev. Lett. **97**, 018105 (2006).
38. S. Ptasiska, S. Denifl, S. Gohlke, P. Scheier, E. Illenberger and T.D. M, Angew. Chem. Int. Ed. **45**, 1893 (2006).
39. J. Gu, Y. Xie and H.F. Schaefer, J. Am. Chem. Soc. **128**, 1250 (2006).
40. P. Swiderek, Angew. Chem. Int. Ed. **45**, 4056 (2006).
41. W. Wang, M. Yan, D. Becker and M.D. Sevilla, Radiat. Res. **137**, 2 (1994).
42. D. Becker, A. Bryant-Friedrich, C. Trzasko and M.D. Sevilla, Radiat. Res. **160**, 174 (2003).
43. L.I. Shukla, R. Pazdro, J. Huang, C. DeVreugd, D. Becker, and M.D. Sevilla, Radiat. Res. **161**, 582 (2004).

44. A. Adhikary, A.Y.S. Malkhasian, S. Collins, J. Koppen, D. Becker, and M. D. Sevilla, Nucleic Acids Res. **33**, 5553 (2005).
45. A. Adhikary, D. Becker, S. Collins, J. Koppen, and M.D. Sevilla, Nucleic Acids Res. **34**, 1501 (2006).
46. A. Adhikary, A. Kumar and M.D. Sevilla, Radiat. Res. **165**, 479 (2006).
47. A. Kumar and M.D. Sevilla, J. Phys. Chem. B **110**, 24181 (2006).
48. A. Messer, K. Carpenter, K. Forzley, J. Buchanan, S. Yang, Y. Razskazovkii, Z. Cai and M.D. Sevilla, J. Phys. Chem. B **104**, 1128 (2000).
49. Z. Cai and M.D. Sevilla, J. Phys. Chem. B **104**, 6942 (2000).
50. Z. Cai, Z. Gu and M.D. Sevilla, J. Phys. Chem. B **104**, 10406 (2000).
51. Z. Cai, Z. Gu and M.D. Sevilla, J. Phys. Chem. B **105**, 6031 (2001).
52. Z. Cai, X. Li and M.D. Sevilla, J. Phys. Chem. B **106**, 2755 (2002).
53. C. Pal and J. Hüttermann, J. Phys. Chem. B **110**, 14976 (2006).
54. M.D. Sevilla, J. Phys. Chem. B **110**, 25122 (2006).
55. W. Wang, D. Becker and M.D. Sevilla, Radiat. Res. **135**, 146 (1993).
56. M.D. Sevilla and D. Becker, in *Royal Society of Chemistry Specialist Periodical Report* edited by B.C. Gilbert, M.J. Davies, D.M. Murphy; Electron Spin Resonance. **14**, 130 (1994).
57. M. Yan, D. Becker, S.R. Summerfield, P. Renke and M.D. Sevilla, J. Phys. Chem. **96**, 1983 (1992).
58. D. Becker, T. La Vere and M.D. Sevilla, Radiat. Res. **140**, 123 (1994).
59. T. La Vere, D. Becker and M.D. Sevilla, Radiat. Res. **145**, 673 (1996).
60. J.F. Ward, Int. J. Radiat. Biol. **66**, 427 (1994).
61. J.R. Milligan, J.A. Aguilera, R.A. Paglinawan, K.J. Nguyen, and J.F. Ward, Int. J. Radiat. Biol. **78**, 733 (2002).
62. M.D. Sevilla, D. Becker, M. Yan and S.R. Summerfield, J. Phys. Chem. **95**, 3409 (1991).
63. D. Becker, Y. Razskazovskii, M.U. Callaghan, M.D. Sevilla, Radiat. Res. **146**, 361 (1996).
64. L.I. Shukla, R. Pazdro, D. Becker and M.D. Sevilla, Radiat. Res. **163**, 591 (2005).
65. S.D. Wetmore, R.J. Boyd, and L.A. Eriksson, Chem. Phys. Lett. **322**, 129 (2000).
66. X. Li, Z. Cai and M.D. Sevilla, J. Phys. Chem. B **105**, 10115 (2001).
67. X. Li, Z. Cai and M.D. Sevilla, J. Phys. Chem. A **106**, 1596 (2002).
68. X. Li, Z. Cai and M.D. Sevilla, J. Phys. Chem A **106**, 9345 (2002).
69. J. C. Rienstra-Kiracofe, G.S. Tschumper, H.F. Schaefer III, S. Nandi and G.B. Ellison, Chem. Rev. **102**, 231 (2002).
70. J. Reynisson and S. Steenken, Phys. Chem. Chem. Phys. **4**, 527 (2002).
71. D. Sovzil, P. Jungwirth, Z. Havlas, Collect. Czech. Chem. Commun. **69**, 1395 (2004).
72. A. Kumar, M. Knapp-Mohammady, P.C. Mishra and S. Suhai, J. Comput. Chem. **25**, 1047 (2004).
73. D.M. Close, J. Phys. Chem. A **108**, 10376 (2004).
74. C.E. Crespo-Hernez, R. Arce, Y. Ishikawa, L. Gorb, J. Leszczynski and D.M. Close, J. Phys. Chem. A **108**, 6373 (2004).
75. D.M. Close, C.E. Crespo-Hernez, L. Gorb and J. Leszczynski, J. Phys. Chem. A **109**, 9279 (2005).
76. T. Caruso, M. Carutenuto, E. Vasca and A. Peluso, J. Am. Chem. Soc. **127**, 15040 (2005).

77. H.-A. Wagenknecht, Angew. Chem. Int. Ed. **42**, 2454 (2003).
78. S.G. Swarts, M.D. Sevilla, D. Becker, C.J. Tokar and K.T. Wheeler, Radiat. Res., **129**, 333 (1992).
79. S.G. Swarts, D. Becker, M.D. Sevilla and K.T. Wheeler, Radiat. Res. **145**, 304 (1996).
80. M.G. Debije, M.T. Milano and W.A. Bernhard, Angew. Chem. Int. Ed. **38**, 2752 (1999).
81. M.G. Debije and W.A. Bernhard, Radiat. Res. **152**, 583 (1999).
82. M.T. Milano and W.A. Bernhard Radiat. Res. **151**,39 (1999).
83. F.D. Lewis, Photochemistry and Photobiology. **81**, 65 (2005).
84. V.B.E. Sartor and G.B. Schuster, J. Phys. Chem. A **105**, 11057 (2001).
85. C. Wan, T. Fiebig, O. Schiemann, J.K. Barton and A.H. Zewail, Proc. Natl. Acad. Sci. USA **97**, 14052 (2000).
86. M.G. Debije and W.A. Bernhard, J. Phys. Chem. B **104**, 7845 (2000).
87. G.D. Reid, D.J. Whittaker, J.A. Day, D.A. Turton, V. Kayser, J.M. Kelly and G.S. Beddard J. Am. Chem. Soc. **124**, 5518 (2002).
88. R.F. Anderson, and G.A. Wright Phys. Chem. Chem. Phys. **1**, 4827 (1999).
89. T.J. Meade and J.F. Kayyem, Angew. Chem. Int. Ed. **34**, 352 (1995).
90. P.F. Barbara and E.J.C. Olson, Adv. Chem. Phys. **107**, 647 (1999).
91. M. Bixon and J. Jortner, J. Phys. Chem. B **104**, 3906 (2000).
92. A. Nitzan, J. Jortner, J. Wilkie, A.L. Burin and M.A. Ratner J. Phys. Chem. B **104**, 5661 (2000).
93. A.A. Voityuk, N. R, M. Bixon and J. Jortner, J. Phys. Chem. B **104**, 9740 (2000).
94. Y.A. Berlin, A.L. Burin and M.A. Ratner, J. Am. Chem. Soc. **123**, 260 (2001).
95. Y.A. Berlin, A.L. Burin, L.D.A. Siebbeles and M.A. Ratner, J. Phys. Chem. A **105**, 5666 (2001).
96. M.H. Baik, J.S. Silverman, I.V. Yang, P.A. Ropp, V.A. Szalai, W. Yang and H.H. Thorp, J. Phys. Chem. A **105**, 6437 (2001).
97. D.M.A. Smith and L. Adamowicz, J. Phys. Chem. B **105**, 9345 (2001).
98. A.A. Voityuk and N. R, J. Phys. Chem. B **106**, 3013 (2002).
99. G. S.M. Tong, I.V. Kurnikov and D.N. Beratan, J. Phys. Chem. B **106**, 2381 (2002).
100. A. Pezeshk, M.C.R. Symons, J.D. McClymont, J. Phys. Chem. **100**, 18562 (1996).
101. B. Weiland and J. Hüttermann, Int. J. Radiat. Biol. **76**, 1075 (2000).
102. P.M. Cullis, G.D.D.Jones, M.C.R. Symons and J.S. Lea, Nature. **330**, 773 (1987).
103. W.H. Nelson, E. Sagstuen, E.O. Hole and D.M. Close, Radiat. Res. **131**, 10 (1992).
104. W.H. Nelson, E. Sagstuen, E.O. Hole and D.M. Close, Radiat. Res. **149**, 75 (1998).
105. B. Giese and S. Wessely, Chem. Commun. **20**, 2108 (2001).
106. V. Shafirovich, A. Dourandin and N.E. Geacintov, J. Phys. Chem. B **105**, 8431 (2001).
107. Y. Razskazovskii, S.G. Swarts, J.M. Falcone, C. Taylor and M. D. Sevilla, J. Phys. Chem. B **101**, 1460 (1997).
108. L.I. Shukla, A. Adhikary, R. Pazdro, D. Becker and M.D. Sevilla, Nucleic Acids Res. **32**, 6565 (2004).

109. A. Adhikary, A. Kumar, D. Becker and M.D. Sevilla, J. Phys. Chem. B **110**, 24171 (2006).
110. A.-O. Colson and M.D. Sevilla, Int. J. Radiat. Biol. **67**, 627 (1995).
111. A.-O. Colson and M.D. Sevilla, J. Phys. Chem. **99**, 3867 (1995).
112. A. Joy, A.K. Ghosh and G.B. Schuster, J. Am. Chem. Soc. **128**, 5346 (2006).
113. Y. Razskazovskiy, M.G. Debije and W.A. Bernhard, Radiat. Res. **159**, 663 (2003).

8 DNA-Based Thermoelectric Nanodevices: A Theoretical Perspective

Enrique Maciá

Departamento de Física de Materiales, Facultad Ciencias Físicas, Universidad Complutense de Madrid, E-28040, Madrid, Spain
emaciaba@fis.ucm.es

8.1 Introduction

The measurement of an appreciable thermoelectric power ($S = +18\,\mu\text{V}\,\text{K}^{-1}$ at room temperature) over guanine molecules adsorbed on a graphite substrate using a STM tip [1], opened novel perspectives for the possible use of organic molecules in the design of nanoscale thermoelectric devices. Firstly, from the study of thermoelectric voltage over a molecule attached to two metallic leads, one can gain valuable information regarding the location of the Fermi energy relative to the molecular levels. Subsequently, one could manage to shift the Fermi level position in order to optimize the thermoelectric performance of a given molecular arrangement. Following this line of reasoning, the thermoelectric response of phenyldithiol organic molecules chemisorbed on gold surfaces was theoretically analyzed, reporting on Seebeck coefficient values comparable to those obtained in Poler's experiment [2]. Similar values ($S = +22\,\mu\text{V}\,\text{K}^{-1}$ at room temperature) have been recently reported for a sample of $FeCl_3$-doped polythiophene [3]. Though these figures are too small to be of interest for most current thermoelectric applications, it is reasonable to expect that they may be significantly enhanced by a proper choice of materials composing the thermoelectric nano-cell. In fact, the extreme sensitivity of thermopower to finer details in the electronic structure suggests that one could optimize the device's thermoelectric performance by properly engineering its electronic structure. Thus, the thermoelectric potential of some conducting polymers, like polythiophene and polyaminosquarine, has been recently reviewed on the basis of their electronic band structures [3]. Also, the thermoelectric properties of nanocontacts made of single-wall carbon nanotubes have been studied numerically, concluding that doped semiconducting nanotubes may exhibit very high figures of thermoelectric merit [4]. These results naturally raise the question regarding the possible use of suitable organic molecules to design novel thermoelectric devices.

With the aim of exploring such a possibility, a systematic theoretical study of the thermoelectric properties of DNA nucleobases guanine (G), cytosine (C), adenine (A), and thymine (T) of increasing complexity – i.e., either as single units or forming dimers or codon trimers of biological relevance – was performed [5–7]. The obtained results showed that relatively

178 Maciá

large thermopower values can indeed be obtained by properly locating the system's Fermi level. In addition, the thermoelectric response of trimer nucleobases shows two resonant features exhibiting large thermopower values ($S = 200 - 400\,\mu\mathrm{V\,K^{-1}}$ at room temperature). Such a behavior closely resembles recently reported thermopower curves of silicon based atomic junctions [8]. Since both the location and the magnitude of the resonance peaks sensitively depend on the energetics of the considered trimer, one may think of introducing a thermoelectric signature for different codons of biological interest [7], in analogy with the electronic signature recently proposed for fast sequencing of single-stranded DNA (ssDNA) chains [9, 10].

Subsequent studies have focused on the Seebeck coefficient and thermoelectric power factor ($S^2\sigma$, where σ is the electrical conductivity) of more realistic double-stranded DNA (dsDNA) chains, in order to estimate the potential use of synthetic DNA chains (such as poly(dG)-poly(dC) and poly(dA)-poly(dT)) as thermoelectric materials. The choice of synthetic, instead of biological DNA (where thousands to millions of base pairs (bps), including four different nucleotides, are aperiodically distributed [11]) is twofold: (i) synthetic DNA strands can be polymerized at will in order to fit any prescribed design; and (ii) quantum chemical calculations show the existence of convenient charge channels in periodic dsDNA chains. Thus, charge transfer mainly proceeds via hole (electron) propagation through the purine (pyrimidine) bases, where the HOMO (LUMO) carriers are respectively located in polyG-polyC (polyA-polyT) chains [12, 13]. In fact, experimental current-voltage curves show that double-stranded poly(dA)-poly(dT) chains behave as n-type semiconductors, whereas poly(dG)-poly(dC) ones behave as p-type semiconductors [14]. Accordingly, these synthetic DNAs may provide the basic building blocks necessary to construct a nanoscale thermoelectric

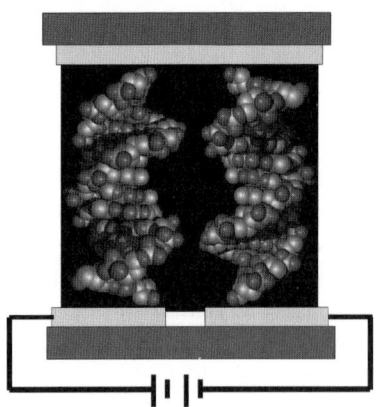

Fig. 8.1. Sketch illustrating the basic features of a nanoscale DNA-based Peltier cell. PolyA-polyT (polyG-polyC) oligonucleotides, playing the role of n-type, *left* (p-type, *right*) semiconductor legs, are connected to organic wires (*light boxes*) deposited onto ceramic heat sinks (*dark boxes*)

cell, where DNA chains will play the role of semiconducting legs in standard Peltier cells, as illustrated in Fig. 8.1.

In this chapter we will consider the possible use of DNA-based nanoscale thermoelectric devices at room temperature in the context of the current search for novel thermoelectric materials. To this end, in Sect. 8.2 we briefly introduce the DNA models and the mathematical tools used in this study, paying special attention to the effective Hamiltonian model parameters choice. In Sect. 8.3, we present the main results obtained from the analytical study of single-stranded oligonucleotides. This approach paves the way to the study of more realistic duplex DNA chains presented in Sect. 8.4, where several transport properties, like the conductance, the Seebeck coefficient and the thermoelectric power factor are analyzed in detail. The role of environmental effects, stemming from the presence of counterions and a hydration shell close to the sugar-phosphate DNA backbone, is addressed in Sect. 8.5. Finally, in Sect. 8.6, we summarize the results and provide some estimations for the thermoelectric figure of merit of synthetic DNA chains.

8.2 DNA Models and the Mathematical Approach

8.2.1 Effective Hamiltonian for Single-Stranded Chains

As a first approximation, ssDNA chains are usually described in terms of a linear chain with an orbital per site, where each lattice site represents a nucleotide. The ends of the chain are connected to leads modeled as semi-infinite one-dimensional chains of atoms with one orbital per site. Thus, the lead-molecule-lead junction is described in terms of three non-interacting subsystems according to the tight-binding Hamiltonian [15]

$$
\mathcal{H} = \left(\sum_{n=1}^{N} \varepsilon_n c_n^\dagger c_n - \sum_{n=1}^{N-1} t_{n,n+1} c_n^\dagger c_{n+1} + h.c. \right)
$$
$$
- \tau \left(c_0^\dagger c_1 + c_N^\dagger c_{N+1} + h.c. \right) + \sum_{k=0}^{-\infty} \left(\varepsilon_M c_k^\dagger c_k - t_M c_{k-1}^\dagger c_k + h.c. \right) \quad (8.1)
$$
$$
+ \sum_{k=N+1}^{+\infty} \left(\varepsilon_M c_l^\dagger c_l - t_M c_l^\dagger c_{l+1} + h.c. \right) .
$$

The first term in (8.1) describes the DNA chain, the second term describes the DNA-metal contact, and the last two terms describe the contacts at both sides of the DNA chain, where N is the number of nucleotides, $c_j^\dagger$ (c_j) is the creation (annihilation) operator for a charge at the jth site in the chain, ε_n are the on-site energies of the G, A, C, and T bases, respectively, $t_{n,n+1}$ is the hopping term between the bases, τ measures the coupling strength between the leads and the end nucleobases, ε_M is the leads on-site energy and t_M ($> \tau$) is their hopping term.

8.2.2 Effective Hamiltonian for Double-Stranded Chains

In the case of realistic dsDNA chains, the electronic energetics should take
into account three different contributions stemming from the nucleobase system, the backbone system and the environment [15, 16]. The energy scale
of the environmental effects, related to the presence of counterions and water molecules, interacting with the nucleobases and the backbone by means
of hydration, solvation and charge transfer processes, is about one order of
magnitude larger than the coupling between the complementary bases, and
about two orders of magnitude larger than the base stacking energies. Quantum mechanical studies show that the hydrogen bonding interaction gives rise
to a spatial separation of the HOMO and LUMO in the nucleobase system,
so that the hole (electron) transfer proceeds through the purine (pyrimidine) bases, where the HOMO (LUMO) carriers are located in polyG-polyC
(polyA-polyT), respectively [12, 13]. Accordingly, we shall consider that the
charge transfer mainly proceeds through the aromatic base stack.

   In order to obtain a simple mathematical description while retaining
most of the relevant physical information, we map the dsDNA chain, including four different nucleotides (which can be arranged either periodically
or aperiodically), into an equivalent binary lattice, where the renormalized
"atoms" correspond to the Watson-Crick complementary pairs in the DNA
molecule [15, 17, 18]. This mapping allows one to write the corresponding
Hamiltonian in a form completely analogous to that of (8.1)

$$
\mathcal{H} = \left(\sum_{n=1}^{N} \tilde{\varepsilon}_n(E) c_n^\dagger c_n - t_0 \sum_{n=1}^{N-1} c_n^\dagger c_{n+1} + h.c. \right) - \tau \left(c_0^\dagger c_1 + c_N^\dagger c_{N+1} + h.c. \right)
$$
$$
+ \sum_{k=0}^{-\infty} \left(\varepsilon_M c_k^\dagger c_k - t_M c_{k-1}^\dagger c_k + h.c. \right) \tag{8.2}
$$
$$
+ \sum_{k=N+1}^{+\infty} \left(\varepsilon_M c_k^\dagger c_k - t_M c_k^\dagger c_{k+1} + h.c. \right) .
$$

This expression properly generalizes the ssDNA Hamiltonian given by (8.1),
where the first term now accounts for the charge carrier propagation through
the DNA chain in terms of the hopping integral t_0, describing the aromatic
π–π base stacking between adjacent nucleotides, and the renormalized on-site
energies $\tilde{\varepsilon}_n(E) = \{\alpha(E), \beta(E)\}$, are given by [17,18]

$$
\alpha, \beta = t_{\alpha,\beta} + \frac{\tau_{G,A}^2 (E - \gamma_{C,T}) + \tau_{C,T}^2 (E - \gamma_{G,A})}{(E - \gamma_{G,A})(E - \gamma_{C,T})} , \tag{8.3}
$$

where $t_\alpha \equiv t_{CG}$ and $t_\beta \equiv t_{AT}$, respectively describe the hydrogen bonding between complementary bases, $\tau_k = t_k + \varepsilon_k(E - \gamma_k)/t_k$, where t_k is the
hopping integral between the backbone state and the base state, ε_k, with

$k = \{G, C, A, T\}$, are the on-site energies of the bases, and γ_k is the on-site backbone energy, which, in general, will depend on the nature of the neighboring base as well as the presence of water molecules and/or counterions attached to the backbone. In this way, this approach provides a realistic description, including 15 physical parameters, $\{\varepsilon_j, t_j, \gamma_j, t_{GC}, t_{AT}, t_0\}$, fully describing the energetics of the DNA molecule in terms of just three variables (i.e., α, β, t_0) in a unified way.

8.2.3 Analytical Tools

To evaluate the thermoelectric power we make use of the approach introduced by Paulsson and Datta relating the Seebeck coefficient to the transmission coefficient as a function of energy by means of the expression [2]

$$S_N(T) = -|e|L_0 \left(\frac{\partial \ln \mathcal{T}_N(E)}{\partial E} \right)_{E_F} T \,, \qquad (8.4)$$

where e is the electron charge, $L_0 = \pi^2 k_B^2/3\,e^2 = 2.44 \times 10^{-8} \; V^2 K^{-2}$ is the Lorenz number, T is the temperature, $\mathcal{T}_N(E)$ is the zero bias transmission coefficient describing the fraction of charge carriers transmitted through a DNA chain of length N in the absence of any applied voltage, and E_F is the Fermi energy. Within the transfer matrix framework, and considering nearest-neighbors interactions only [19], the Schrödinger equation corresponding to the Hamiltonians (8.1) and (8.2) can be expressed in the form

$$\begin{pmatrix} \psi_{N+1} \\ \psi_N \end{pmatrix} = T_{N+1} T_N .... T_1 T_0 \begin{pmatrix} \psi_0 \\ \psi_{-1} \end{pmatrix} \,, \qquad (8.5)$$

where ψ_n is the wavefunction amplitude for the energy E at site n and

$$T_n(E) = \begin{pmatrix} \dfrac{E - \varepsilon_n}{t_{n,n+1}} & -\dfrac{t_{n,n-1}}{t_{n,n+1}} \\ 1 & 0 \end{pmatrix} \,, \qquad (8.6)$$

is the local transfer matrix. $\mathcal{T}_N(E)$ can then be obtained from the knowledge of the leads dispersion relation, $E = \varepsilon_M + 2t_M \cos k$, and the matrix elements of the metal-DNA-metal global transfer matrix $M(E) \equiv \prod_{n=N}^1 T_n(E)$, by means of the relationship [20]

$$\mathcal{T}_N(E) = \frac{4 \sin^2 k}{[M_{12} - M_{21} + (M_{11} - M_{22}) \cos k]^2 + (M_{11} + M_{22})^2 \sin^2 k} \,. \qquad (8.7)$$

From the knowledge of the transmission coefficient given by (8.7), the conductance through the lead-DNA-lead is determined using the Landauer formula [21]

$$G_N(E_F) = G_0 \mathcal{T}_N(E_F) \,, \qquad (8.8)$$

where $G_0 \equiv 2e^2/h \simeq 1/12906\,\Omega^{-1}$ is the conductance quantum. Finally, the thermoelectric quality of a material is expressed in terms of the dimensionless figure of merit

$$ZT = \frac{S^2(T)\sigma(T)}{\kappa(T)}T = \frac{P}{\kappa}T\,, \qquad (8.9)$$

where κ is the thermal conductivity and $P(T) \equiv S^2(T)\sigma(T)$ is referred to as the thermoelectric power factor.

8.2.4 Model Parameters

Single-Stranded Oligonucleotides

In order to obtain the quantitative results, we evaluate (8.4) and (8.8) at room temperature taking $\varepsilon_G = -7.75\,\text{eV}$, $\varepsilon_A = -7.95\,\text{eV}$ and $\varepsilon_T = \varepsilon_C = -8.30\,\text{eV}$ [16]. Depending on the DNA sequence composition, its length and the temperature, the effective value of the hopping integral t can vary over a relatively broad range. In our study we will adopt the values $t_{CC} = 0.3\,\text{eV}$, $t_{GG} = 0.25\,\text{eV}$, $t_{TT} = 0.13\,\text{eV}$ and $t_{AA} = 0.035\,\text{eV}$ in the case of homopolymers, and $t_{GT} = 0.083\,\text{eV}$, $t_{TG} = 0.26\,\text{eV}$, $t_{AC} = 0.11\,\text{eV}$ and $t_{CA} = 0.37\,\text{eV}$ in the study of heteropolymers. These values were derived from ab-initio calculations for 5'-XY-3' intrastrand stacked pairs [22]. In this way, we expect to provide a more realistic description for codon triplets of biological interest.

The base-lead electronic coupling strongly depends on the contact geometry between the molecule and the lead at the junction. In the thermoelectric experiments by Poler and co-workers, guanine molecules were adsorbed on a graphite substrate [1]. The reported adsorption energy of simple organic molecules, such as methane or ethane on a graphite surface is 0.126 eV and 0.17–0.19 eV, respectively [23]. The adsorption energies of the same organics on single-walled nanotubes are reported to be 0.23 eV and 0.30 eV, respectively [24]. Quite interestingly, similar figures within the range 0.2–0.5 eV were measured for the $E_F - E_{HOMO}$ shift, recently reported for oligothiophene derivatives adsorbed on gold [25]. Since nucleobases contain more atoms than methane or ethane, their adsorption energies are expected to be larger on similar substrates. Accordingly, we will consider values within the range $\tau = 0.1$–$0.5\,\text{eV}$ for the base-metal electronic coupling in order to compare with the experimental results.

Finally, we have considered two different contact parameters of technological interest. On the one hand, that corresponding to a molecule connected to an open edge of a graphene sheet at both sides. In this way, the spectral window is approximately given by the graphite π bandwidth $[-6.8, 0]\,\text{eV}$, corresponding to the tight-binding parameters $t_M = 1.7\,\text{eV}$, and $\varepsilon_G = -3.4\,\text{eV}$. On the other hand, we consider platinum contacts corresponding to the tight-binding parameters $t_M = 2.2\,\text{eV}$, and $\varepsilon_M = -5.4\,\text{eV}$, determining the allowed spectral window $[-9.8, -1.0]\,\text{eV}$.

Double-Stranded DNA

The on-site energies for the different nucleobases are chosen as the ionic potentials of their N-methylated forms [22]. On the basis of a recent study about Hückel parameters for biomolecules we adopt the value $\gamma = 12.25$ eV for the backbone phosphate group on-site energy and the value $t = 1.5$ eV for the resonance integral between the nucleobases and the sugar moiety [26]. The precise nature of the hydrogen bonding in Watson-Crick bps has been the subject of a number of quantum chemistry studies indicating that the orbital interaction accounts for about 40% and the electrostatic attraction about 60% of all attractive forces [27]. Our adopted values are taken from the ab initio calculations considering a B-DNA fragment [22]. A broad collection of possible values for the hopping integral between the stacked bases can be found in the literature, ranging from $t_0 = 0.01$ to $t_0 = 0.4$ eV [12,28–33]. Our adopted value for the double-stranded DNA is based on (i) quantum chemistry calculations yielding $t_0 = 0.14$–0.22 eV for poly(dG)-poly(dC) and poly(dA)-poly(dT) duplexes in B-DNA geometry [34], (ii) first principles calculations for a four base pair G-C stacking arranged in B-DNA configuration ($t_0 - 0.115$ cV) [35], and (iii) previous works where some experimental I–V curves for polyG-polyC chains were correctly reproduced by using $t_0 = 0.17$ eV [36]. In Table 8.1, we list the dsDNA model parameters adopted in (8.2). As we can see, their values span over three orders of magnitude, ranging from high energy values related to the sugar-phosphate and nucleobases on-site energies, to the aromatic base stacking low energies.

Finally, in order to reasonably fulfill the transmission resonance condition $\tau = t_M = t_0$ [37], we consider a DNA chain connected to guanine wires at both sides. In this way, the spectral window is given by the energy interval $[-0.3, 0.3]$ eV, where the origin of energies is set at the guanine contact level (i.e., $\varepsilon_M = \varepsilon_G \equiv 0$). Note that the resulting contact bandwidth ($4t_M = 0.6$ eV) compares well with the HOMO bandwidths reported for periodic guanosine stacked ribbons from first-principle studies [38].

Table 8.1. Parameters adopted for the dsDNA effective Hamiltonian considered in this work arranged by decreasing energies in order to illustrate the different energy scales of relevance in the DNA system

Model Hamiltonian parameters (eV)	
$\gamma = 12.27$	
$\varepsilon_A = 8.25$	$\varepsilon_T = 9.13$
$\varepsilon_G = 7.77$	$\varepsilon_C = 8.87$
$t = 1.5$	
$t_{GC} = 0.90$	$t_{AT} = 0.34$
$t_0 = 0.15$	

8.3 Thermopower of Single-Stranded Oligonucleotides

8.3.1 Analytical Expressions

Single Nucleobase

In the case of a single nucleobase the transmission coefficient reads [6],

$$\mathcal{T}_v(E) = \left[1 + \zeta^2 W^{-1}(E - \xi_v)^2\right]^{-1} , \tag{8.10}$$

where $v = \{G, A, C, T\}$ labels the considered nucleobase, $\zeta \equiv (t_M^2 - \tau^2)/\tau^2$ is a coupling factor which vanishes when $\tau = t_M$, $W \equiv (E - E_-)(E_+ - E)$, with $E_{\pm} = \varepsilon_M \pm 2t_M$, defines the allowed spectral window as determined by the leads bandwidth, and

$$\xi_v \equiv \frac{\varepsilon_v \lambda^{-2} - \varepsilon_M}{\zeta} , \tag{8.11}$$

with $\lambda \equiv \tau/t_M$, is a base-dependent resonance energy. According to (8.10), the transmission amplitude is modulated by the lead-nucleobase coupling strength through the factor ζ. The particular case $\zeta = 0$ ($\mathcal{T}_v(E) = 1$) corresponds to metallic conduction over the molecule. In that case, a very small thermoelectric voltage is expected after (8.4). Consequently, we will consider the general case $\zeta \neq 0$. In this case (8.11) defines a resonance energy satisfying the full transmission property $\mathcal{T}_v(\xi_v) = 1$. Making use of (8.10) into (8.4) we obtain [6],

$$S_v(T) = 2|e|L_0 T \left(1 + \frac{ba_v}{cd}\right) \frac{a_v \zeta^2}{cd + a_v^2 \zeta^2} , \tag{8.12}$$

where $a_v \equiv E_F - \xi_v$, $b \equiv E_F - \varepsilon_M$, $c \equiv E_+ - E_F$, and $d = E_F - E_-$. As expected, in the cases $\zeta = 0$ or $E_F = \xi_v$ (i.e., $a_v = 0$) we have a vanishing thermopower. Conversely, the Seebeck coefficient asymptotically diverges as E_F approaches the band edges (i.e., $c = 0$ or $d = 0$). Therefore, very large thermopower values can be eventually reached by properly shifting the Fermi level through the electronic structure of the system.

Dimer Nucleobase

In the case of a dimer oligonucleotide we have nucleobases of energies ε_1^v and ε_2^v, respectively coupled with a hopping term t between them. By following a similar procedure as above one obtains [6],

$$\mathcal{T}_v(E) = \left[1 + q_v^2 + \Omega^2 W^{-1}(E - \xi_2^v)^2(E - \xi_1^v)^2\right]^{-1} , \tag{8.13}$$

where ξ_n^v are base-dependent resonance energies, $\Omega \equiv t_{\mathrm{M}}^2/(t\tau^2)$, and

$$q_v \equiv \frac{\varepsilon_2^v - \varepsilon_1^v}{2t} , \tag{8.14}$$

is a measure of the chemical diversity of the dimer. Thus, in the case of homobases (i.e., GG, AA, CC or TT) we have $q_v = 0$ and, according to (8.13), we get full transmission ($\mathcal{T}_v = 1$) under resonance conditions satisfying $E = \xi_n^v$. Conversely, when both bases are different in nature we get $\mathcal{T}_{\max} = (1 + q_v^2)^{-1} < 1$, under the same conditions. Therefore, a direct consequence of the chemical diversity of heterodimers is a conductance reduction. Consequently, a corresponding enhancement of the thermoelectric voltage (as compared to that exhibited in the single base case) can be expected from general principles. For a given q_v the transmission amplitude is modulated by the Ω ratio factor, involving the different hopping terms. At variance with the single base case, in the dimer case it is not possible to get $\Omega = 0$, so that the transmission through a dimer base will be in general, lower than that corresponding to a single base.

Making use of (8.13) into (8.4) we get the Seebeck coefficient corresponding to the dimer case as

$$S_v(T) = 2|e|L_0T \left[1 - (1 + q_v^2)\mathcal{T}_v\right] \left(a_v^{-1} + \frac{b}{cd}\right) , \tag{8.15}$$

where $a_v^{-1} \equiv a_{1v}^{-1} + a_{2v}^{-1}$, with $a_{nv} \equiv E_{\mathrm{F}} - \xi_n^v$, is a reduced mean value.

Trimer Nucleobase

In the case of a trimer oligonucleotide we have three nucleobases of energies $\varepsilon_1^v, \varepsilon_2^v$ and ε_3^v, respectively, coupled with hopping terms t and ηt, and the transmission coefficient takes the form [6],

$$\mathcal{T}_v(E) = \left[(4fK_v - Q_v)^2 + 4W^{-1}t_{\mathrm{M}}^2 P_v^2\right]^{-1} , \tag{8.16}$$

where $K_v = 4x_1^v x_2^v x_3^v - x_1^v \eta^2 \mu^2 - x_3^v \mu^2$, $Q_v = \lambda^2 (J_v + H_v)/2$, with $J_v \equiv 4x_2^v x_1^v - \mu^2$ and $H_v = 4x_3^v x_2^v - \eta^2 \mu^2$, $P_v = \lambda^4 x_2^v + K_v - fQ_v$, $\lambda \equiv \tau/t_{\mathrm{M}}$, measures the coupling strength between the nucleobase and the leads, and $\mu \equiv t/t_{\mathrm{M}}$ measures the coupling strength between the bases in units of the lead bandwidth, and we have introduced the auxiliary variables $2x_v = (E - \varepsilon_v)/t_{\mathrm{M}}$ and $2 \cos k \equiv 2f = (E - \varepsilon_{\mathrm{M}})/t_{\mathrm{M}}$. In this case we have several resonance conditions given by the relationships $P_v = 0$ and $Q_v = 4K_v \cos k$, respectively.

On the other hand, by plugging Eq. (8.16) into Eq. (8.4) we obtain

$$S_v(T) = 2|e|L_0 T \mathcal{T}_v \mathcal{F}_v , \tag{8.17}$$

where

$$\mathcal{F}_v \equiv t_m^{-1}(4fK_v - Q_v)(8f\tilde{K}_v + 2K_v - \tilde{Q}_v)$$
$$+ P_v \frac{2t_{\mathrm{M}}}{cd} \left(\lambda^4 + 4\tilde{K}_v - 2f\tilde{Q}_v - Q_v + \frac{4t_{\mathrm{M}}^2}{cd}fP_v\right) , \tag{8.18}$$

with $\tilde{K}_v = x_1^v x_2^v + x_1^v x_3^v + x_2^v x_3^v - \mu^2(1+\eta^2)/4$, and $\tilde{Q}_v = \lambda^2(x_1^v + 2x_2^v + x_3^v)$.

8.3.2 Transport Curves

Single Nucleobase Case

The resulting conductance and thermopower curves are very similar for the four bases considered. Thus, we get $\Delta S/S_G$ less than 10% for the different nucleobases over the entire spectral window, the general trend being that both transport curves increase as the base on-site energy increases.

For the sake of illustration, in the main frame of Fig. 8.2 we show the guanine thermopower curve for $\tau = 0.5\,\text{eV}$. We have checked that the thermopower curves are rather insensitive to the adopted coupling strength value. Thus, we get $\Delta S/S_G$ less than 1% within the range $\tau = 0.1$–0.5, the general trend being that thermopower increases as τ decreases. Clearly, a significantly large value of the thermopower can be reached when the Fermi level is located close to the band edges. In the left (right) inset of Fig. 8.2, we compare the conductance (thermopower) of different nucleobases as a function of the Fermi level position. The curves are very similar in shape, exhibiting a well defined maximum at the energy $E_v = \varepsilon_M + 4\zeta\tau^2/(\varepsilon_v - \varepsilon_M)$ in the conductance case. The $G_v(E_v)$ curves peak in the interval $G = 4$–$7 \times 10^{-7}\ \Omega^{-1}$, in reasonable agreement with experimental outcomes previously reported for metal-molecule-metal junctions [39].

By combining the obtained curves we can determine the magnitude S^2G, closely related to the so-called thermoelectric power factor. In Fig. 8.3 we compare the obtained graphite-G-graphite curves for two different values of

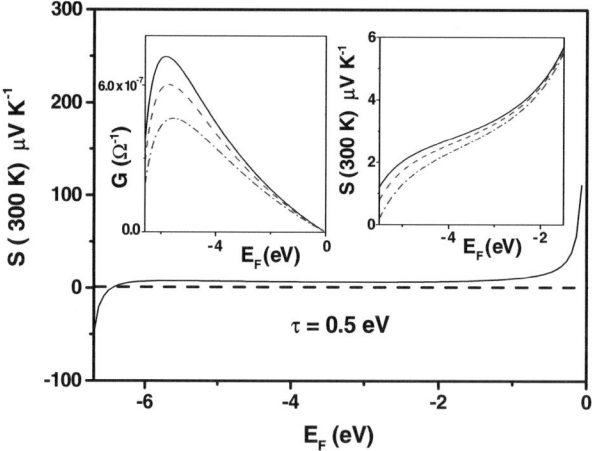

Fig. 8.2. Room temperature dependence of the Seebeck coefficient as a function of the Fermi level energy for a G nucleobase with $\tau = 0.5\,\text{eV}$, $t_M = 1.7\,\text{eV}$, and $\varepsilon_M = -3.4\,\text{eV}$. (*Left inset*) The Landauer conductance as a function of the Fermi level energy for G (*solid lines*), A (*dashed line*), C,T (*dot dashed line*) bases. (*Right inset*) Comparison of the thermopower corresponding to G (*solid line*), A (*dashed line*) and C,T (*dot dashed line*) nucleobases

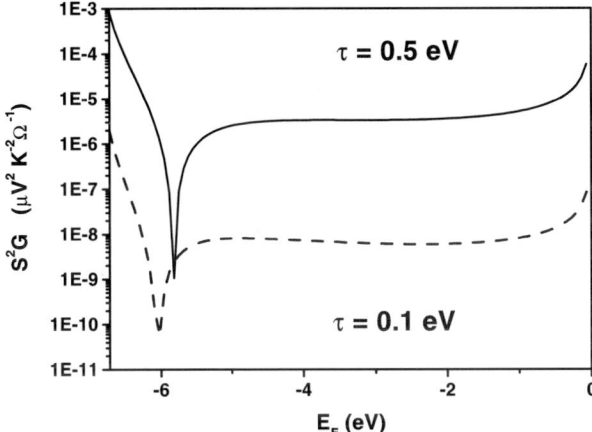

Fig. 8.3. Room temperature dependence of the thermoelectric power factor as a function of the Fermi level energy in a graphite-G-graphite junction for two different values of the molecule-lead coupling strength

the coupling strength τ. The overall shape of the power factor is mainly determined by the energy dependence of the Seebeck coefficient shown in Fig. 8.2. However, at variance with the behavior observed for the Seebeck coefficient itself, the power factor is extremely sensitive to minor variations in the molecule-lead electronic coupling. In fact, S^2G is enhanced by about three orders of magnitude by changing the coupling parameter from $\tau = 0.1$ eV to $\tau = 0.5$ eV, due to the conductance modulation. This result indicates that, in principle, we could optimize the $S^2\sigma$ output by properly controlling the main contact features at the interface.

Dimer Nucleobase Case

In this case we are mainly interested in the *dimerization effects* on the transport properties. In the main frame of Fig. 8.4 we compare the monomer (dashed line) and dimer (solid line) guanine thermopower curves for $\tau = 0.5$ eV. An overall increase of the thermopower stemming from dimerization effects is clearly appreciated. This effect exhibits a systematic behavior, increasing as the Fermi level shifts towards lower energies. We note that the guanine dimers exhibit a p-type behavior over a broader range of energies, as compared to the monomer case. In the left inset of Fig. 8.4, we compare the conductance of G and C monomers and GG and CC dimers for two different values of the t parameter, respectively. For a given t value, the dimerization effects in general decrease the conductance of dimers as compared to that observed in monomers, and this effect is significantly enhanced by decreasing the interbase hopping term. Conversely, the thermopower curve of GG dimers is only slightly modified when the t parameter value is changed by a factor of 40. This result is illustrated in the right inset of Fig. 8.4.

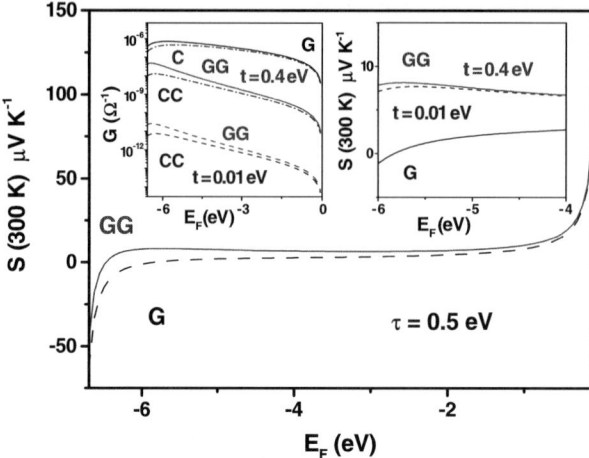

Fig. 8.4. Room temperature dependence of the Seebeck coefficient as a function of the Fermi level energy for GG (*solid line*) and G (*dashed line*) nucleobases with $\tau = 0.5\,\mathrm{eV}$, $t_m = 1.7\,\mathrm{eV}$, and $\varepsilon_m = -3.4\,\mathrm{eV}$. (*Left inset*) Landauer conductance as a function of the Fermi level energy for G and GG (*solid lines*) as compared to C and CC nucleobases with $t = 0.4\,\mathrm{eV}$ (*dot dashed lines*) and $t = 0.01\,\mathrm{eV}$ (*dashed lines*). (*Right inset*) Energy dependence of the thermopower curve for two GG dimers with $t = 0.4\,\mathrm{eV}$ (*solid line*) and $t = 0.01\,\mathrm{eV}$ (*dashed line*) as compared to that of a single G nucleobase

A similar qualitative and quantitative behavior is observed for the other homodimers (AA,CC, TT) and heterodimers (GC,GT,GA,AC,AT,TA) as well. The general trend is that both the conductance and thermopower curves of a given heterodimer, say XY, are intermediate between those corresponding to the homodimers XX and YY, respectively. The subsidiary influence due to the chemical nature of the bases on the dimer transport properties can be understood on the basis of the relative weight of the chemical diversity parameter q_v versus the coupling ratio factor Ω in (8.13). Thus, in the most favorable case, which corresponds to the GC dimer ($q_{GC} = 0.6875$), we get $\Omega/q_{GC} = 42.036$ (for $t = 0.4\,\mathrm{eV}$), so that the contribution of the q_v^2 term is almost negligible as compared to that coming from the last term in (8.13).

Trimer Nucleobase Case

Homonucleotide Codons

In the first place we shall consider the transport properties corresponding to GGG, AAA, CCC, and TTT codon trimers, respectively, codifying for glycine, lysine, proline and phenylalanine amino acids in the homo sapiens genetic code. In this case we have $\eta = 1$ in (8.16) and (8.17). In the main frame of Fig. 8.5 we show the CCC thermopower curve for $\tau = 0.5\,\mathrm{eV}$. This curve is

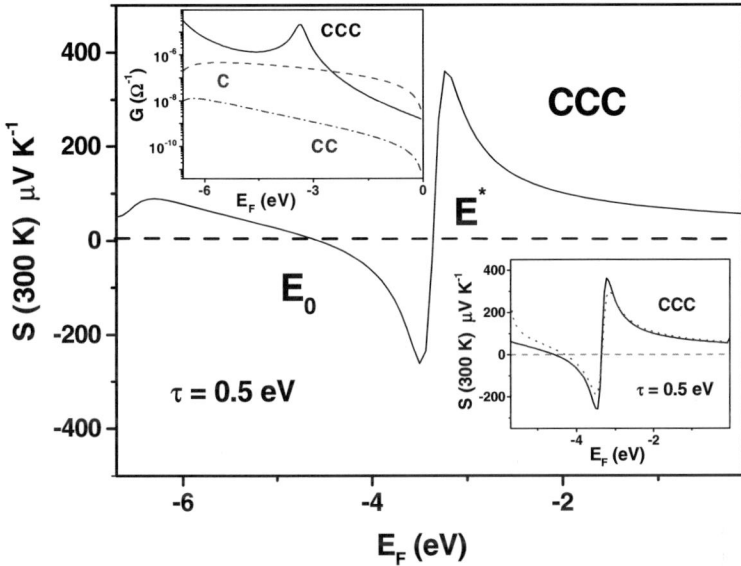

Fig. 8.5. Room temperature dependence of the Seebeck coefficient as a function of the Fermi level energy for a CCC trimer with $\tau = 0.5\,\mathrm{eV}$, $t_m = 1.7\,\mathrm{eV}$, and $\varepsilon_m = -3.4\,\mathrm{eV}$. (*Upper inset*) Landauer conductance as a function of the Fermi level energy for CCC (*solid line*), CC (*dot-dashed line*) and C (*dashed line*) nucleobases. (*Lower inset*) Environmental effects in the thermopower due to the presence of backbone counterions giving rise to nucleobase energy shifts within the range $\Delta E = -1\,\mathrm{eV}$ (*dotted line*) and $\Delta E = +1\,\mathrm{eV}$ (*dashed line*)

characterized by three peaks and two crossing points E_0 and E^*, respectively, defining three different regimes exhibiting p-type or n-type thermopower alternatively. The thermopower values attained at the peaks is significantly high, and compares well with the values reported for benchmark thermoelectric materials. In the upper inset of Fig. 8.5, we compare the conductance of a C monomer, a CC dimer and CCC trimer as a function of the Fermi energy. While the dimer conductance is degraded as a consequence of dimerization effects, we observe that the trimer conductance is significantly enhanced as compared to that corresponding to both C and CC bases over a broad energy range. In addition, a well defined, narrow resonance peak is located at about $E^* = -3.35\,\mathrm{eV}$, flanked by a shallow conductance minimum at about $E_0 = -4.57\,\mathrm{eV}$. Taking into account (8.4) the origin of the crossing points in the thermopower curve can be properly traced back to these topological features.

Heteronucleotide Codons

Now we consider the transport properties corresponding to TGT, CAC and TTG codon trimers, respectively codifying for cysteine, histidine and leucine amino acids in the homo sapiens genetic code. The aim is to compare their

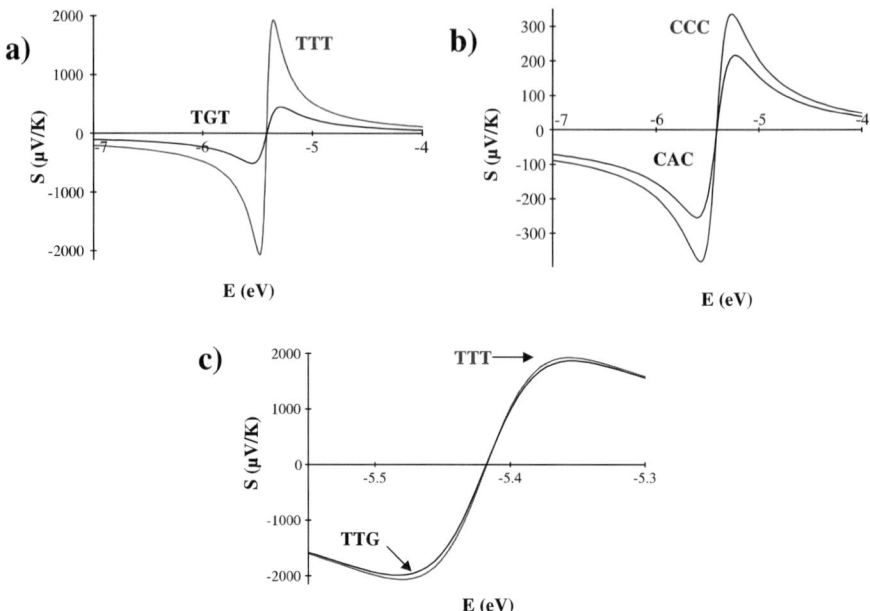

Fig. 8.6. Dependence of the room temperature thermopower as a function of the Fermi level energy for **a** TTT and TGT codons; **b** CCC and CAC codons; and **c** TTT and TTG codons connected to platinum leads with $\tau = 0.5\,\mathrm{eV}$, $t_m = 2.2\,\mathrm{eV}$, and $\varepsilon_m = -5.4\,\mathrm{eV}$

properties with those previously obtained for the TTT and CCC trimers in order to see the effects stemming from the change of one of their original nucleobases ($\eta \neq 1$ in this case). In Fig. 8.6a and b we compare the thermopower curves for the TTT/TGT and CCC/CAC trimers respectively as a function of the Fermi energy. We can see that the thermopower is substantially reduced upon the substitution of the central nucleobases in both cases. This is a general feature of the codons obeying the formula XYX. This result strikingly contrasts with the small effect associated to the substitution of an end nucleobase instead, as it is illustrated in Fig. 8.6c. Accordingly, the thermoelectric properties of codons obeying the general formula XXY are quite similar to those observed for the corresponding homonucleotides.

8.4 Thermopower of Double-stranded DNA Chains

8.4.1 Analytical Expressions

PolyG-PolyC and PolyA-PolyT Chains

Within the transfer matrix framework, considering nearest-neighbors interactions only, the Schrödinger equation corresponding to the Hamiltonian (8.2) can be expressed in the form [17, 40]

$$\begin{pmatrix} \psi_{N+1} \\ \psi_N \end{pmatrix} = L_N Q_\alpha^{N-2} L_1 \begin{pmatrix} \psi_1 \\ \psi_0 \end{pmatrix} \equiv M_N(E) \begin{pmatrix} \psi_1 \\ \psi_0 \end{pmatrix} , \qquad (8.19)$$

where ψ_n is the wavefunction amplitude for the energy E at site n,

$$Q_\alpha \equiv \begin{pmatrix} 2x & -1 \\ 1 & 0 \end{pmatrix} , \qquad (8.20)$$

where $2x \equiv (E - \alpha)/t_0$ describes the DNA energetics related to the GC bps, and the contact matrices

$$L_N \equiv \begin{pmatrix} 2x\Lambda^{-1} & -\Lambda^{-1} \\ 1 & 0 \end{pmatrix} , L_1 \equiv \begin{pmatrix} 2x & -\Lambda \\ 1 & 0 \end{pmatrix}$$

measure the DNA-lead coupling strength measured in terms of the π–π bonding value from the ratio $\Lambda \equiv \tau/t_0$ (completely analogous expressions are obtained for the polyA-polyT chain by simply replacing $\alpha \to \beta$ in (8.20)). Note that we are considering a chain of arbitrary length, composed of N bps, rather than short oligonucleotides in (8.19). Consequently, in order to obtain the global transfer matrix $M_N(E)$ we must evaluate the power matrix Q^{N-2}. To this end, we will make use of the Cayley-Hamilton theorem for unimodular matrices [41]. The transmission coefficient at zero bias as a function of energy is then given by [6]

$$T_N(E) = \left\{ 1 + W^{-1} \left[(E - \varepsilon_M) U_{N-1} - \Omega_0 (U_N + \Lambda^2 U_{N-2}) \right]^2 \right\}^{-1} , \qquad (8.21)$$

where $\Omega_0 \equiv t_M/\Lambda$, and $U_k(x) \equiv \sin[(k+1)\theta]/\sin\theta$, with $x \equiv \cos\theta$, are Chebyshev polynomials of the second kind. By inspecting (8.21) we realize that the transmission coefficient in general does not reach the full transmission condition $T_N = 1$. This transmission degradation stems from contact effects [37]. In fact, even in the most favorable conditions for charge transport (i.e., $E = \varepsilon_M$) we get $T_N(\varepsilon_M) = \left[1 + (\Lambda U_{N-2}^* + \Lambda^{-1} U_N^*)^2/4 \right]^{-1} < 1$, where $U_k^* \equiv U_k(x_M)$, and $2x_M = (\varepsilon_M - \alpha(\varepsilon_M))/t_0$.

Making use of (8.21) into (8.4) and (8.8) one gets,

$$S_N(E_F, T) =$$

$$\tilde{S}_0(T)\Delta G \left\{ B(E_F) + \left(\frac{\partial \ln\left[(E - \varepsilon_M) U_{N-1} - \Omega_0 (U_N + \Lambda^2 U_{N-2}) \right]}{\partial E} \right)_{E_F} \right\} , \qquad (8.22)$$

where $\tilde{S}_0(T) = 2|e|L_0 T$, $\Delta G \equiv 1 - G_N/G_0$, and $B(E_F) \equiv (E_F - \varepsilon_M)/W(E_F)$. The Seebeck coefficient is then expressed as a product involving three contributions. The factor $\tilde{S}_0$ sets the thermovoltage scale (in $\mu VK^{-1}eV$ units) and accounts for the linear temperature dependence of S_N [42]. The factor ΔG links the thermopower magnitude to the conductance properties of the chain so that the Seebeck coefficient progressively decreases (increases) as the conductance increases (decreases), vanishing when $T_N = 1$, as expected from basic transport theory. The last factor in (8.22) depends on two additive contributions in turn. The value of $B(E_F)$ depends on the relative position of the Fermi level with respect to both the center, ε_M, and the band edges, $E_\pm$, of the contacts. Thus, its contribution vanishes when $E_F \rightarrow \varepsilon_M$, whereas B (and consequently S_N) asymptotically diverges as the Fermi level approaches the spectral window edges (i.e., $E_F \rightarrow E_\pm$). Finally, the logarithmic derivative term in (8.22) contains most physically relevant information, accounting for (i) the contact effects (related to the coupling constants Λ and Ω_0), (ii) the size effects (described by the N parameter dependence), and (iii) the resonance effects related to the DNA energetics by means of the Chebyshev polynomials' argument. In order to focus on the *intrinsic* transport properties of DNA chains, we will minimize the contact effects by adopting $t_M = t_0 = \tau$ henceforth, so that $\Lambda = 1$ and $\Omega_0 = t_0$. Thus, taking into account the recurrence relationship $U_{k+1} - 2xU_k + U_{k-1} = 0$, we can rewrite (8.8) and (8.22) in the form

$$G_N(E_F) = \frac{G_0}{1 + C(E_F)U_{N-1}^2} , \qquad (8.23)$$

where $C(E_F) \equiv [\alpha(E_F) - \varepsilon_m]^2 / W(E_F)$, and

$$S_N(E_F, T) = \tilde{S}_0(T)\left[1 - T_N(E_F)\right]\left[B(E_F) + \frac{P_2(E_F)}{E_F - \gamma} + \left(\frac{\partial \ln U_{N-1}}{\partial E}\right)_{E_F}\right], \qquad (8.24)$$

where

$$P_2(E_F) \equiv \frac{a_1(E_F - \gamma)^2 - 2t^2}{a_1(E_F - \gamma)^2 + (a_0 - \varepsilon_M)(E_F - \gamma) + 2t^2} , \qquad (8.25)$$

with $a_0 \equiv t_{GC(AT)} + 2(\varepsilon_{G(A)} + \varepsilon_{C(T)})$, and $a_1 \equiv (\varepsilon_{G(A)}^2 + \varepsilon_{C(T)}^2)/t^2$. By comparing (8.22) and (8.24) we see that the logarithmic derivative in (8.22) has been split into two separate contributions. The first one includes sugar-phosphate backbone effects through the γ parameter dependence. In particular, since $P_2(\gamma) = -1$, we realize that S_N asymptotically diverges as the Fermi level approaches the backbone on-site energy (i.e., $E_F \rightarrow \gamma$). In general, the γ value will depend on the chemical nature of the nucleotides, as well as the possible presence of water molecules and/or counterions attached to the backbone. Accordingly, this resonant enhancement of the thermoelectric power strongly depends on environmental conditions affecting the DNA electronic structure. Finally, the Chebyshev polynomial logarithmic derivative appearing in (8.24) describes possible size effects in the thermoelectric response for DNA chains of different length.

GACT-CTGA Chain

In this case, the effective Hamiltonian given by (8.2) describes a binary lattice, so that the global transfer matrix is expressed as $M_N(E) = L_N(Q_\alpha Q_\beta)^{m-1} L_1$, where $m \equiv N/2$, and the contact matrices

$$L_1 = \begin{pmatrix} 2x & -\Lambda \\ 1 & 0 \end{pmatrix}, \qquad L_N = \Lambda^{-1} \begin{pmatrix} 2y & -1 \\ \Lambda & 0 \end{pmatrix}$$

describe the coupling between the DNA and the metallic leads in terms of the coupling strength Λ. Thus, one gets [18]

$$M_N(E) = \begin{pmatrix} \Lambda^{-1}(U_m + U_{m-1}) & -2yU_{m-1} \\ 2xU_{m-1} & -\Lambda(U_{m-1} + U_{m-2}) \end{pmatrix}.$$

Making use of the relationship $U_{m-1}^2 - U_m U_{m-2} = 1$, it is easy to check that $\det[M_N(E)] = 1$. Taking into account the relationship $U_m^2 + U_{m-1}^2 - 2zU_m U_{m-1} = 1$, after some algebra one gets the following expression for the Landauer conductance [18]

$$G_m(E) =$$
$$G_0 \left\{ 1 + (x - y)^2 U_{m-1}^2 + t_M^2 W^{-1} [f_\Lambda(E, U_m) - 2(x + y)U_{m-1} \cos k]^2 \right\}^{-1},$$
$$(8.26)$$

where the auxiliary function $f_\Lambda(E, U_m) \equiv \Lambda^{-1}(U_{m-1} + U_m) + \Lambda(U_{m-2} + U_{m-1})$, describes contact effects. The term $(x - y)^2 U_{m-1}^2$ in (8.26) accounts

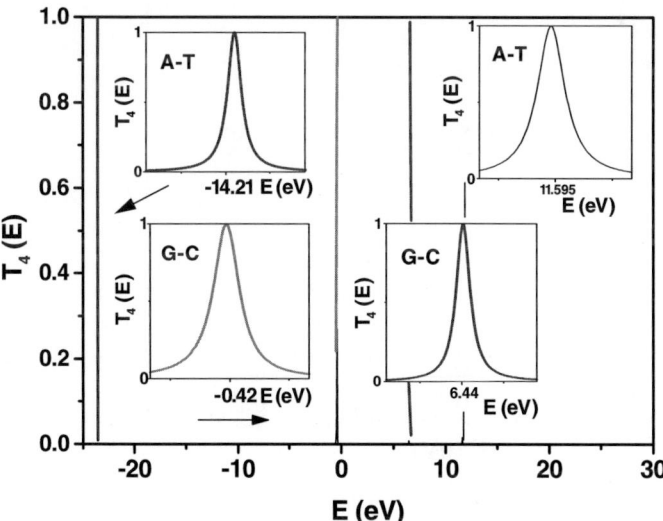

Fig. 8.7. Transmission coefficient as a function of the energy for a periodic polyGACT-polyCTGA chain with $N = 4$ bps

for the chemical diversity of a polyGACT-polyCTGA chain as compared to either polyG-polyC or polyA-polyT chains, and its main physical effect is to reduce the overall conductance of the former with respect to that obtained for the simpler ones.

In Fig. 8.7, the energy dependence of the transmission coefficient is shown as a function of the injected charges energy at zero bias. In the insets the transmission band profile is magnified. As we see, the full transmission condition is fulfilled for all four bands, indicating the extended nature of their eigenstates. By taking $y = x$ in (8.26) we can obtain the transmission spectra for the polyG-polyC chain (the expression for polyA-polyT is then obtained by simply replacing $x \rightarrow y$). In this way, we can assign the central bands in the energy spectrum to GC bps, while the edge bands in the spectrum are related to the AT bps.

8.4.2 Transport Curves

In Fig. 8.8, we plot the thermopower and electrical conductance curves as a function of the Fermi energy obtained from (8.23) and (8.24) for either G-C or A-T complementary pairs ($N = 1$). The $S(E)$ curves exhibit typically metallic values ($1 - 10\,\mu\mathrm{V\,K^{-1}}$) over a broad energy interval around the guanine energy level and then suddenly grow (in absolute value) as E_F

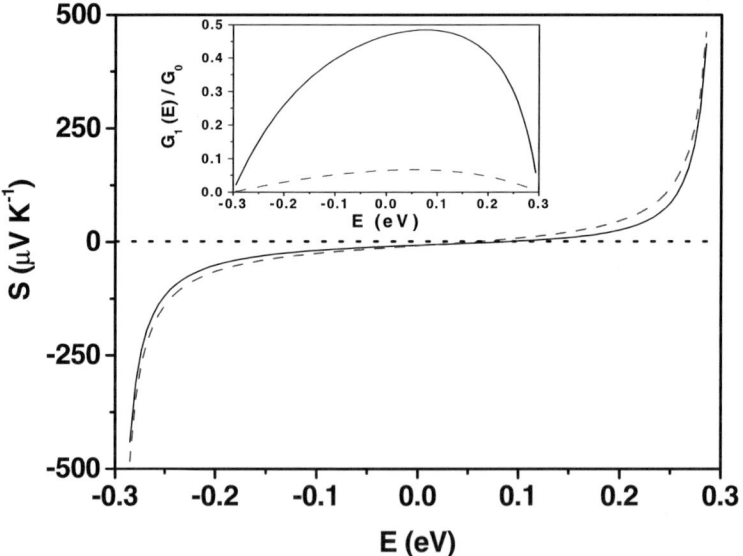

Fig. 8.8. Room temperature dependence of the Seebeck coefficient as a function of the Fermi level energy for a G-C (*solid curve*) and A-T (*dashed curve*) Watson-Crick bps. (*Inset:*) The Landauer conductance as a function of the Fermi level energy for the same bps

approaches the band edges (due to the $B(E_F)$ contribution). Clearly, the thermoelectric response is very similar for both kinds of Watson-Crick pairs, though the Seebeck coefficient is somewhat larger for the A-T one, due to its smaller conductance value (shown in the inset). In this case ($U_0 = 1$) the transmission coefficient reduces to $T_1 = (1 + C)^{-1}$ and the corresponding conductance curves attain the maximum $G_1 \simeq 3.8 \times 10^{-5}$ ($G_1 = 5.1827 \times 10^{-6}$) Ω^{-1} at the resonance energy $E_1^* = 8.638 \times 10^{-2}$ ($E_1^* = 5.501 \times 10^{-2}$) eV for G-C (A-T) bp, respectively. These conductance values are remarkably large (in particular, the G-C bp value is about one order of magnitude larger than the values usually reported for organic molecular junctions [43,44]) accounting for the small values of the Seebeck coefficient in the energy interval $-0.2 \lesssim E \lesssim 0.2$, as prescribed by the ΔG factor in (8.22).

As the number of bps composing the DNA chain is progressively increased, several topological features (i.e, maxima, minima, and crossing points) appear in the thermopower curves of the polyG-polyC chains, as illustrated in Figs. 8.9 and 8.12 for the case $N = 5$. The Seebeck coefficient is clearly characterized by the presence of two peaks around a crossing point located at the energy $E_0 = -0.116$ eV. The thermopower values attained at the peaks are significantly high, and compare well with the values reported for bench-

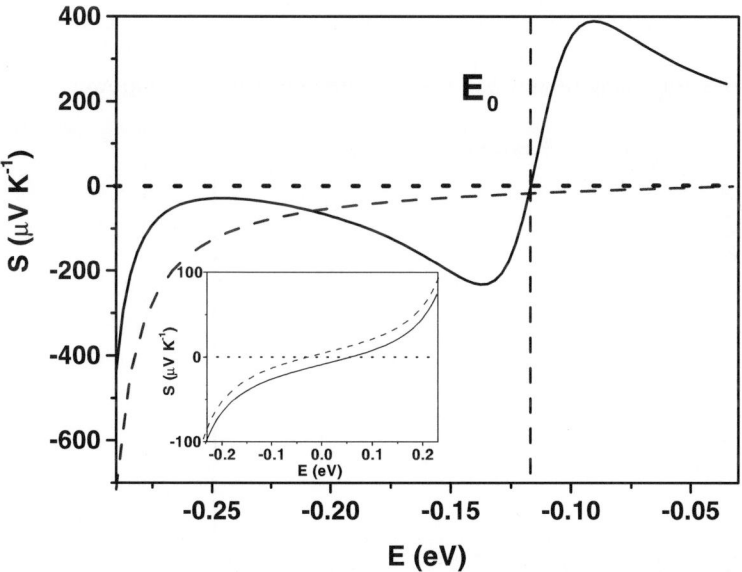

Fig. 8.9. Seebeck coefficient as a function of the Fermi level energy for a polyG-polyC (*solid curve*) and a polyApolyT (*dashed curve*) oligomer with $N = 5$ bps. The *vertical dashed line* separates the energy regions exhibiting the n-type and p-type thermopowers, respectively. (*Inset:*) The Seebeck coefficient as a function of the Fermi level energy for an A-T Watson-Crick bp (*solid line*) is compared to that correspoding to a polyA-polyT oligomer with $N = 5$ (*dashed line*)

mark thermoelectric materials. Nevertheless, as the Fermi level shifts away from the resonance energy, the Seebeck coefficient significantly decreases, illustrating the fine tuning capabilities of the thermopower measurements. On the contrary, the thermoelectric response of the polyA-polyT chain is rather insensitive to the chain length. This is illustrated in the inset of Fig. 8.9, where we compare the thermoelectric curves of a single A-T bp and a $N = 5$ polyA-polyT oligomer.

This contrasting behavior can be understood by inspecting the conductance curves shown in Fig. 8.10 for different N values. Obviously, the overall topology of the polyA-polyT $G_N(E)$ curves do not substantially change as we progressively increase their length, although the conductance peak ratio G_2/G_5 significantly reduces by more than five orders of magnitude. This degradation of the charge transport efficiency is related to the fact that both adenine and thymine energy levels are far above the contact Fermi level, meanwhile the guanine level is just aligned to the contact, one in the polyG-polyC chain. In that case, a pronounced resonance peak (saturating at the quantum conductance value G_0) appears in the conductance curve, shown in the inset of Fig. 8.10. On the other hand, according to (8.24) the main features of the polyG-polyC Seebeck coefficient shown in Fig. 8.9 can be properly accounted for in terms of the conductance curve shown in this inset. In fact, when the Fermi level is located at the left (right) of the conductance peak, the slope of the transmission coefficient curve $T_N(E)$ is positive (negative) leading to the n-type (p-type) thermopower, respectively. In addition, the steeper the conductance curve the higher the thermopower value

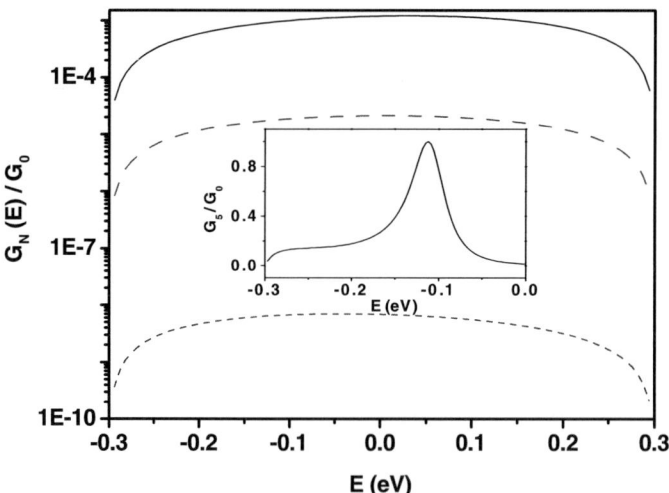

Fig. 8.10. Landauer conductance as a function of the Fermi level energy for polyA-polyT oligomers with $N = 2$ (*solid curve*); $N = 3$ (*dasehd curve*), and $N = 5$ (*short dashed curve*). (*Inset:*) Landauer conductance as a function of the Fermi level energy for a polyG-polyC oligomer with $N = 5$

close to the resonance energy, as it can be readily seen by comparing Fig. 8.9 and Fig. 8.10. Finally, we note that the crossover energy E_0 defines the two different regimes where the polyG-polyC oligomer alternatively exhibits the n-type or p-type thermopower. In this regard, it is worth mentioning that when the Fermi level is located above E_0, the Seebeck coefficient of each DNA chain exhibits contrary signs, so that the polyG-polyC chain behaves as a p-type material, while the polyA-polyT chain behaves like a n-type one, in agreement with previous experimental results [14].

By properly combining the previous results, and making use of the typical values $L_N = 0.34 \times N$ nm for the length, and $R = 1$ nm for the radius of B-form DNA, we can determine the magnitude of the thermoelectric power factor $P_N = \sigma_N S_N^2 = G_N L_N S_N^2 / (\pi R^2)$ for the considered samples. In Fig. 8.11 we plot the power factors of polyG-polyC (solid line) and polyA-polyT (dashed line) chains as a function of the energy for $N = 1$ (main frame) and $N = 5$ (inset). The overall shape of the power factor is mainly determined by the energy dependence of the Seebeck coefficient. In fact, in the case $N = 1$ the power factor takes on relatively small values over a broad range of energies located around the conductance peak, but it significantly increases as the Fermi level approaches the band edges, as it was previously discussed. In the case $N = 5$, in addition to this general behavior (not shown), we observe that the power factor also attains significantly large values *close* to the resonance energy of the polyG-polyC chain due to the presence of the above mentioned Seebeck coefficient peaks. The values of the power factor maxima attained in this case ($P_5 = 1.5–3 \times 10^{-3} \, \mathrm{Wm^{-1} K^{-2}}$) nicely fit with those reported for benchmark thermoelectric materials ($P = 2.5–3.5 \times 10^{-3} \, \mathrm{Wm^{-1} K^{-2}}$) at

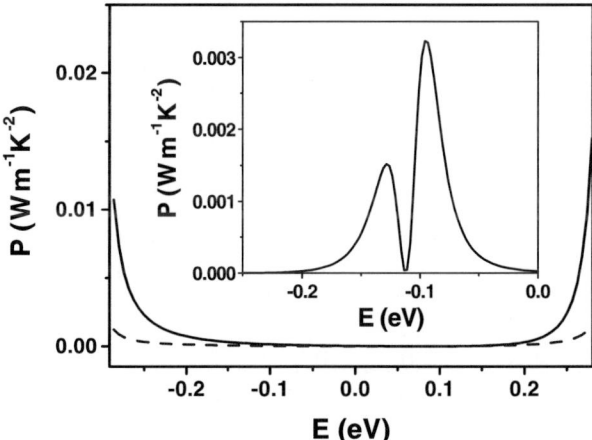

Fig. 8.11. Room temperature thermoelectric power factor as a function of the Fermi level energy for G-C (*solid curve*) and A-T (*dashed curve*) Watson-Crick bps. (*Inset:*) Room temperature thermoelectric power factor as a function of the Fermi level energy for a polyG-polyC oligomer with $N = 5$

high temperatures [45, 46]. On the contrary, the power factor is completely negligible for polyA-polyT oligonucleotides.

8.5 Environmental Effects

The sensitivity of thermopower to backbone effects should be considered in any realistic treatment, for the presence of a number of counterions located along the DNA sugar-phosphate backbone (mainly in the vicinity of negatively charged phosphates) as well as the grooves of the DNA helix (mainly near the nitrogen electronegative atoms of guanine and adenine) are expected [47]. These ions are compensating for the negative charge of the backbone and may play a significant role in the effective charge distribution along the DNA chain. Therefore, it seems pertinent to briefly consider the possible effect of energy fluctuations, stemming from local charge transfer among the ions and the nucleobases, on the previously obtained thermopower curves. To this end, we shall express the on-site nucleotide energies in the form $\varepsilon'_n = \varepsilon_n \pm \Delta E$ in either (8.1) and (8.2), where ΔE measures the magnitude of the perturbation in the diagonal terms of the model Hamiltonian. On the basis of ab-initio calculations for polyG chains [48], we have evaluated the

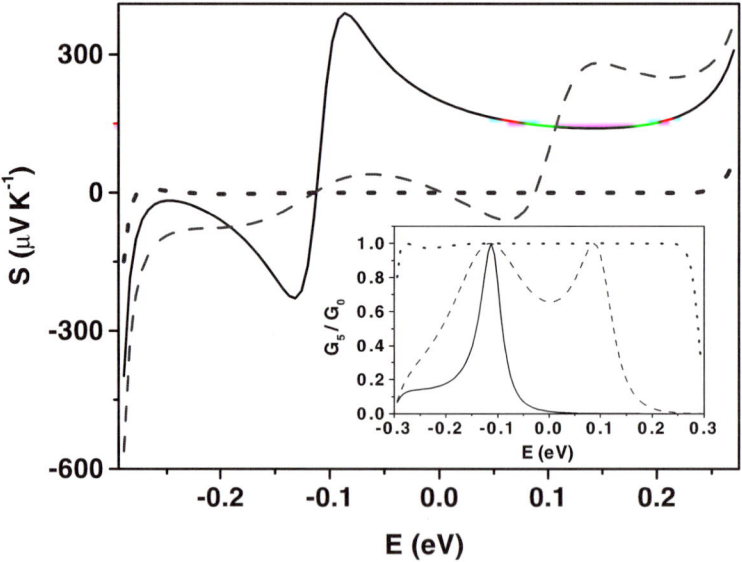

Fig. 8.12. Seebeck coefficient as a function of the Fermi level energy for a polyG-polyC oligomer with $N = 5$ bps and $\gamma = 12.25\,\text{eV}$ (*solid curve*), $\gamma = 11.75\,\text{eV}$ (dashed curve), and $\gamma = 10.0\,\text{eV}$ (*dotted curve*) with $\tau = t_\text{M} = 0.15\,\text{eV}$, and $\varepsilon_\text{M} = 0\,\text{eV}$. (*Inset:*) Landauer conductance as a function of the Fermi level energy for the same samples shown in the main frame

corresponding thermopower curves allowing for on-site energies fluctuations within the range $\Delta E = \pm 1\,\mathrm{eV}$.

In the case of the trimer nucleobases the obtained results are shown in the lower inset of Fig. 8.5. As it can be seen, the resonance peak in the thermopower curve is quite robust under the influence of local charge transfer due to environmental effects.

In the double-stranded DNA case a crude estimation regarding the influence of cations on the unperturbed backbone on-site energies yields γ values within the range $9.75 \leq \gamma \leq 10.75\,\mathrm{eV}$ in some realistic situations [6]. In Fig. 8.12, we compare the Seebeck coefficient as a function of the energy for different γ values for a polyG-polyC chain with $N = 5$. By inspecting this plot we realize the remarkable role played by the environmental effects on thermopower in double-stranded chains. In fact, by systematically varying the on-site energy parameter from $\gamma = 12.25\,\mathrm{eV}$ (no environmental effects) to $\gamma = 10.75\,\mathrm{eV}$, the thermoelectric response of the DNA chain can be modulated from typically semiconducting values to typically metallic ones. As expected from basic theory (see (8.22)), the degradation of the thermopower is related to a progressive enhancement of the DNA conductance. This result is shown in the inset of Fig. 8.12, where we plot the systematic variation of the polyG-polyC oligomer conductance as γ is progressively decreased.

8.6 Outlook and Perspectives

In this chapter, we have reviewed some recent theoretical results concerning the thermoelectric properties of both single-stranded and double-stranded DNA chains. From the results reported in Sect. 8.3, we conclude that the main features of the electrical conductance and thermoelectric response are quite similar for both single nucleobases and dimer nucleobases, irrespective of the chemical nature of their constituent bases. In fact, by comparing their transport curves we observe that the main dimerization effect is an appreciable decrease in the conductance, accompanied by a slight increase in the thermopower. In comparison, the trimer nucleobases exhibit quite different transport curves, characterized by a significant enhancement of both electrical conductance and thermoelectric power due to the presence of resonance peaks. In particular, the thermoelectric response of trimer codons connected between two metallic leads strongly depends on the relative position between the metal Fermi level and the trimer molecular levels. Therefore, we can efficiently optimize the power factor by properly shifting the Fermi level position, suggesting that DNA-based molecular junctions may be of interest for thermoelectric applications. In addition, by comparing the transport curves corresponding to different types of trimers, we have shown that a characteristic thermoelectric signature can be used to identify the XYX type codons from XXX homonucleotide ones on the basis of their different thermoelectric responses. Since the coding properties of DNA introns are closely related to the codon triplet associations, our preliminary result may enclose some biological relevance well deserving a more detailed study by means of more realistic modelling of both the electronic structure of nucleotides and the codon-lead bonding geometry.

From the results reported in Sect. 8.4, we conclude that the thermoelectric response of short duplex DNA chains strongly depends on (i) the chemical nature of the considered DNA chain and the contacts, and (ii) the relative position between the contacts Fermi level and the DNA molecular levels. Thus, while the thermoelectric power of the polyA-polyT oligomers is quite insensitive to the number of bps composing the chain, polyG-polyC oligomers exhibit a strong dependence on the chain length. Accordingly, we can efficiently optimize the power factor of polyG-polyC chains by properly shifting the Fermi level position close to the resonance energy, which plays the role of a tuning parameter. On the other hand, depending on the Fermi level position, both n-type and p-type thermoelectric responses can be simultaneously obtained for polyA-polyT and polyG-polyC DNA chains, respectively. This is a very convenient feature in order to design DNA-based thermoelectric devices, where both oligomers would play the role that semiconducting materials legs usually play in standard Peltier cells. To this end, the relatively low value of the polyA-polyT chain Seebeck coefficient could be significantly improved by connecting it to adenine wires, rather than the guanine ones, in order to get a proper alignment between the contacts Fermi level and the DNA molecular levels.

The thermoelectric quality of a material is expressed in terms of the dimensionless figure of merit given by (8.9). We have seen that the relatively large values of the power factor (appearing at the numerator of ZT) can be obtained by properly locating the Fermi level close to some of the characteristic resonances of the lead-DNA-lead junction. Therefore, the potential of DNA oligomers as thermoelectric materials will ultimately depend on their thermal transport properties which, to the best of my knowledge, have not yet been fully analyzed. Nevertheless, we can make a rough estimation of ZT by assuming that the thermal transport properties recently reported for a series of simple organic semiconductors (e.g. pentacene) are representative of more complex biomolecules as well. In particular, it seems reasonable to expect that the thermal conduction is dominated by phonon transport in these organic compounds, leading to small thermal conductivities in general. In fact, room temperature thermal conductivity values in the range $\kappa = 0.25$–$0.50\,\mathrm{Wm^{-1}\,K^{-1}}$ were measured for different organic films [49]. It is well known that the thermal conductivity of low dimensional systems is usually lower than the bulk, accounting for the higher thermoelectric performance reported for multilayers and nanowires [50]. Accordingly, bulk values provide an upper limit to the expected thermal conductivity. A suitable estimation of thermal conductivity for ideal coupling between a ballistic thermal conductor and the reservoirs relies on the quantum of thermal conductance $g_0 = \pi^2 k_B^2 T/(3h) = 9.46 \times 10^{-13}\,\mathrm{T\,WK^{-1}}$, which represents the maximum possible value of the energy transported per phonon mode [51]. In the regime of low temperatures, four main modes arising from dilatational, torsional and flexural degrees of freedom are expected for a quantum wire [52]. Therefore, the thermal conductivity of a DNA oligomer of length $L_N = 0.34N$ nm will be given by $\kappa_N \simeq 4g_0 L_N/(\pi R^2) = 0.02\,\mathrm{Wm^{-1}K^{-1}}$ (at $T = 10$ K) and $\kappa_N \simeq 0.6\,\mathrm{Wm^{-1}K^{-1}}$ (at room temperature) in optimal conditions. By taking $\kappa \simeq 0.1\,\mathrm{Wm^{-1}K^{-1}}$ as a suitable reference value, along with the power factor values previously obtained, we get $ZT \simeq 4.5$–9.0 for polyG-polyC chains with five bps at room temperature (well above the usual highest $ZT \simeq 1$ for conventional bulk materials) [53].

These remarkably high figure of merit values (comparable to those exhibited by the best thermoelectric materials [54]), must be properly balanced with the significant role played by the unavoidable environmental effects, stemming from the presence of a cation/water molecules atmosphere around the DNA chain, on the actual thermoelectric efficiency of DNA-based nanocells, as it has been indicated in Sect. 5. The inclusion of the phonon degrees of freedom, following the approaches introduced in some recent works [55,56], would be then pertinent in order to obtain more accurate estimations on the feasibility of this proposal. In particular, the role of polarons (whose formation is a very common process for organic polymers with a flexible backbone such as DNA) in the electrical transport efficiency will deserve a closer scrutiny [14,57–60]. Broadly speaking, the on-site interaction of the charge carrier with phonon modes tends to localize it, leading to charge transfer rates within the range $T = 5$–$75\,\mathrm{ps}$, as reported by experiments [61]. These values are much larger than the charge transfer rates related to coherent tunneling (the dominant process assumed in our approach), which are given by $T \simeq t_0/h \simeq 0.03\,\mathrm{ps}$. Accordingly, one reasonably expects that the presence of polarons gives rise to a degradation of the charge transfer efficiency, as compared to that corresponding to coherent transport conditions. From the basic principles one knows that a decrease in the charge transfer efficiency is generally accompanied by an enhancement of the Seebeck coefficient in most samples. On this basis, one could then expect that the inclusion of polaronic effects would lead to further improvement in the thermoelectric properties of DNA chains.

In summary, this theoretical prospective study on the thermoelectric properties of synthetic DNA oligonucleotides suggests that these materials are suitable candidates to be considered in the design of highly-performing, nanoscale sized thermoelectric cells. Experimental work aimed to test the actual capabilities of DNA based thermoelectric devices under different environmental conditions as well as to accurately determine the thermal transport properties of synthetic DNA samples would be then very appealing.

Acknowledgement. I warmly thank Emilio Artacho, Giovanni Cuniberti, Rafael Gutiérrez, Daniel Porath, Stephan Roche, and Eugene B. Starikov for enlightening conversations and for sharing useful information. I acknowledge M.V. Hernández for a critical reading of the manuscript. This work has been supported by the Universidad Complutense de Madrid through project PR27/05-14014-BSCH.

References

1. J.C. Poler, R.M. Zimmermann, and E.C. Cox, Langmuir **11**, 2689 (1995).
2. M. Paulsson and S. Datta, Phys. Rev. B **67**, 241403(R) (2003).
3. X. Gao, K. Uehara, D.D. Klug, S. Patchkovskii, J.S. Tse, and T.M. Tritt, Phys. Rev. B **72**, 125202 (2005).
4. K. Esfarjani, M. Zebarjadi, and Y. Kawazoe, Phys. Rev. B **73**, 2006 (2006).
5. S. Roche, and E. Maciá, Modern Phys. Lett. B **18**, 847 (2004).
6. E. Maciá, Nanotechnology **16**, S254 (2005).
7. E. Maciá, Rev. Adv. Mater. Sci. **10**, 166 (2005).
8. X. Zheng, W. Zheng, Y. Wei, Z. Zeng, and J.J. Wang, Chem Phys. **121**, 8537 (2004).
9. J. Lagerqvist, M. Zwolak, and M. Di Ventra, Nano Lett. **6**, 779 (2006).
10. M. Zwolak, and M. Di Ventra, Nano Lett. **5**, 421 (2005).
11. E. Maciá, Rep. Prog. Phys. **69**, 397 (2006).
12. E. Artacho, M. Machado, D. Sánchez-Portal, P. Ordejn, and J.M. Soler, Molecular Phys. **101**, 1587 (2003).
13. E.B. Starikov, Phil. Mag. **85**, 3435 (2005).
14. K.H. Yoo, D.H. Ha, J.O. Lee, J.W. Park, J. Kim, J.J. Kim, H.Y. Lee, T. Kawai, and H.Y. Choi, Phys. Rev. Lett. **87**, 198102 (2001).
15. G. Cuniberti, E. Maciá, and R. Römer, *this volume*
16. R.G. Endres, D.L. Cox, and R.R.P. Singh, Rev. Mod. Phys. **76**, 195 (2004).
17. E. Maciá, and S. Roche, Nanotechnology **17**, 3002 (2006).
18. E. Maciá, Phys. Rev. B **74**, 245105 (2006).
19. A comparison between density functional calculations and a tight-binding model, reported in Ref. [12], indicates that second nearest neighbor interactions are one order of magnitude smaller than nearest-neighbor ones. A detailed numerical study on the interbase coupling in ds B-DNA structures containing 4-8 bps concluded that the hopping integrals beyond second nearest-neighbor bases are negligible. H. Mehrez, and M.P. Anantram, Phys. Rev. B **71**, 115405 (2005).
20. E. Maciá and F. Domínguez-Adame, *Electrons, Phonons ans Excitons in low dimensional aperiodic systems* (Colección Línea 300, Ed. Complutense, Madrid, 2000).
21. M. Büttiker, Y. Imry, R. Landauer, and S. Pinhas, Phys. Rev. B **31**, 6207 (1985).
22. Y.J. Yan, and H. Zhang, J. Theor. Comp. Chem. **1**, 225 (2002).
23. G. Vidali, G. Ihm, H.-Y. Kim, and M.W. Cole, Surf. Sci. Rep. **12**, 133 (1991).
24. A. Kleinhammes, S.-H. Mao, X.-J. Yang, X.-P. Tang, H. Shimoda, J.P. Lu, O. Zhou, and Y. Wu, Phys. Rev. B **68**, 075418 (2003).
25. L. Patrone, S. Palacin, J. Charlier, F. Armand, J.P. Bourgoin, H. Tang, and S. Gauthier, Phys. Rev. Lett. **91**, 096802 (2003).
26. K. Iguchi, J. Phys. Soc. Jpn. **70**, 593 (2001); Int. J. Mod. Phys. B **11**, 2405 (1997); K. Iguchi, Int. J. Mod. Phys. B **18**, 1845 (2004).

27. C. Fonseca Guerra, F.M. Bickelhaupt, and E.J. Baerends, Cryst. Growth Design. **2**, 239 (2002).
28. P.J. de Pablo, F. Moreno-Herrero, J. Colchero, J. Gomez-Herrero, P. Herrero, A.M. Baro, P. Ordejón, J.M. Soler, and E. Artacho, Phys. Rev. Lett. **85**, 4992 (2000).
29. R. Di Felice, A. Calzolari, E. Molinari, and A. Garbesi, Phys. Rev. B **65**, 045104 (2002).
30. H. Wang, J.P. Lewis, and O.F. Sankey, Phys. Rev. Lett. **93**, 016401 (2004).
31. Y.A. Berlin, A.L. Burin, and M.A. Ratner, Superlattices and Microstruc. **28**, 241 (2000).
32. H. Sugiyama, and I. Saito, J. Am. Chem. Soc. **118**, 7063 (1996).
33. A.A. Voityuk, J. Jortner, M. Bixon, and N. Rösch, J. Chem. Phys. **114**, 5614 (2001).
34. E.B. Starikov, Phil. Mag. Lett. **83**, 699 (2003).
35. H. Mehrez, and M. P. Anantram, Phys. Rev. B **71**, 115405 (2005).
36. G. Cuniberti, L. Graco, D. Porath, and C. Dekker, Phys. Rev. B **65**, 241314(R) (2002).
37. E. Maciá, F. Triozon, and S. Roche, Phys. Rev. B **71**, 113106 (2005).
38. R. Di Felice, A. Calzoları, and H. Zhang, Nanotechnology **15**, 1256 (2004); A. Calzolari, R. Di Felice, E. Molinari, and A. Garbesi, J. Phys. Chem B **108**, 2509 (2004); A. Calzolari, R. Di Felice, and E. Molinari, Appl. Phys. Lett. **80**, 3331 (2002); A. Calzolari, R. Di Felice, E. Molinari, and A. Garbesi, Physica E **13**, 1236 (2002).
39. C. Kergueris, J.P. Bourgoin, S. Palacin, D. Esteve, C. Urbina, M. Magoga, and C. Joachim, Phys. Rev. B **59**, 12505 (1999).
40. E. Maciá, Phys. Rev. B **75**, 035130 (2007).
41. Let $\mathbf{A}$ be a matrix belonging to the SL(2,$\mathbb{R}$) group. Then, $\mathbf{A}^n = U_{n-1}(z)\mathbf{A} - U_{n-2}(z)\mathbf{I}$, where $U_k(z)$ is a Chebyshev polynomial of the second kind, and $z \equiv \mathrm{tr}\mathbf{A}/2$. For an elegant derivation of this result the reader is referred to D.J. Griffiths, and C.A. Steinke, Am. J. Phys. **69**, 137 (2001).
42. Essentially describing a diffusive behaviour of the charge carriers. Possible electron-phonon interactions, giving rise to phonon-drag effects at low temperatures, are not included in our treatment. According to photoinduced infrared spectroscopy measurements a relatively low hole-vibrational coupling constant ($\lambda_{\mathrm{ph}} \approx 0.2$) is expected for biological dsDNA samples, as it was reported by A. Omerzu, N. Licer, T. Mertelj, V.V. Kabanov, and D. Mihailovic, Phys. Rev. Lett. **93**, 218101 (2004).
43. C. Kergueris, J.P. Bourgoin, S. Palacin, D. Esteve, C. Urbina, M. Magoga, and C. Joachim, Phys. Rev. B **59**, 12505 (1999).
44. M. Tsutsui, Y. Teramae, S. Kurokawa, and A. Sakai, Appl. Phys. Lett. **89**, 163111 (2006).
45. S. Yamaguchi, Y. Nagawa, N. Kaiwa, and A. Yamamoto, Appl. Phys. Lett. **86**, 153504 (2005).
46. Y. Kimura, and A. Zama, Appl. Phys. Lett. **89**, 172110 (2006).
47. R.N. Barnett, C.L. Cleveland, A. Joy, U. Landman, and G.B. Schuster, Science **294**, 567 (2001).
48. Ch. Adessi, S. Walch, and M.P. Anantram, Phys. Rev. B **67**, 081405 (2003).
49. N. Kim, B. Domercq, S. Yoo, A. Christensen, B. Kippelen, and S. Grahm, Appl. Phys. Lett. **87**, 241908 (2005).

50. G. S. Nolas, J. Sharp, and H. J. Goldsmid, *Thermoelectrics Basic Principles and New Materials Developments*, (Springer Series in Materials Science, vol. **45**, Springer, Berlin, 2001), p. 235.
51. K. Schwab, E.A. Henriksen, J.M. Worlock, and M.L. Roukes, Nature (London) **404**, 974 (2000).
52. N. Nishiguchi, Phys. Rev. B **52**, 5279 (1995); N. Nishiguchi, Y. Ando, and M.N. Wybourne, J. Phys. Condens. Matter **9**, 5751 (1997).
53. T.C. Harman, P.J. Taylor, M.P. Walsh, and B.E. LaForge, Science **297**, 2229 (2002).
54. *Semiconductors and Semimetals; Recent Trends in Thermoelectric Materials Research III*, (ed. by T.M. Tritt, Academic, San Diego, CA, 2001).
55. D. Segal, Phys. Rev. B **72**, 165426 (2005).
56. J. Koch, F. von Oppen, Y. Oreg, and E. Sela, Phys. Rev. B **70**, 195107 (2004).
57. E.M. Conwell and S.V. Rakhmanova, Proc. Natl. Acad. Sci. U.S.A **97**, 4556 (2000).
58. W. Zhang, A.O. Govorov, and S.E. Ulloa, Phys. Rev. B **66**, 060303(R) (2002).
59. S.S. Alexandre, E. Artacho, J.M. Soler, and H. Chacham, Phys. Rev. Lett. **91**, 108105 (2003).
60. H. Yamada, E.B. Starikov, D. Hennig and J.F.R. Archilla, Eur. Phys. J. E **17**, 149 (2005).
61. C. Wan, T. Fiebig, S.O. Kelley, C.R. Treadway, J.K. Barton, and A.H. Zewail, Proc. Natl. Acad. Sci. U.S.A **96**, 6014 (1999).

9 Transverse Electronic Signature of DNA for Electronic Sequencing

Mingsheng Xu[1], Robert G. Endres[2], and Yasuhiko Arakawa[1]

[1] Nanoelectronics Collaborative Research Center, IIS & RCAST, The University of Tokyo, 4-6-1 Komaba, Meguro-ku, Tokyo 153-8505, Japan
[2] Department of Molecular Biology, Princeton University, Princeton, NJ 08544-1014, USA
    msxu@iis.u-tokyo.ac.jp (MSX), rendres@Princeton.EDU (RGE)

9.1 Introduction

Previous investigations into the charge transport properties of DNA along the DNA helix were aimed at the DNA's potential application in molecular electronics instead of electronic DNA sequencing. Those studies led to controversial results [1,2], and the charge transport mechanism along the DNA helix is still unclear [1,3,4]. Carefully controlled experiments with short DNA molecules by scanning probe microscopy technology recently demonstrated that the charge can migrate *along* the DNA helix [5–9]. However, very few experimental studies have as yet, looked into the charge transport properties of DNA in the *transverse direction* [10,11]. No experiment has been carried out to study the current-voltage characteristics of the individual nucleotides or nucleosides [12,13] for potential electronic DNA sequencing.

In recent years, proliferation of the large-scale DNA-sequencing projects for applications in clinical medicine and health care has driven the search for alternative methods to reduce time and cost [14–22]. The commonly used Sanger sequencing method relies on the chemistry to read the bases G, C, A, and T in DNA and is still far too slow and costly for reading the personal genetic codes. It costs an estimated 10–25 million US-dollars to sequence a single human genome and $20,000–$50,000 to sequence a microbial genome [21]. New technologies for low-cost genome sequencing [17] must be evaluated not only by their accuracy but also by the length of the genome fragments which can be sequenced at once.

In this chapter, we shall report a potential physical alternative for direct and potentially rapid DNA sequencing via detection of the unique transverse electronic signatures of DNA bases and base-pairs [10,18] (see Fig. 9.1). Previously, there has been a lot of attempts to sequence DNA by directly visualizing the nucleotide composition of the DNA molecules with scanning tunnelling microscopy (STM), but such effects have not been successful [11,12,23,24]. In contrast, the technology discussed here is based on the ultrahigh vacuum (UHV) STM to directly sense the molecular levels of the single DNA bases. Combined with a first-principles study, we demonstrate the characteristic electronic signatures of the poly(AT) and poly(GC)

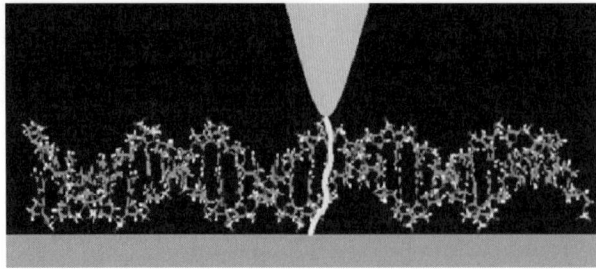

Fig. 9.1. Schematic picture of an idealized dsDNA on a conducting surface with a STM tip and the tunnelling current through a single basepair

double-stranded (ds) DNA in the transverse direction and of the four DNA bases (nucleosides). By exploiting these electronic signatures, it would become possible to sequence DNA by the STM-based technology without DNA modification and obvious length limitations.

In the following sections, we discuss the STM-based sequencing method, the transverse electronic signatures of poly(GC) and poly(AT) dsDNA molecules and DNA nucleosides (bases with ribose), as well as first-principles calculations of single methylated bases (bases with the methyl-groups replacing ribose) on a Au surface. Based on the experimental signatures of the nucleosides, Bayes' theorem is used to estimate the accuracy of predicting bases correctly for sequencing by STM/STS. Finally, we discuss the remaining questions regarding realistic sequencing of single-stranded (ss) DNA molecules based on STM-based technology.

9.2 Characterization of DNA on Au(111) Surface

9.2.1 Sample Fabrication and STM Measurements

The studied samples were fabricated by immersing the Au(111) substrate into the DNA or the nucleoside solution for about $10 \sim 15$ minutes. The DNA or the nucleoside powder was dissolved in the 1 MK_2HPO_4+TE buffer solution (pH 7.0) [6,10]. Immediately after preparation, the DNA or the nucleoside sample was loaded into the UHV treatment chamber (below 2.2×10^{-7} Pa), and transferred to the UHV/STM chamber (below 1.5×10^{-8} Pa). Imaging and spectroscopy measurements were made using an etched tungsten tip with a bias voltage applied to the samples at room temperatures [6,10]. A same tip was used for measuring poly(GC) and poly(AT) dsDNA; another tip was used for the measurements of the four nucleosides. The tips were regularly cleaned. All the I–V curves presented here are averages over about five individual I–V curves obtained from the same location to reduce noise. Each I–V data point of an individual I–V curve is an average of thirty-two repeated measurements collected by the STM control electronics. From the average I–V curves, we calculated the normalized conductances defined as $(\mathrm{d}I/\mathrm{d}V)/((I/V)^2 + c^2)^{\frac{1}{2}}$. Introducing the small constant c avoids dividing by zero, but shifts the intensities of the small peaks in the band gap. We choose c to be equal to about 2.0% of the maximal current of the I–V curve throughout this work. Based on the normalized conductance, we assign HOMO and LUMO energies according to the following criterion. The first peak or the shoulder larger than 2.0 to the left (right) of the band gap is the putative HOMO (LUMO) [25].

9.2.2 Chemical Characterization

X-ray photoelectron spectroscopy (XPS) was used to chemically characterize DNA on the Au(111) surface. Figure 9.2 shows the XPS patterns for C_{1s} and N_{1s} from a sample where a 25 basepair-long dsDNA was immobilized on the Au(111) surface [10]. The appearance of the strong N_{1s} and C_{1s} peaks indicates the presence of DNA on the Au(111) surface [26]. In contrast, residue of the buffer solution was not traced by XPS. The XPS results confirm the presence of DNA on the Au(111) surface. Note that compared with the binding

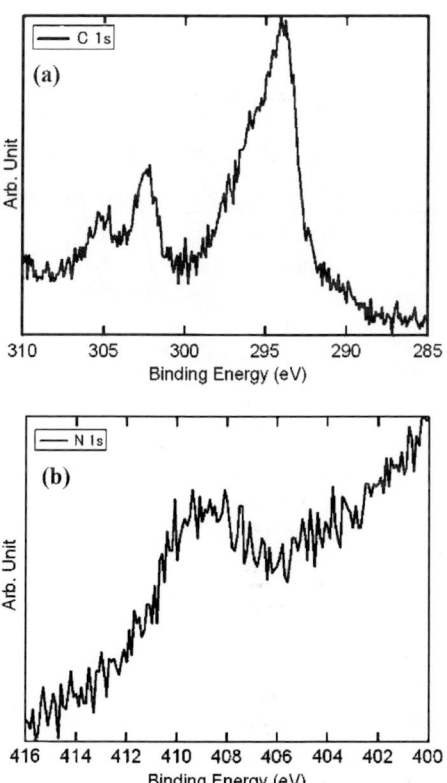

Fig. 9.2. X-ray photoelectron spectroscopy (XPS) characterization of DNA on the Au surface: **a** C_{1s} pattern, **b** N_{1s} pattern

energy of the $Au_{4f7/2}$, we find that the whole spectrum is shifted by $+9.2\,eV$, which may be due to the charge effect.

9.2.3 Control Experiments

To examine the effects of the K_2HPO_4+TE buffer solution, which was used to dissolve the dsDNA or the nucleosides, the Au(111) substrate was immersed into the buffer solution without containing dsDNA or nucleosides [10]. Figure 9.3a shows a typical UHV/STM image of the buffer-treated Au(111) surface. No contamination is visible. Figure 9.3b compares the normalized conductances of the clean Au(111) substrate, of the buffer-treated Au(111) substrate, of the poly(GC)-dsDNA sample, and of the thymidine sample. Although the normalized conductance spectrum obtained from the buffer-treated Au(111) surface, compared to that obtained from the clean Au(111) surface, shows additional peaks, none of these peaks match the dominant peaks of the DNA or of the nucleoside spectra. Furthermore, no insulating gap is observed for the buffer-treated Au(111) surface [10]. Therefore, we conclude that our buffer-treated Au(111) surfaces are suitable for studying dsDNA and nucleosides.

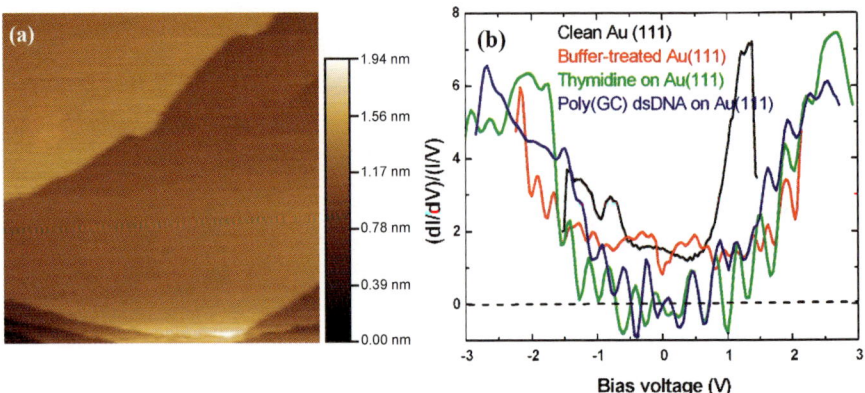

Fig. 9.3. a UHV/STM image (150 nm×150 nm) of a buffer–treated Au(111) surface acquired at preset tunnelling condition 1.5 V and 0.2 nA. **b** Comparison of the normalized conductances of the clean Au(111) substrate, of buffer–treated Au(111), of thymidine on Au(111) and of poly(GC) dsDNA on Au(111). No vertical offset is applied

9.3 Transverse Electronic Signature of DNA

9.3.1 Poly(GC) and Poly(AT) Double-Stranded DNA

We first demonstrate that dsDNA of two different homogeneous sequences can be differentiated electronically by scanning tunnelling spectroscopy

(STS) [6, 10]. Here, we study 40 basepair-long poly(GC) and poly(AT) ds-DNA molecules by using the same etched tungsten tip for both measurements, since the density of states (DOS) of the STM tip affects the tunnelling current. In Fig. 9.4a, as confirmed by the tapping mode AFM in air, poly(AT) dsDNA forms networks of 1–2 nm thick bundles, and hence, our DNA molecules most likely lie flat on the substrate with the helix parallel to substrate surface instead of standing upright [27]. Consequently, it is feasible to locate the STM tip above the molecules to study the charge transport in the transverse direction (see illustration in Fig. 9.1). Both the poly(GC) and the poly(AT) dsDNA form networks on the Au(111) surfaces, but with different patterns (not shown here). Figure 9.4b shows a typical UHV/STM image of 40 basepair-long poly(AT) dsDNA on a Au(111) surface. We observe, somewhat depending on the sample region, a difference between poly(GC) and poly(AT) DNA molecules. Poly(GC) dsDNA frequently exhibits positive (light) features, whereas poly(AT) dsDNA shows depressive (dark) features. In addition to such contrasting features, we also observe the positive images of isolated molecules and films of both poly(GC) and poly(AT) samples, which indicates the difficulty of sequencing DNA by STM imaging.

Sequencing by imaging the chemical composition of DNA by STM remains elusive. Despite the obvious contrast between poly(AT) and poly(GC) dsDNA molecules [11] and the high (base)-resolution images [11, 28], identification of the chemical composition of the four bases is extremely difficult. To improve the sample-tip contact, Ohshiro et al. [12] have shown facilitated electron tunnelling between nucleobases and a complementary nucleobase-modified STM tip, which demonstrates the ability to identify the chemicals due to selective chemical interactions. For sequencing by chemical-selective STM, however, a reliable preparation of readable samples and nucleobase-modified tips would be challenging. For this reason, sequencing by spectroscopy (STS) is a realistic possibility instead.

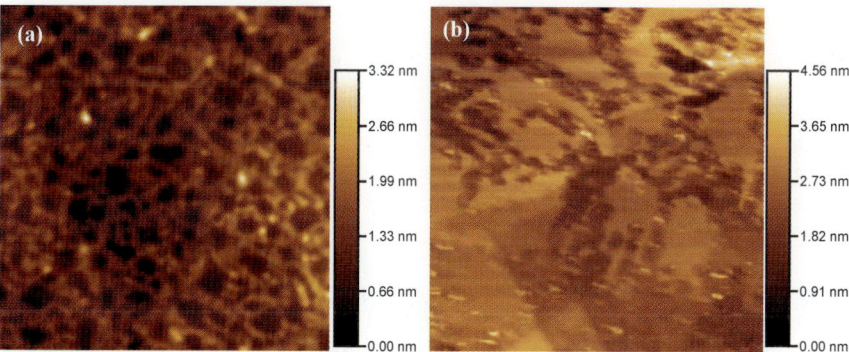

Fig. 9.4. Structural properties of poly(AT) dsDNA on Au(111). **a** Tapping mode AFM image (600 nm × 600 nm) obtained in air. **b** UHV/STM image (200 nm × 200 nm) acquired at 2.5 V and 30 pA

Figure 9.5a shows representative I–V characteristics of the poly(GC) and poly(AT) dsDNA obtained by the STS in the *transverse direction*, along with the corresponding normalized conductances (Fig. 9.5b) calculated from the digital I–V data [29]. The current is essentially zero up to a threshold voltage of a few volts. Beyond the threshold, the current rises sharply. The current through G–C basepairs tends to be larger than the current through A–T basepairs. The normalized conductance, which can be interpreted as the local density of states [30], exhibits pronounced peaks. The first peak or shoulder to the left and to the right of the band gap with a height above a certain threshold correspond to the HOMO (highest occupied molecular orbitals) and the LUMO (lowest unoccupied molecular orbitals), respectively. The results suggest that the charge can transversely transport through the basepairs beyond a threshold voltage analogous to a solid-state semiconductor.

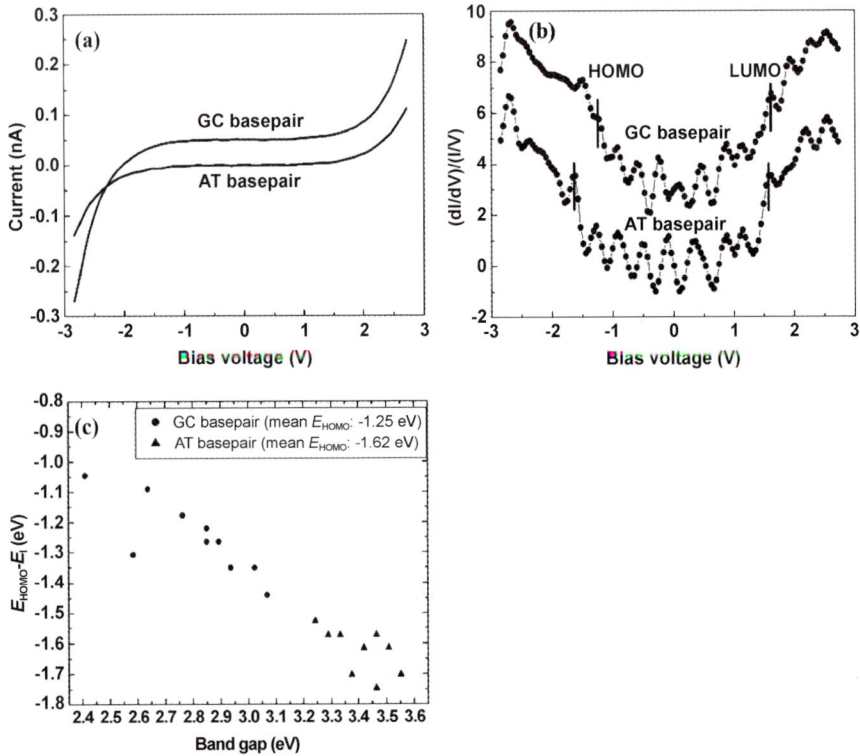

Fig. 9.5. Transverse electronic signatures of poly(GC) and poly(AT) dsDNA on Au(111). **a** Representative I V curves acquired at preset tunnelling condition 2.0 V and 30 pA. **b** The corresponding normalized conductances calculated from the I–V data. For clarity, the curves are offset vertically. **c** Scatter plot of HOMO energies (relative to the Fermi level of Au(111)) versus the band gaps of poly(GC) and poly(AT) dsDNA

Figure 9.5c shows the HOMO energies and the corresponding band gaps, and provides information about the electronic differences between G–C and the A–T basepairs in dsDNA. While the data of poly(GC) and poly(AT) dsDNA cluster separately, the data generally fluctuate depending on the surface location, which may be due to a variation of the local DNA conformation and of the environment [10]. The averages and the standard deviations of the HOMO energies of poly(GC) and of poly(AT) dsDNA are $-1.25 \pm 0.12\,\mathrm{eV}$ and $-1.62 \pm 0.07\,\mathrm{eV}$, respectively. (Energies are relative to the Fermi level of Au(111)). Similar differences of the HOMO energies were recently observed by STM [31], where the difference of the electronic structure between A–T and the G–C base pairs was used to address the charge-transfer mechanism along the DNA helix. The separation of the average HOMO energies ($0.37\,\mathrm{eV}$) is larger than the sum of the standard deviations, and hence, is statistically significant to differentiate (to sequence:) G–C and A–T basepairs along double-stranded DNA, but contains no information about distinguishing G–C from C–G and A–T from T–A base pairs.

9.3.2 DNA Bases

Having demonstrated crude sequencing of dsDNA via detection of the HOMO-energy differences between two sequences, we further demonstrate that the four bases can be identified individually based on their electronic signatures. For that purpose, we measure the four types of nucleosides, i.e., guanosine (G), adenosine (A), cytosine (C), and thytosine (T), on Au(111) separately. This proof-of-principle study is directly related to sequencing of ssDNA. Here, we only show typical UHV/STM images of nucleosides on Au(111), not describing them in full detail in the current work. Briefly, we observed the well-ordered patterns of adenosine (Fig. 9.6a) and cytidine (Fig. 9.6b) on Au(111). We confirm that the well-ordered pattern is indeed adenosine (or cytidine) rather than Au(111) reconstruction through the nonlinearity in the small bias range of the I–V curves. Similar mimetic reconstruction stripes were previously observed and possible formation mechanisms of such nucleoside patterns were proposed [24, 32, 33]. For all the four nucleosides we observed the isolated molecules (or small clusters) (Fig. 9.6c and d) and uniform films on the Au(111) as well.

Figure 9.7a (bottom-left axes) and Fig. 9.7b show representative I–V curves of the four nucleosides and the corresponding normalized conductances, respectively. Figure 9.7a (top-right axes) shows the corresponding current ratios I_X/I_T between the current of the nucleoside X (= A, G, or C) and the current of the nucleoside T for different biases. For negative bias voltages below the band gap threshold, the nucleoside T exhibits the smallest current among the four nucleosides (Fig. 9.7d), as predicted theoretically [18]. However, the current ratios have a complicated dependence on the bias voltage in the range from $-3.0\,\mathrm{V}$ to $3.0\,\mathrm{V}$. Hence, se-

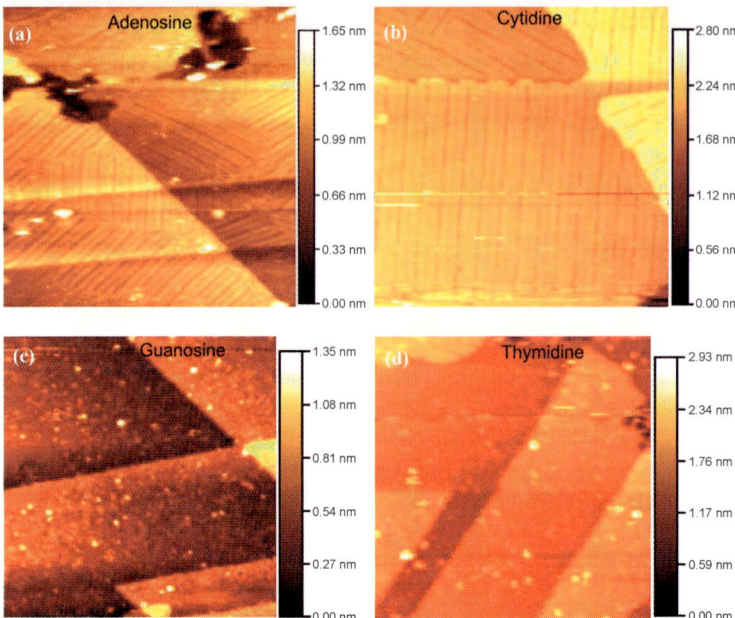

Fig. 9.6. The UHV/STM images (100 nm × 100 nm) of the four nucleosides on Au(111): **a** Adenosine and **b** Cytidine show well-ordered patterns. **c** Guanosine; spots may indicate single or very few molecules. **d** Thymidine. Preset tunnelling condition: 2.0 V and 30 pA

quenching of DNA based on the tunnelling current alone may be complicated due to the exponential dependence of the current on the tip-sample separation and the applied bias [30]. In contrast, the exponential dependence is removed by calculating the normalized conductance from the digital I–V data [30]. Thus for sequencing by STM, we highlight the significant differences between the HOMO (or LUMO) energies among the four nucleosides, which are not only accessible via the present electronic probe, but also via potential nanoscale optical probes [19]. For each nucleoside, however, we find that the HOMO energies fluctuate (Fig. 9.7c). This may be due to the change of the contact geometry of the nucleoside molecules on Au(111), which influences the charge injection barrier. The resultant averages and standard deviations of the HOMO energies are -1.39 ± 0.07 (G), -1.56 ± 0.06 (A), -1.69 ± 0.06 (C) and -1.74 ± 0.08 (T) eV (relative to the Fermi energy E_F of Au(111)), respectively. The relative order of the four averaged HOMO energies is consistent with redox-potential measurements, i.e., -7.926, -8.503, -8.968, and -9.516 eV for bases G, A, C, and T [34].

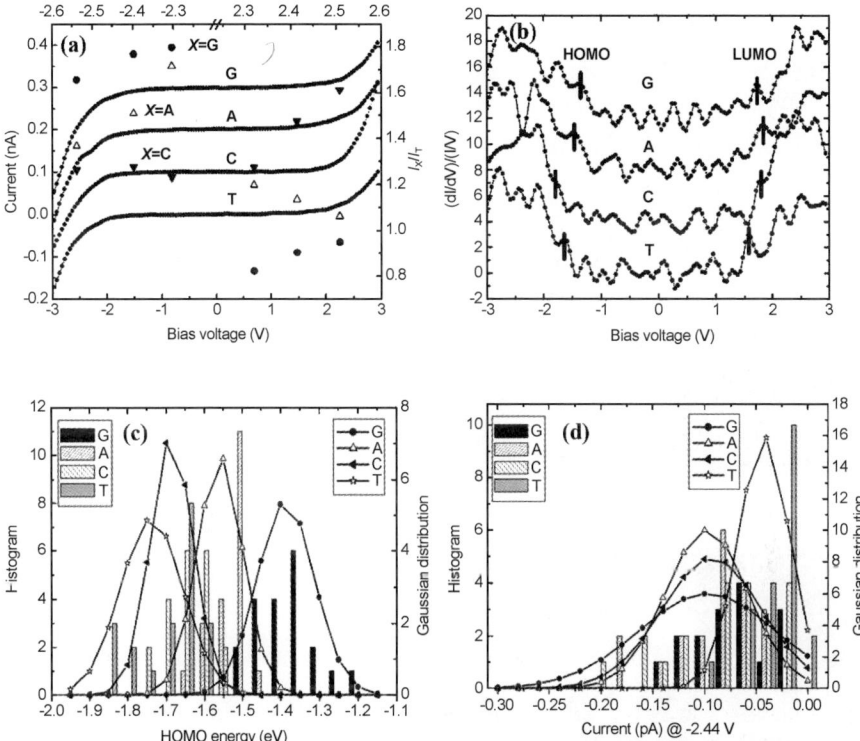

Fig. 9.7. Electronic signatures of the four nucleosides. **a** Representative I–V curves (*bottom-left axes*) acquired at 2.3 V and 30 pA, as well as corresponding current ratios of X (X = A, G or C) relative to T (*top-right axes*) at fixed biases. **b** Normalized conductances obtained from the I–V curves in panel **a**. **c** Distribution of the HOMO energies of the four nucleosides. HOMO energies are given with respect to Fermi energy E_F of Au(111). **d** Current distributions of the four nucleosides at -2.44 V. *Bars* indicate histograms of experimental data, *lines* show ideal Gaussian distributions based on the experimental averages and standard deviations

9.4 First-Principles Calculations

In order to support the experimental finding of the base-specific electronic signatures, we performed first-principles calculations of the electronic structures of single methylated bases (ribose replaced by a methyl-group, see Fig. 9.8a and b as examples) on a periodically repeated Au (111) slab.

All structures were relaxed to the nearest local minimum on the energy landscape, leading predominantly to a parallel absorption of the bases with respect to the surface. The density of states (DOS) of the methylated bases on a Au (111) slab was obtained with the SIESTA package [35, 36]. SIESTA

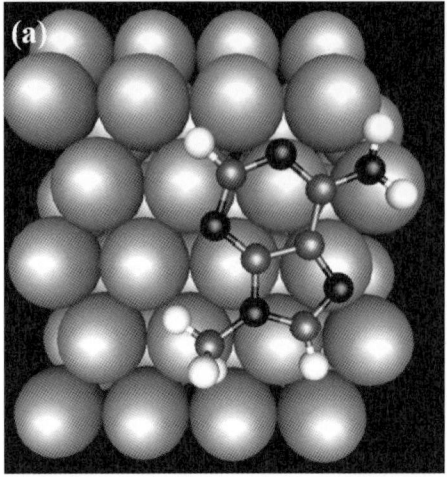

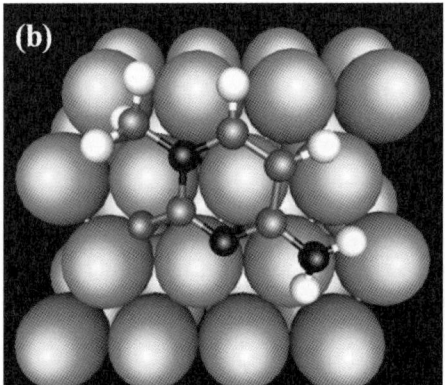

Fig. 9.8. Relaxed structure of methylated bases on a Au(111) slab from first-principles calculations. Methylated A (**a**) and C (**b**)

is based on the *ab initio* density functional theory, and uses the generalized gradient approximation, as well as pseudo potentials for the core electrons and a localized atomic orbital-like basis set for high efficiency. We used a double-zeta basis with additional polarization orbitals. The basis set of Au includes the 4f electrons explicitly, while the 5d and 6s electrons are included in the pseudo-potential. The Au surface was modelled with a periodically repeated 3 layer slab of 4 times 5 (4) Au atoms in each layer for a purine (pyrimidine) calculation using the experimental lattice constant 4.08 Å [37]. The atoms of the base and the top surface layer were relaxed until the forces on atoms were below 0.04 eV/Å. For each base, the structural relaxation was started from two slightly different initial structures concerning the lateral base position on the surface, but with the same parallel distance to the surface (2 Å). For comparison, we also conducted calculations of each base separated by 6 Å from the surface. At this separation, there is essentially no interaction between the bases and the surface. To check whether three surface layers with only the atoms of the top layer being able to move are sufficient, we used the relaxed output structure of a cytosine calculation, added a forth layer of Au on the bottom of the slab, and restarted the relaxation calculation while additionally allowing the Au atoms of the second from the top surface layer to move. The initial forces of the restarted relaxation on the atoms of the base in the presence of the extra layer of Au were below 0.1 eV/Å and hence remained small. The relaxed structure and the density of states were essentially indistinguishable from before.

Figures 9.9a–d show the density of states projected onto the contributions of the four bases. While bases absorbed at about 2.9 Å separation from the surface (upper two curves in each panel), we randomly observed both physical (solid line) and chemical absorption (dashed line) depending on small differences in the initial lateral orientation of the bases. The chemically absorbed bases always lost hydrogens of a methyl-group and bound to the surface with the unsaturated carbon. While the chemical absorption smears out the characteristic electronic features, the ground state energy of the chemically absorbed structures ($E_{0,\mathrm{Che}}$) was always higher, and hence more unfavorable, than the ground state energy of the physically absorbed structures ($E_{0,\mathrm{Che}}$), i.e. $\Delta E = E_{0,\mathrm{Che}} - E_{0,\mathrm{Phy}} > 0$. For comparison, we also calculated the density of states for the non-interacting situation where each base was separated by 6 Å, from the surface (lower curve in each panel). The calculated HOMO energies of G, A, C, and T are -4.86, -5.13, -5.21, and -5.40 eV, respectively. We also note that absorption leads to energetic downward shifts of the molecular states, and that the density of states of C and T is smeared out and shows very little features – even when physically absorbed. The similarity between C and T, as well as the order of the HOMO energies (best visible for the non-interacting bases at 6 Å separation) are in line with our experimental findings.

9.5 Sequencing

To sequence ssDNA from the electronic signatures of the bases, one needs to correctly predict the sequence of the bases A, C, G, and T from the measurements. We first use the HOMO energies of the four nucleosides alone for prediction, but additional electronic features can also be included. As described in the appendix, we apply Bayes' theorem to calculate the probability

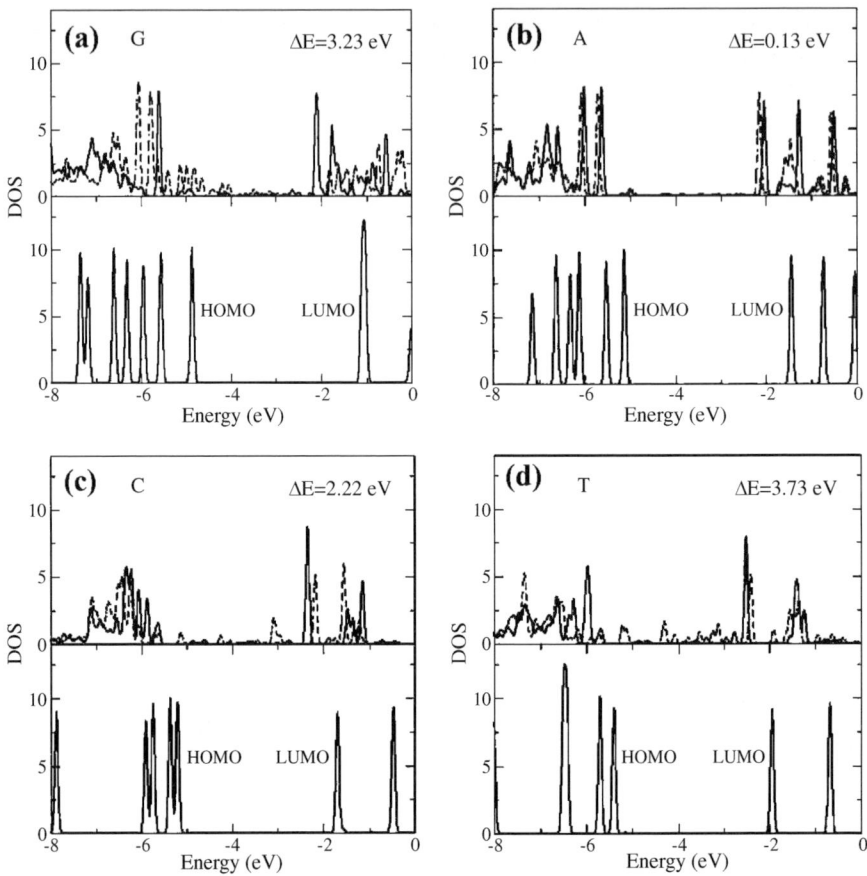

Fig. 9.9. First–principles calculations of the methylated DNA bases on Au(111) slab. The density of states (DOS) of methylated guanine (G), adenine (A), cytosine (C), and thymine (T), shown in panels **a**, **b**, **c**, and **d**, respectively, were obtained with the density functional theory code SIESTA. For clarity, only the projection onto the p-orbitals of the elements C, N and O are shown. In each panel, the two *upper curves* show a representative physically (*solid line*) and a chemically (*dashed line*) absorbed base. ΔE is the energy difference between the chemically and the physically absorbed bases. Physical absorption is always energetically favorable for the cases studied here ($\Delta E > 0$). The binding distances between the DNA bases and the top layer of the Au(111) slab range from 2.6 Å to 3.1 Å; the angle between the normal vectors of the DNA bases and Au(111) ranges from 4° and 22°, indicating that the bases are rather flat on the surface. For comparison, the *lower curve* in each panel shows the non-interacting situation, where the DNA bases have a larger ($\sim$ 6 Å) separation from the surface. From the lower curves, the HOMO energies of the G, A, C, and T are $-4.86, -5.13, -5.21$ and -5.40 eV, respectively

of predicting the correct base, given the data. For this purpose, we simplify the experimental distributions of the HOMO energies by Gaussian distributions based on the averages and the standard deviations (see Fig. 9.7c). Consequently, the expected probabilities of predicting G, C, A, and T correctly are 0.86, 0.48, 0.62, and 0.55, respectively. This is significantly larger than 0.25 when guessing, but much lower than the consensus accuracy 0.9999 of the Sanger sequencer [16].

Our method of sequencing based on the HOMO energies can be improved by including additional electronic features. We consider two such cases separately (for details see appendix). First, if the sequencing is very fast and inexpensive, sequencing both strands based on the HOMO energies leads to the expected probability 0.87 of predicting G–C and C–G basepairs correctly, and 0.82 for predicting A–T and T–A basepairs correctly. Second, currents through the bases can be used in addition to the HOMO energies when sequencing a single DNA strand. The current would help to identify T since this base or nucleoside has the smallest current of all four at negative bias. This leads to the expected probabilities 0.86, 0.59, 0.64, and 0.69 of predicting G, C, A, and T correctly, respectively (see the statistical analysis below) The combination of the two cases will lead to a further improvement. Note that we have shown electronic signatures of the isolated nucleosides, while the potential electronic sequencing of the bases in ssDNA may suffer from interference with neighboring bases. However, Zwolak et al. [13], have recently shown through simulations that the relative current through the bases in ssDNA is independent of the nearest-neighbors as long as the electrode width is of nanometer size. Hence the development of our method to identify the electronic signatures of single nucleosides should facilitate sequencing of real ssDNA.

9.6 Outlook and Perspectives

With the development of an ultrafast STM technology, we can envision a futuristic STM probe with the single-base resolution gliding along ssDNA on a solid surface with a scan speed of 15,000 nm/s [38] and reading out HOMO energies with a sequencing speed of up to estimated 40,000 bases per second. Note that the scan speed of 15,000 nm/s [38] was achieved on a flat metal and a semiconductor surface. The situation may be different when obtaining spectra of single molecules with their higher corrugation. The present STM-based method does not require imaging the internal atomic structure of the bases but requires single-base resolution, so that the STM tip can be located above each base individually. The genomic DNA (without obvious length limitations) needs only to be isolated and separated into ssDNA, not requiring complicated DNA modifications such as cloning of DNA fragments with subsequent amplification and fluorescent labelling. The question remains how to get suitable samples of ssDNA with the bases regularly aligned for accurate sequencing, since the electronic properties of DNA and the bases are sensitive to the conformational variation. Stretching of anchored ssDNA by molecular combing [39] or flow may constitute an inexpensive possibility. The current most promising idea is threading ssDNA through solid-state nanopores [40] and simultaneously detecting differences in HOMO energies and currents at a fixed bias by STM-based technologies. The nanopore method will especially support rapid sequencing, but has to cope with the additional solution and base rotation effects [22] in contrast to measuring ssDNA on a surface by UHV/STM–based technology.

9.7 Appendix: Accuracy of Sequencing

We want to know the probability of predicting the correct base b, e.g., G, given a measurement B, such as the HOMO energy of the base b. According to Bayes' theorem, this posterior probability is given by [41]

$$P(b|B) = \frac{P(B|b)P(b)}{P(B)} = \frac{P(B|b)P(b)}{\sum_b P(B|b)P(b)} , \tag{9.1}$$

where $P(b)$ is the *a priory* probability of encountering base b, and $P(B|b)$ is the probability distribution of data B for base b. In order to obtain the expected prediction $P(b|B)$, we average over the distribution of all possible measurements $P(B|b)$, i.e.,

$$\langle P(b|B)\rangle = \int P(b|\tilde{B})P(\tilde{B}|b)d\tilde{B} . \tag{9.2}$$

Assuming $P(b) = 1/4$ for all the bases and using (9.1), (9.2) simplifies to

$$\langle P(b|B)\rangle = \int \frac{P(\tilde{B}|b)d\tilde{B}}{1 + P(\tilde{B}|b)^{-1}[P(\tilde{B}|b_2) + P(\tilde{B}|b_3) + P(\tilde{B}|b_4)]} , \tag{9.3}$$

where b_2, b_3, and b_4 are the other three possible bases whose measurement may also lead to value B. Using idealized Gaussian distributions of the HOMO energies from Fig. 9.7c, the expected prediction of the four bases is $P(G) = 0.86$, $P(C) = 0.48$, $P(A) = 0.62$, and $P(T) = 0.55$ based on the numerical integration of (9.3) with a cutoff plus/minus five standard deviations of the distribution $P(B|b)$.

The prediction can be improved by including additional data. If the data C, e.g., current, is available (in addition to B), then $P(b|B)$ can be replaced by $P(b|B, C)$ leading to an extra dimension and further separation of the base–specific electronic features. Assuming the independence of data B and C, we obtain the double integral

$$\langle P(b|B, C)\rangle = \tag{9.4}$$

$$\int \frac{P(\tilde{B}|b)P(\tilde{C}|b)d\tilde{B}d\tilde{C}}{1 + P(\tilde{B}|b)^{-1}P(\tilde{C}|b)^{-1}[P(\tilde{B}|b_2)P(\tilde{C}|b_2) + ... + P(\tilde{B}|b_4)P(\tilde{C}|b_4)]} .$$

Using idealized Gaussian distributions of the currents from Fig. 9.7d for data C, the expected prediction of the four bases is $P(G) = 0.86$, $P(C) = 0.59$, $P(A) = 0.64$, and $P(T) = 0.69$.

Further improvement can be achieved, if both strands are sequenced. Considering only HOMO energies for simplicity (without the current data), the probability of correctly predicting the basepair is

$$P(b_{11}b_{21}|B_1B_2) = \frac{P(B_1|b_{11})P(B_2|b_{21})}{P(B_1|b_{11})P(B_2|b_{21}) + ... + P(B_1|b_{14})P(B_2|b_{24})} , \quad (9.5)$$

where b_{ij} stands for one of the four bases j on strand i. The denominator describes that the measurement of the four possible basepairs (two basepairs with two orientations, e.g., G–C or C–G) can principally lead to the data tuple B_1B_2. In deriving (9.5), we assumed that the probability of encountering any of the four basepairs is $P(b_{1j}b_{2j}) = 1/4$. The expected prediction is obtained from the double integral

$$\langle P(b_{11}b_{21}|B_1B_2) \rangle = \quad (9.6)$$

$$\int \frac{P(\tilde{B}_1|b_{11})P(\tilde{B}_2|b_{21})d\tilde{B}_1 d\tilde{B}_2}{1 + P(\tilde{B}_1|b_{11})^{-1}P(\tilde{B}_2|b_{21})^{-1}[P(\tilde{B}_1|b_{12})P(\tilde{B}_2|b_{22}) + \cdots + P(\tilde{B}_1|b_{14})P(\tilde{B}_2|b_{24})]} .$$

As a consequence, the expected predictions for the basepairs are $P(G-C) = P(C-G) = 0.87$ and $P(A-T) = P(T-A) = 0.82$. All the integrals are evaluated with the adaptive Simpson quadrature method.

Acknowledgement. This work was partially supported by the IT program of the Ministry of Education, Culture, Sports, Science and Technology of the Japanese Government.

References

1. R.G. Endres, D.L. Cox and R.R.P. Singh, Rev. Mod. Phys. **76**, 195 (2004).
2. M. Taniguchi and T. Kawai, Physica E **33**, 1 (2006).
3. M. di Ventra and M. Zwolak, in *Encyclopedia of Nanoscience and Nanotaechnology*, edited by H.Singh–Nalwa (American Scientific Publishers, 2004).
4. E. Maciá and S. Roche, Nanotechnology, **17**, 3002 (2006).
5. B. Xu, P.M. Zhang, X.L. Li and N.J. Tao, Nano Lett. **4**, 1105 (2004).
6. M.S. Xu, S. Tsukamoto, S. Ishida, M. Kitamura, Y. Arakawa, R.G. Endres and M. Shimoda, Appl. Phys. Lett. **87**, 083902 (2005).
7. H. Cohen, C. Nogues, R. Naaman and D. Porath, Proc. Natl. Acad. Sci. USA **102**, 11589 (2005).
8. S.M. Lindsay, Y. Li, J. Pan, T. Thundat, L.A. Nagahara, P. Oden, J.A. Derose, U. Knipping and J.W. White, J. Vac. Sci. & Technol. B **9**, 1096 (1991).
9. H. van Zalinge, D.J. Schiffrin, A.D. Bates, W. Haiss, J. Ulstrup and R.J. Nicho, Chem. Phys. Chem **7**, 94 (2006).
10. M.S. Xu, R.G. Endres, S. Tsukamoto, M. Kitamura, S. Ishida and Y. Arakawa, Small **1** 1168 (2005).

11. H. Tanaka and T. Kawai, Surf. Sci. **539**, L531 (2003).
12. T. Ohshiro and Y. Umezawa, Proc. Natl. Acad. Sci. USA **103**, 10 (2006).
13. Q. Chen, D.J. Frankel and N.V. Richardson, Langmuir **18**, 3219 (2002).
14. A. Marziali and M. Akeson, Annu. Rev. Biomed. Eng. **3**, 195 (2001).
15. C. Stroh, H. Wang, R. Bash, B. Ashcroft, J. Nelson, H. Gruber, D. Lohr, S.M. Lindsay and P. Hinterdorfer, Proc. Natl. Acad. Sci. USA **101**, 12503 (2004).
16. M. Margulies *et al.*, Nature **437**, 376 (2005).
17. R.G. Blazej, P. Kumaresan and R.A. Mathies, Proc. Natl. Acad. Sci. USA **103**, 7240 (2006).
18. M. Zwolak and M. Di Ventra, Nano Lett. **5**, 421 (2005).
19. A. Hübsch, R.G. Endres, D.L. Cox and R.R.P. Singh, Phys. Rev. Lett. **94**, 178102 20 (2005).
20. J. Shendure, G.J. Porreca, N.B. Reppas, X. Lin, J.P. McCutcheon, A.M. Rosenbaum, M.D. Wang, K. Zhang, R.D. Mitra and G.M. Church, Science **309**, 1728 (2005).
21. Y.H. Rogers and J.C. Venter, Nature **437**, 326 (2005).
22. J. Lagerqvist, M. Zwolak and M. Di Ventra, Nano Lett. **6**, 799 (2006).
23. D.J. Driscoll, M.G. Youngquist and J.D. Baldeschwieler, Nature **346**, 294 (1990).
24. N.J. Tao, J.A. Derose and S.M. Lindsay, J. Phys. Chem. **97**, 910 (1993).
25. M.S. Xu, R.G. Endres and Y. Arakawa, Appl. Phys. Lett. (submitted).
26. D.Y. Petrovykn, H. Kimura–Suda, L.J. Whiteman and M.J. Tarlov, J. Am. Chem. Soc. **125**, 5219 (2003).
27. A.B. Steel, R.L. Levicky, T.M. Herne and M.J. Tarlov, Biophys. J. **79**, 975 (2002).
28. E. Shapir, H. Cohen, N. Borovok, A.B. Kotlyar and D. Porath, J. Phys. Chem. **110**, 4430 (2006).
29. M. Prietsch, A. Samsavar and R. Ludeke, Phys. Rev. B **43**, 11850 (1991).
30. *Scanning Tunneling Microscopy and Spectroscopy*, edited by R. Wiesendanger (Cambridge University Press, Cambridge, 1994).
31. M. Iijima, T. Kato, S. Nakanishi, H. Watanabe, K. Kimura, K. Suzuki and Y. Maruyama, Chem. Lett. **34**, 1084 (2005).
32. M. Edelwirth, J. Freund, S.J. Sowerby and W.M. Heckl, Surface Science **417**, S.201 (1998).
33. M. Furukawa, H. Tanaka and T. Kawai, Chem. Phys. **115**, 3419 (2001).
34. C.A.M. Seidel, A. Schule and M.H.M. Sauer, J. Phys. Chem. **100**, 5541 (1996).
35. P.J. de Pablo, F. Moreno-Herrero, J. Colchero, J. Gómez Herrero, P. Herrero, A.M. Bar, P. Ordejón, J.M. Soler and E. Artacho, Phys. Rev. Lett. **85**, 4992 (2002).
36. P. Ordejón, E. Artacho and J.M. Soler, Phys. Rev. B. **53**, 10441 (1996).
37. N.W. Ashcroft and N.D. Mermin, *Solid State Physics*, (Harcourt Brace College Publishers, New York, 1976).
38. R. Curtis, T. Mitsui and E. Ganz, Rev. Sci. Instrum. **68**, 2790 (1997).
39. D. Bensimon, A.J. Simon, V. Croquette and A. Bensimon, Phys. Rev. Lett. **74**, 4754 (1995).
40. J. Li, M. Gershow, D. Stein, E. Brandin and J. Golovchenko, Nature Materials **2**, 611 (2003).
41. *Biological Sequence Analysis*, edited by R. Durbin, S. Eddy, A. Krogh and G. Mitchison (Cambridge Unversity Press, Cambridge 1998).

10 DNA Photonics –
Probing Light-Induced Dynamics in DNA on the Femtosecond Timescale

Qiang Wang and Torsten Fiebig

Eugene F. Merkert Chemistry Center, Boston College, Chestnut Hill, MA 02467, USA
Fiebig@bc.edu

10.1 Introduction

The electronic properties of DNA have continued to capture the interest of many research groups for the past four decades. In 1962, Eley and Spivey first suggested that the DNA π-stack structure might provide an efficient pathway for electron transport [1]. Thirty years later, the pioneering work by Jacqueline Barton on photoinduced electron transfer in DNA [2] lead to the foundation of a new field surrounding DNA charge transfer chemistry [3, 4]. After a decade of intense research the initial controversy about potential "wire-like behavior" has faded and a broad consensus about the conduction properties seems to emerge: DNA is not a molecular wire, but it still conducts charges better than proteins. Femtosecond time-resolved studies of DNA charge migration in the late 1990s revealed conformational gating, i.e. DNA base pairs must undergo structural rearrangements in order to facilitate efficient charge migration [5], particularly over longer distances. Hence, the internally flexible helical base stack cannot be considered as a static charge donor-bridge-acceptor system [6]. Structural dynamics, particularly on the femtosecond to nanosecond time scale, are critical and their characterization becomes an indispensable component of present and future charge transfer experiments.

While the initial controversies regarding the long-range conductivity properties and wire-type behavior of DNA have been settled, a new field, **DNA photonics**, has emerged around the photophysics of nucleic acids [7]. The contributions that can be expected from future studies in DNA photonics will likely answer the question whether – and to what extent – DNA can be used as a functional building block in molecular nanoscale devices [8]. They will also be focused on the complex interactions between structural and electronic properties of DNA which are profound for applications in nanobiotechnology [9, 10]. Thus, DNA photonics is not solely focused on charge transfer phenomena but includes all possible photophysical processes and their potentially complex interplays. For example, DNA ultrafast electronic energy dissipation, delocalization after initial electronic excitation, excimer and exciton formations are also topics for DNA photonics studies.

10.1.1 Electron Transfer versus Hole Transfer

Although the distinction between electron transfer (ET) and hole transfer (HT) appears semantic in nature, it is an important one to make. The difference becomes clear when both processes are viewed in the orbital picture (see Fig. 10.1). In ET processes, the electron donor is excited and an electron is transferred from the LUMO (lowest unoccupied molecular orbital) of the donor to the LUMO of the acceptor. The corresponding one-electron coupling matrix element V involves both LUMO wave functions [11]. In contrast, the hole transfer occurs between the HOMO (highest occupied molecular orbital) of the donor and the HOMO of the acceptor, after the excitation of the acceptor has created an electron hole in the HOMO level. Hence, the resulting one-electron coupling matrix element involves both HOMO wave functions.

Both the energetical and the topological structure of the "HOMO-band" and the "LUMO-band" differ very much in DNA. Until recently, most studies on DNA charge transfer were targeting (oxidative) hole transfer reactions. In the early 2000s, DNA assemblies were designed specifically to study excess electrons in DNA [4]. Based on the aforementioned orbital argument, one would expect very different electronic coupling pathways for electron migrations involving the LUMO band. Although differences between oxidative HT and reductive ET were found, there are many similarities as well. In Sect. 10.4, we present a comparative study of HT and ET, where many similarities between these two types of charge transfer are demonstrated.

In this chapter, we review some of the real-time dynamical studies that were carried out in our group over the past two years. The work was aimed at probing the electronic properties of DNA on the ultrafast time scale. Some

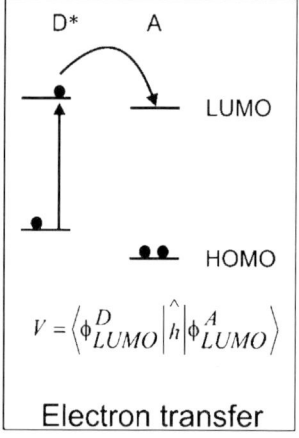
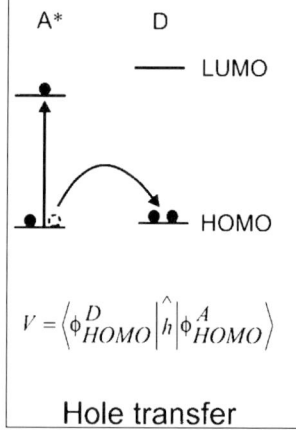

Fig. 10.1. Orbital schemes for photoinduced electron transfer (*left*) and hole transfer (*right*). V is the electronic coupling matrix element within the one-electron approximation [11]

of the presented projects were carried out in collaboration with the groups of Frederick Lewis (Northwestern University, Evanston IL) and Hans-Achim Wagenknecht (University of Regensburg, Germany). First, we discuss our experimental methodology, femtosecond broadband pump-probe spectroscopy. In the subsequent sections, we present results obtained on various natural and chemically-modified DNA assemblies that enabled us to gain new mechanistic insight into DNA photonics.

10.2 Femtosecond Broadband Pump-Probe Spectroscopy

Real-time information about the dynamics of photoinduced charge transfer processes can be obtained from optical pump-probe experiments using ultra-short laser pulses. Figure 10.2 shows a schematic of the femtosecond broadband pump-probe setup that has been described in detail elsewhere [12]. The changes in optical density were probed by a femtosecond white-light continuum (WLC) generated by tight focusing of a small fraction of the output of a commercial Ti:Sa based pump laser (CPA-2010, Clark-MXR) into a 3 mm thick CaF_2 plate.

The obtained WLC provides a usable probe source that covers the UV-VIS-NIR spectral range (from 250 to 1160 nm). The WLC is split into two beams (probe and reference) and focused into the sample using reflective optics. After passing through the sample, both probe and reference are spectrally dispersed and simultaneously detected on a CCD sensor. The change in optical density is calculated using the standard formalism for pump-probe spectroscopy [12]. The pump pulse is typically-generated by frequency doubling (optional) of the compressed output of a home-built non-collinear optical parametric amplifier (NOPA) system. To compensate for group velocity dispersion in the UV-pulse, we used an additional prism compressor. Independent measurements of the chirp of the WLC were carried out to correct the pump-probe spectra for time-zero differences. The overall time resolution has been estimated to be $100 - 120$ fs and a spectral resolution of $7 - 10$ nm is typically obtained. Measurements are typically performed with magic angle geometry ($54.7°$) for the polarization of pump and probe pulses to avoid contributions from orientation relaxation.

Compared with the conventional (two-color) pump-probe technique, broadband pump-probe spectroscopy can (in principle) capture and resolve reactant, intermediate and product states simultaneously. By measuring the pump-probe spectra as a function of time, one does not only obtain "kinetic traces at multiple wavelengths" but the complex spectral evolution which includes detailed information about spectral shifts, lineshapes and linewidths [12]. There are up to three contributions to the pump-probe spectra. Depending on the spectral range of interest one can observe (a) induced excited state absorption (ESA), (b) stimulated emission from excited states,

of several 100 ps (Fig. 10.15c). The observed spectral dynamics can be illustrated in a simple energy level diagram shown in Fig. 10.16. Depending upon the local environment of each adenine, one may simplistically distinguish two categories of chromophores:

Well stacked base domains: UV absorption leads to delocalized exciton states where the electronic wave functions are spread over more than a single base. Following optical excitation, the initially excited states undergo electronic relaxation within the exciton state manifold, characterized by a pronounced spectral blue shift (from 380 to 330 nm, see Fig. 10.15b). Entirely consistent with this internal conversion process is the observed change (i.e. decrease) in oscillator strength. The ultrafast electronic relaxation (sub-100 fs, in agreement with [46]) is followed by a slower nuclear relaxation process (several picoseconds) which may involve modes from the base stack and from the surrounding medium and which manifests itself in minor changes of the 330 nm band shape.

A comparison of the short-wavelength ESA bands (< 380 nm) in AMP and in $(dA)_n$ (see Fig. 10.15a,b) suggests that the underlying electronic transition is localized and thus insensitive to electronic delocalization. The broad ESA band > 400 nm results from interchromophoric interactions and reflects

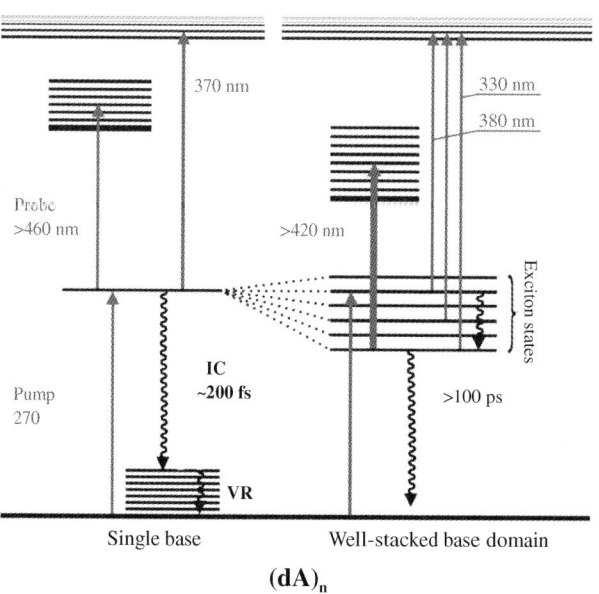

Fig. 10.16. Energy-level diagram for the energy dissipation pathways in homoadenine sequences $(dA)_n$. In addition to poorly stacked single bases that show very similar behavior as AMP, there are well stacked base domains that can be excited cooperatively by UV light, forming delocalized exciton states. Energy dissipation in these delocalized states involves internal conversion (IC) and can be traced spectroscopically (Copyright 2007, National Academy of Sciences, USA)

electronic delocalization in the system (see below). The electronically relaxed exciton state has a lifetime of several 100 ps, in accordance with results from previously reported transient absorption measurements [36, 47].

Poorly stacked single bases: Parallel to the optical excitation of well stacked base domains, monomer like excitations of single adenine bases occur at locations in the sequence where local static and dynamic disorder disrupts the electronic coupling between neighboring bases. These localized adenine states resemble similarity with AMP which undergoes sub-200 fs internal conversion (IC) to its vibrationally hot ground state. The typical spectral fingerprint of hot ground states in all $(dA)_n$ systems (as seen in AMP, Fig. 10.15a) is superimposed to the $330-380$ nm ESA bands.

10.5.2 Exciton Delocalization

In conclusion, there are two types of excited state processes that occur simultaneously in $(dA)_n$ oligonucleotides: (i) A sub-200 fs internal conversion from the initially excited state forming the hot ground state (similar as in AMP) and, (ii) electronic relaxation (IC) of delocalized exciton states. Figure 10.15d displays the ESA spectra of $(dA)_2$, $(dA)_6$, and $(dA)_{18}$ 5 ps after excitation. Clearly, all spectra contain the same 330 nm ESA band. However, the long-wavelength part (above 400 nm) of the spectrum depends on the length of the base stack n. This critical observation indicates that the electronic composition of the excited base states varies with the number of individual bases in the stack, and is a manifestation for exciton delocalization. The intensity ratio ρ_{435} – between the long-wavelength part of the ESA spectrum, dominated by interchromophoric transitions centered around 435 nm, and the local transition at 330 nm – is proportional to the oscillator strength for exciton absorption. Thus measuring ρ_{435} as a function of n will provide information about the spatial extent of the exciton. For instance, if the exciton were only spread over two adjacent bases, ρ_{435} would not change with increasing length of the stack because the effective number of stacked adenine dimers is approximately constant in all samples. On the other hand, ρ_{435} would continuously increase with increasing n if the exciton were completely delocalized over all bases in the stack. Figure 10.17a shows ρ_{435} after 3 ps for all $(dA)_n$ systems and for three A·T DNA duplexes.

10.5.3 Exciton Dynamics

There are two characteristics about the exciton absorption intensity ρ_{435}:

– ρ_{435} increases monotonically with increasing length of the base stack
– ρ_{435} is not constant over the excited state lifetime of the bases but decays with a characteristic time constant of 8 to 10 ps in all DNA systems studied, except for $(dAT)_9 \cdot (dAT)_9$ oligonucleotides (Fig. 10.17b).

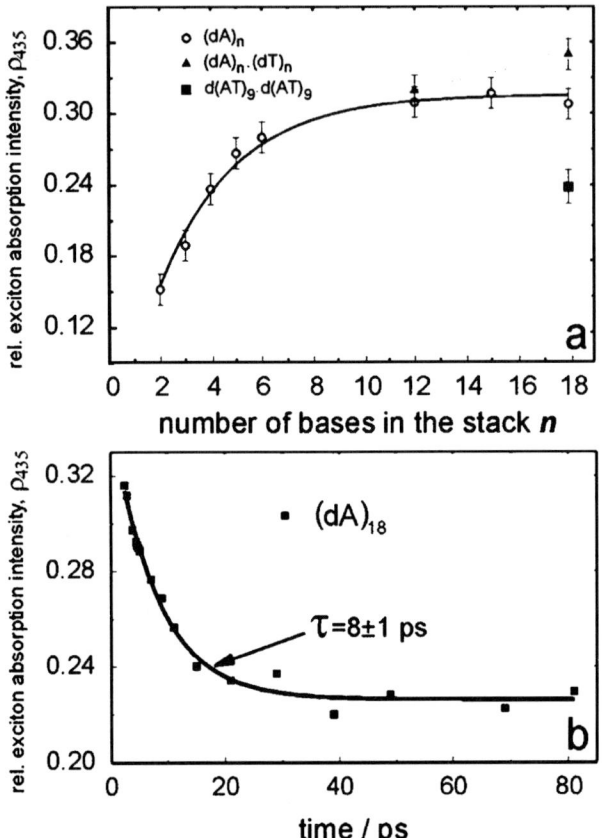

Fig. 10.17. a: Spectral intensity ratio between the exciton absorption (in a 30-nm interval, centered at 435 nm) and the monomer absorption at 330 nm (ρ_{435}), 3 ps after excitation, as a function of the stack length n. For the $(dA)_n$ series, a single exponential fit (*solid curve*) was used to extract the "1/e delocalization length" d. The ρ_{435} values for the AT duplexes are connected with a *dotted line*. **b:** Time dependence of ρ_{435} for $(dA)_{18}$. The decay of ρ_{435} is characterized by time constants of $8-10$ ps in all DNA systems studied, except for $d(AT)_9 \cdot d(AT)_9$ where no decay of ρ_{435} was observed. (Modified with permission from [38], Copyright 2007, National Academy of Sciences, USA)

Structural conversions in nucleic acids such as local transitions from A to B-DNA are known to occur on time scales several orders of magnitude slower than the observed 10 ps [48]. Therefore, one must conclude that changes in the relative spectral intensities during the excited state lifetime of the bases are caused by small amplitude fluctuations in the base stack and/or the surrounding environment (metal ions, solvent) [49].

Figure 10.17a shows larger changes of $\rho_{435}(n)$ at short stack lengths and a more gradual evolution for large n. The obtained base stack dependence of ρ_{435} reflects an exponential decay of the exciton delocalization length in DNA and reveals a "$1/e$ delocalization length" d for $(dA)_n$ of 3.3 ± 0.5 base pairs. Since shorter duplexes are not thermodynamically stable at room temperature, the delocalization length in A·T duplexes can only be estimated by comparing $(dA)_{12} \cdot (dT)_{12}$ and $(dA)_{18} \cdot (dT)_{18}$. Although ρ_{435} has not reached saturation at $n = 12$ and continues to increase towards $n = 18$, the relative change is small compared to the increase of ρ_{435} at shorter stack lengths (i.e. between $(dA)_2$ and $(dA)_5$). Given the structural similarities with respect to stacking in single-stranded homo adenine sequences and A·T duplexes, these results suggest a similar length distribution of the delocalized domains in both types on nucleic acids, albeit A·T duplexes appear to have a slightly larger fraction of more extended ($d > 4$) delocalized domains. The overall larger values of ρ_{435} reflect stronger exciton absorption and thus larger electronic coupling due to shorter base-base distances and/or a more rigid stack structure in duplexes as opposed to single-stranded sequences.

It is interesting to note that the alternating sequence $d(AT)_9 \cdot d(AT)_9$ reveals an intermediate ρ_{435} value that corresponds to the degree of delocalization found in $(dA)_4$. In contrast to all the other nucleic acid systems with vertically stacked adenines (in the same strand), ρ_{435} in $d(AT)_9 \cdot d(AT)_9$ does not change with time. This observation may simply reflect the different electronic character of intrastrand excited A-T-A ... complexes, compared to A-A-A ... domains. For instance, one would expect the former complexes to have substantial charge transfer contributions which would not participate in the exciton state wavefunctions formed in homadenines. Alternatively, the observation could indicate that interstrand and intrastrand electronic coupling pathways are affected inherently differently by ultrafast structural dynamics.

Guided by the temporal evolution of the ESA spectra shown in Fig. 10.15, the following conclusions emerge:

- The electronic coupling between stacked bases leads to the formation of delocalized exciton states upon UV absorption. This possibility has been widely discussed by theoreticians in the past. The experimental data presented here, directly support this exciton model.
- In single-stranded homoadenine sequences, the typical "$1/e$ delocalization length" is 3–4 bases. However, given the conformational inhomogeneiety of these flexible biopolymers, more extended delocalization is likely to be present in some molecules. Ensembles of A·T duplexes have a larger fraction of more extended delocalized domains. This is evidenced by the significant increase in ρ_{435} from $(dA)_{12} \cdot (dT)_{12}$ to $(dA)_{18} \cdot (dT)_{18}$.
- The electronic exciton structure is dynamic, i.e. the delocalization length changes during the lifetime of the excited base stack.
- A substantial fraction of excited DNA molecules undergo ultrafast internal conversion to the hot ground state, similar as in single bases [36, 47].

It is reasonable to assume that the optical excitation in this molecular subensemble remains localized due to static and dynamic disorder in the stack.

The last point provides an answer to the introductory question about the intrinsic DNA protection mechanism. Random DNA sequences (containing both A·T and G·C base pairs) are even more likely to yield localized "monomer type" electronic states. The effective competition of the monomer type photophysical pathway with the exciton formation is critical from an evolutionary view point because excess energy is funnelled to the ground state in times shorter than needed to make and break chemical bonds. The amount of disorder in the genome will define what fraction of stacked bases can avoid irreversible photo damage by eliminating electronic excess energy in the same fashion as single DNA bases.

10.6 Competition Between Energy Delocalization and Charge Transfer

The study of natural DNA base sequences presented in the previous section revealed important details about energy delocalization, relaxation and dissipation in sequences containing identical bases or base pairs. However, natural bases are not very suitable for studying charge transfer interactions between non-identical bases. Small extinction coefficients for excited state absorption of neutral bases and their radical ions exacerbate spectroscopic studies of natural DNA base stacks. An experimental alternative is the selective replacement of adenine by its fluorescent analog 2-aminopurine (Ap, see Fig. 10.18). Because of its favorable spectroscopic properties, its ability to form hydrogen bonds with thymine, and the fact that Ap introduces only minor structural distortions, it has been widely used as a fluorescent base analog and site specific probe in DNA assemblies [50–52]. In addition, Ap has also been employed as excited state charge donor for all four natural bases [53,54]. Due to its complex redox activity in DNA, multiple pathways for fluorescence quenching can lead to highly non-exponential decays of the Ap fluorescence.

Recently, we presented our first time-resolved study on Ap dimers in the trinucleotide 5'-d(**ApApC**)-3' (**ApApC**) [55]. For comparison, we also studied 5'-d(**ApCC**)-3' (**ApCC**) which contains only a single Ap moiety. After photoexcitation of **ApApC**, Ap* can either donate an electron to cytosine, or form an excimer with the adjacent Ap (i.e. delocalization of excess electronic energy) [56]. The dynamics of the ET process has been studied previously using conventional two-color transient absorption spectroscopy [53,54]. Here, the aim was to design a model DNA conjugate where energy delocalization (between two identical chromophores) can compete with interbase electron transfer.

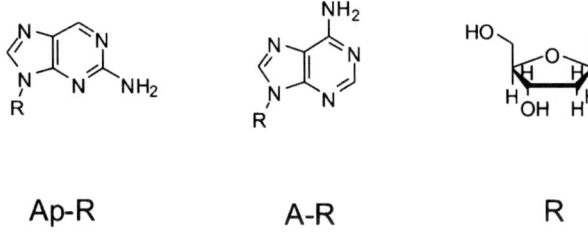

Ap-R   A-R   R

Fig. 10.18. Ribose-substituted 2-aminopurine (Ap-R), ribose-substituted adenine (A-R) and ribose(R)

Figure 10.19 indicates the formation of excimers in **ApApC**. The temporal evolution of the pump-probe spectra of **ApApC** differs significantly from that of the Ap monomer in **ApCC**, particularly below 450 nm. The spectral changes are due to energetic shifts of selected vibronic transitions in the excimer state and have previously been reported for excimer formation in pyrene [57]. The main difference between the two trinucleotides is the overall excited state lifetime: After ~ 2 ns, most **ApApC** molecules have returned to the electronic ground state, whereas the excited state lifetime of **ApCC** is significantly longer. Global fitting analysis reveals that the **ApApC** dynamics are characterized by a fast decay time component of $\sim 4 - 5$ ps which is assigned to excimer formation $(ApAp)^*C$. The excimer state has a ~ 70 ps lifetime, which implies strong vibrational coupling between the excimer and the electronic ground state that lead to a fast internal conversion process. Concomitant competitive electron transfer processes caused by the "ApC" moiety or the asymmetric Ap sites in **ApApC** are possible, however, they should occur with small magnitudes, as evidenced by a residual third lifetime component of 0.9 ns [55]. By contrast, electron transfer from Ap* to the neighboring C is the primary process in ApCC, in accordance with a driving force of approximately -0.45 eV from previous results [54]. The charge separates with a time constant of 90 ps and recombines on a ~ 1.5 ns time scale.

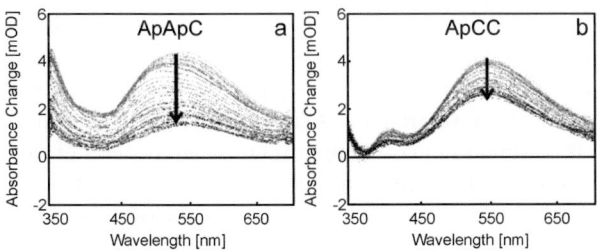

Fig. 10.19. Pump-probe spectra of ApApC and ApCC in the time range of 5 ps to 150 ps after excitation at 316 nm. The *arrows* show the temporal evolution of the spectra from early time to late time

A number of conclusions emerge from this study. First, the time scale for excimer formation in the trinucleotide ($\sim 4\,\text{ps}$) is very fast which indicates that only minor structural rearrangements accompany this process. Second, energy delocalization (i.e. excimer formation) is an order of magnitude faster than exergonic electron transfer ($\Delta G_{\text{ET}} = -0.45\,\text{eV}$) [54]. It is reasonable to assume that the relative yields of each of these processes are dictated by structural parameters, i.e. conformational gating (see Sect. 10.4), although it is not clear whether these findings are relevant for natural DNA, a direct competition between electronic delocalization and electron transfer is expected to occur in random DNA sequences.

10.7 New Experimental Methodology and DNA Photonics Applications

10.7.1 Femtosecond Circular Dichroism (CD) Spectroscopy

Although numerous measurements have been performed on ultrafast photoinduced phenomena in DNA during the past years, the employed experimental methods have been limited to a few variations of pump-probe and fluorescence techniques. As pointed out in the chapter, structural dynamics are strongly linked to electronic properties in DNA. Therefore, one needs to develop new instrumental tools that are sensitive to structural parameters on the ultrafast time scale.

A new approach in this direction is femtosecond time-resolved circular dichroism (CD) spectroscopy. CD spectroscopy has been widely applied as a tool in material sciences, chemistry, biology and physics [59]. Chiral molecules exhibit circular dichroism, i.e. a difference in extinction for left- and right-circularly polarized light $\Delta\varepsilon = \varepsilon_{\text{L}} - \varepsilon_{\text{R}}$. While chirality is necessary for observing a CD spectrum, the molecules (i.e. chromophores) do not have to be chiral to show a CD signal if they are spatially oriented to form a chiral array as typically found in biopolymers like folded proteins or nucleic acids (exciton coupled CD spectroscopy, EC-CD). Furthermore, the EC-CD signal will depend strongly on the relative orientation of the chromophores. If the interaction between the chromophores is time-dependent – due to structural dynamics in the system – the CD spectrum will also be time-dependent. Femtosecond time-resolved CD measurements will be particularly sensitive to deviations from the helical stacking arrangement and hence reveal detailed conformational fluctuations and base-pair dynamics in DNA.

10.7.2 DNA Arrays for Genetic Variation Diagnosis

Future high-impact applications of DNA photonics are expected in the area of nanobiotechnology [9, 10, 60–62]. Currently, considerable efforts are being spent on the development of reliable and sensitive high-throughput DNA assays for the detection of genetic variations, mainly single nucleotide poly-morphisms (SNPs). SNPs are single base-pair variations that are the most common of approximately three million genetic differences in individual human beings. Fluorescent DNA assays have been shown effective to detect genetic variations, but in case of SNPs, the thermodynamic differences between the fully matched and the mishybridized duplexes are often too small to be detectable by a fluorescence difference.

A new research direction in DNA biophotonics is concentrated on π-arrays and clusters of organic chromophores (multichromophores) which possess optical properties that often differ significantly from those of the corresponding monomers. Multichromophores find potential applicability in molecular devices and biosensors based on multistep electronic exciton migration [63]. DNA has been applied as a template for the helical assembly of non-covalently-bound chromophores, e.g. cyanine dyes, in the minor groove [64]. Preliminary results have shown that it is possible to apply the self-assembled and regular structure of duplex DNA as a supramolecular scaffold to allow for more than two organic chromophores to interact with each other. The most remarkable example is a combinatorial approach by the group of Eric Kool that yielded astonishing new fluorescence properties when four different chromophores have been used as C-nucleosidic DNA base substitutes in a completely artificial DNA base stack [65–68]. Alternately, organic chromophores can be attached to the ribofuranosides in RNA. For example, a helical pyrene array along the outside of duplex RNA exhibits a significant pyrene excimer fluorescence enhancement [69]. Recently, Wagenknecht et al. showed that five adjacent pyrene modified uridines form an ordered, helical π-stack that exhibits a significant fluorescence enhancement as a result of the strong electronic interaction between the chromophores [6]. Most importantly, the observed strong emission requires the complete base sequence complementarity with the unmodified counterstrand.

10.8 Final Remarks

Over the past decade many conflicting reports on the electronic conduction properties of DNA appeared in the literature. Many of these conflicts have arisen because inherently different molecular systems (chromophores, sequences, surrounding media, temperature, etc.) were unjustifiably compared with one another. The few examples presented here reflect the complexity of electronic and structural interactions that dictate electronic transfer processes in DNA. Unravelling the details of these interactions will undoubtedly result in a better understanding of **DNA photonics**. Future studies using new spectroscopic techniques that are sensitive to both the structural and the dynamical evolution of complex molecules will strongly assist these efforts. The first generation of experiments on photoinduced electron transfer (ET) in DNA has spawned a basic mechanistic picture from which simple kinetic models were derived. In these models, ET through the base stack has been reduced to a static donor-bridge-acceptor problem. Recent experimental and theoretical results have demonstrated that structural dynamics are critical for a comprehensive mechanistic understanding of the ET process. Finally, we emphasized the importance of the initially prepared electronic state. Ultrafast electronic energy delocalization, dissipation and migration must be included into the theoretical description of light induced dynamics in DNA.

Acknowledgement. This work was generously supported by the NSF (CHE 0645565 and CHE 0521503) and by the Volkswagen foundation (Germany).

References

1. D.D. Eley and D.I. Spivey, Trans. Faraday Soc. **58**, 411 (1962).
2. C.J. Murphy, M.R. Arkin, Y. Jenkins, N.D. Ghatlia, S.H. Bossmann, N.J. Turro and J.K. Barton, Science **262**, 1025 (1993).
3. *Long-Range Charge Transfer in DNA, Vol. I & II*, edited by G.B. Schuster (Springer, 2004).
4. *Charge Transfer in DNA*, edited by H.A. Wagenknecht (Wiley-VCH, Weinheim, 2005).
5. C.Z. Wan, T. Fiebig, S.O. Kelley, C.R. Treadway, J.K. Barton and A.H. Zewail, Proc. Natl. Acad. Sci. U.S.A. **96**, 6014 (1999).
6. P. Kaden, E. Mayer-Enthart, A. Trifonov, T. Fiebig and H.-A. Wagenknecht, Angew. Chem. Int. Ed. Engl. **44**, 1637 (2005).
7. F.D. Lewis, Photochem. Photobiol. **81**, 65 (2005).
8. D. Porath, G. Cuniberti and R.D. Felice, Top. Curr. Chem. **237**, 183 (2004).
9. F.J.M. Hoeben, P. Jonkheijm, E.W. Meijer and A.P.H.J. Schenning, Chem. Rev. **105**, 1491 (2005).
10. A. Ajayaghosh, S.J. George and A.P.H.J. Schenning, Top. Curr. Chem. **258**, 83 (2005).
11. M.D. Newton, Chem. Rev. **91**, 767 (1991).
12. M. Raytchev, E. Pandurski, I. Buchvarov, C. Modrakowski and T. Fiebig, J. Phys. Chem. A **107**, 4592 (2003).
13. R.R. Alfano and S.L. Shapiro, Phys. Rev. Lett. **24**, 592 (1970).
14. *The Supercontinuum Laser Source*, edited by R.R. Alfano (Springer, New York, 1989).
15. A. Brodeur and S.L. Chin, Phys. Rev. Lett. **80**, 4406 (1998).
16. A.L. Gaeta, Phys. Rev. Lett. **84**, 3582 (2000).
17. H. Ward and L. Berge, Phys. Rev. Lett. **90**, art. no. 053901 (2003).
18. I. Buchvarov, A. Trifonov and T. Fiebig, Opt. Lett., submitted (2007).
19. J. Jortner, M. Bixon, T. Langenbacher and M.E. Michel-Beyerle, Proc. Natl. Acad. Sci. U.S.A. **95**, 12759 (1998).
20. B. Giese, Annu. Rev. Biochem. **71**, 51 (2002).
21. Y.A. Berlin, I.V. Kurnikov, D. Beratan, M.A. Ratner and A.L. Burin in *Long-Range Charge Transfer in DNA, II*, edited by G.B. Schuster (Springer, 2004).
22. E.M. Conwell and S.V. Rakhmanova, Proc. Natl. Acad. Sci. USA **97**, 4556 (2000).

23. P.T. Henderson, D. Jones, G. Hampikian, Y.Z. Kan and G.B. Schuster, Proc. Natl. Acad. Sci. U.S.A. **96**, 8353 (1999).
24. F.D. Lewis, H.H. Zhu, P. Daublain, T. Fiebig, M. Raytchev, Q. Wang and V. Shafirovich, J. Am. Chem. Soc. **128**, 791 (2006).
25. F.D. Lewis, T.F. Wu, X.Y. Liu, R.L. Letsinger, S.R. Greenfield, S.E. Miller and M.R. Wasielewski, J. Am. Chem. Soc. **122**, 2889 (2000).
26. F.D. Lewis, X.Y. Liu, S.E. Miller, R.T. Hayes and M.R. Wasielewski, J. Am. Chem. Soc. **124**, 11280 (2002).
27. L. Valis, Q. Wang, M. Raytchev, I. Buchvarov, H.A. Wagenknecth and T. Fiebig, Proc. Natl. Acad. Sci. U.S.A. **103**, 10192 (2006).
28. S.O. Kelley, R.E. Holmlin, E.D.A. Stemp and J.K. Barton, J. Am. Chem. Soc. **119**, 9861 (1997).
29. T. Fiebig, C. Wan, S.O. Kelley, J.K. Barton and A.H. Zewail, Proc. Natl. Acad. Sci. USA **96**, 1187 (1999).
30. R. Huber, N. Amann and H.-A. Wagenknecht, J. Org. Chem. **69**, 744 (2004).
31. L. Valis, N. Amann and H.-A. Wagenknecht, Org. Biomol. Chem. **3**, 36 (2005).
32. S.O. Kelley and J.K. Barton, Chem. Biol. **5**, 413 (1998).
33. G. Kokkinidis and A. Kelaidopoulou, J. Electroanal. Chem. **414**, 197 (1996).
34. C.V. Sonntag *Free-Radical-Induced DNA Damage and Its Repair* (Springer, Berlin, 2006).
35. J. Eisinger and R.G. Shulman, Science **161**, 1311 (1968).
36. C.E. Crespo-Hernandez, B. Cohen and B. Kohler, Nature **436**, 1141 (2005).
37. T. Förster and K. Kasper, Z. Phys. Chem. N.F. **1**, 275 (1954).
38. I. Buchvarov, Q. Wang, M. Raytchev, A. Trifonov and T. Fiebig, Proc. Natl. Acad. Sci. U.S.A. **104**, 4794 (2007).
39. H. Simpkins and E.G. Richards, Biochemistry **6**, 2513 (1967).
40. C.S.M. Olsthoorn, L.J. Bostelaar, J.F.M. De Rooij, J.H. Van Boom and C. Altona, Eur. J. Biochem. **115**, 309 (1981).
41. A.S. Davydov *Theory of Molecular Excitons* (McGraw-Hill Book Company, Inc., New York, 1962).
42. J.M.L. Pecourt, J. Peon and B. Kohler, J. Am. Chem. Soc. **123**, 10370 (2001).
43. D. Markovitsi, A. Sharonov, D. Onidas and T. Gustavsson, ChemPhysChem **4**, 303 (2003).
44. J. Peon and A.H. Zewail, Chem. Phys. Lett. **348**, 255 (2001).
45. R.J. Sension, S.T. Repinec and R.M. Hochstrasser, J. Chem. Phys. **93**, 9185 (1990).
46. D. Markovitsi, D. Onidas, T. Gustavsson, F. Talbot and E. Lazzarotto, J. Am. Chem. Soc. **127**, 17130 (2005).
47. C.E. Crespo-Hernandez, B. Cohen, P.M. Hare and B. Kohler, Chem. Rev. **104**, 1977 (2004).
48. V.A. Bloomfield, D.M. Crothers and J. Ignacio Tinoco, *Nucleic Acids – Structures, Properties, and Functions* (University Science Books, Sausolito, CA, 1999).
49. R.N. Barnett, C.L. Cleveland, A. Joy, U. Landman and G.B. Schuster, Science **294**, 567 (2001).
50. M.A. O'Neill and J.K. Barton, J. Am. Chem. Soc. **124**, 13053 (2002).
51. T.M. Nordlund, S. Andersson, L. Nilsson, R. Rigler, A. Graslund and L.W. McLaughlin, Biochemistry **28**, 9095 (1989).
52. L.W. McLaughlin, T. Leong, F. Benseler and N. Piel, Nucleic Acids Res. **16**, 5631 (1988).

53. C.Z. Wan, T. Fiebig, O. Schiemann, J.K. Barton and A.H. Zewail, Proc. Natl. Acad. Sci. U.S.A. **97**, 14052 (2000).

54. T. Fiebig, C.Z. Wan and A.H. Zewail, ChemPhysChem **3**, 781 (2002).

55. Q. Wang, M. Raytchev and T. Fiebig, Photochem. Photobiol., in press (2007).

56. M. Rist, H.A. Wagenknecht and T. Fiebig, ChemPhysChem **3**, 704 (2002).

57. M.F.M. Post, Langelaa.J and Vanvoors.Jd, Chem. Phys. Lett. **10**, 468 (1971).

58. H.A. Wagenknecth and T. Fiebig in *Charge Transfer in DNA*, edited by H.A. Wagenknecht (Wiley-VCH, Weinheim, 2005).

59. *Circular Dichroism – Principles and Applications*, edited by N. Berova, K. Nakanishi and R.W. Woodyeds. (Wiley-VCH, New York, 2000).

60. T. Carell, C. Behrens and J. Gierlich, Org. Biomol. Chem. **1**, 2221 (2003).

61. P.J. Hrdlicka, B.R. Babu, M.D. Sorensen and J. Wengel, Chem. Commun. 1478 (2004).

62. J.L. Sessler and J. Jayawickramarajah, Chem. Commun., 1939 (2005).

63. P. Tinnefeld, M. Heilemann and M. Sauer, ChemPhysCHem **6**, 217 (2005).

64. B.A Armitage, Top. Curr. Chem. **253**, 55 (2005).

65. J.M. Gao, C. Strassler, D. Tahmassebi and E.T. Kool, J. Am. Chem. Soc. **124**, 11590 (2002).

66. H.J. Gao and Y. Kong, Ann. Rev. Mater. Res. **34**, 123 (2004).

67. A. Cuppoletti, Y. Cho, J.-S. Park, C. Strässler and E.T. Kool, Bioconjugate Chem. **16**, 528 (2005).

68. Y. Cho and E.T. Kool, ChemBioChem **7**, 669 (2006).

69. M. Nakamura, Y. Ohtoshi and K. Yamana, Chem. Commun., 5163 (2005).

11 Vibrons in DNA:
Their Influence on Transport

Benjamin B. Schmidt[1,2], Evgeni B. Starikov[1,2], Matthias H. Hettler[2], and
Wolfgang Wenzel[2]

[1] Institut für Theoretische Festkörperphysik and DFG-Center for Functional
Nanostructures (CFN), Universität Karlsruhe, 76128 Karlsruhe, Germany
[2] Forschungszentrum Karlsruhe, Institut für Nanotechnologie, Postfach 3640,
76021 Karlsruhe, Germany
matthias.hettler@int.fzk.de

11.1 Introduction

Transport measurements on DNA display various types of behavior ranging
from insulating [1] over semi-conducting [2] to quasi-metallic [3], depending
on the measurement setup, environment and the measured DNA molecule.
The variance of the experimental results as well as the quantum-chemical calculations [4] suggest that the environment and its influence on the vibrational
modes (vibrons) of DNA are an important factor for the transport properties
of the DNA wires.

While there were suggestions of the importance of the environment,
the vibrons and the electron-vibron coupling effects on the electron trans-
fer/transport [5–7], only recently [8–11] there have been attempts to describe
the consequences of this coupling on the transport through DNA in a more microscopic way. Typically, the DNA is described within a tight-binding model
approach for the electronic degrees of freedom, where the model parameters
are either motivated by the quantum-chemistry computations [4,12,13] or by
the desire to fit a certain experiment. The variance of qualitatively different
tight-binding models is large, ranging from the involved all-atomic represen-
tations to models where each base pair is represented by only a single orbital
(see also other articles in this book).

On the other hand, the vibrational modes have so far been treated within
very simple models, where only a local electron-vibron coupling has been
taken into account [10]. If the coupling is sufficiently strong this leads to
the formation of the Holstein polarons, i.e. a bound state of electrons with
a local dispersionless vibrational mode (or a distortion). While the inclusion
of a local coupling is sufficient to describe the transition from the elastic
(quasi-ballistic) transport to the inelastic (dissipative), it ignores the fact that
the non-local electron-vibron coupling strength is found to be comparable in
magnitude to the local coupling [4]. Furthermore, as the non-local electron-
vibron coupling leads effectively to a vibron-assisted hopping, the proper
inclusion of this coupling can be important for the transport through the
(in)homogeneous sequences of the 'natural' DNA (see, for example [14,15]).

In this chapter, we study the electron transport through the simple tight-binding models for the double-stranded DNA wires strongly coupled (both locally and non-locally) to vibrational modes of the DNA. The DNA base pairs are represented by a single tight-binding orbital, the energy of which is dependent on whether it is a Guanine-Cytosine (GC) pair or a Adenine-Thymine (AT) pair. The vibrational modes are also coupled to the surrounding environment (water or buffer solution) which we represent by an ohmic bath. The inclusion of the bath allows for the dissipation of the energy and therefore opens the possibility for the "inelastic" transport processes. By applying the equation-of-motion techniques, we address the influence of specific DNA vibrational modes on the transport with parameters motivated by the quantum-chemical calculations [4, 13]. We find that for the homogeneous DNA sequences, such as the poly(dG)-poly(dC) wires, the vibrons strongly enhance the linear conductance at low temperatures, but affects only weakly the conductance at a large bias, which remains dominated by the quasi-ballistic transport through the extended electronic states.

The chapter is organized as follows: in the following we define the model we have considered (Sect. 11.2.1). We then sketch the technique used to obtain the parameters of the vibrational modes and their couplings to the electronic degrees of freedom (Sect. 11.2.2). The equation-of-motion method to obtain the current is briefly outlined in Sect. 11.2.3. In Sect. 11.3, we then present the results of our calculations for the homogeneous DNA wires. We summarize our conclusions in Sect. 11.4.

11.2 Model and Technique

11.2.1 The Model

The quantum chemistry calculations [16, 17] show that the highest occupied molecular orbital (HOMO) of a DNA base pair is (in the conventional cases) located on the Guanine or Adenine, whereas the lowest unoccupied molecular orbital (LUMO) is located on the Thymine and Cytosine. Between the HOMO and the LUMO there is an energetic gap of approximately $2-3\,\mathrm{eV}$. Experimental evidences hint at the prevalence of the hole transport through DNA. Given the energetic and the spatial separation of the HOMO and the LUMO and considering a sufficiently low bias voltage, we can model one base pair by a single tight-binding orbital (site). This is the minimal model for the electronic degrees of freedom for DNA.

We consider a DNA sequence with N base pairs. The first and the last base pair (BP) are coupled to semi-infinite metal electrodes. The chemical details of this coupling are not the focus of our interest, so we assume the simplest possible coupling which leads to a line width Γ_{L} (Γ_{R}) of the adjacent (the left most or the right most) base pair orbital. We further allow for a coupling to (in general multiple) vibrational modes, that can be excited by

local and non-local coupling to the charge carriers on the DNA. These modes
in turn are coupled to the environment. When later performing the numerical
calculations we will restrict ourselves to a single vibrational mode of the
DNA base pair (e.g. the 'stretch' mode, see below). Such a choice is justified
by earlier theoretical findings that, for example, this mode is most strongly
coupled to the electron motions in the DNA duplexes (see, for example [18]).
We then arrive at the Hamiltonian $H = H_{el} + H_{vib} + H_{el-vib} + H_{L/R} + H_{T,L} + H_{T,R} + H_{bath}$ with

$$H_{el} = \sum_i \epsilon_i a_i^\dagger a_i - \sum_i \sum_{j \neq i} t_{ij} a_i^\dagger a_j$$

$$H_{T,L} + H_{T,R} = \sum_r \sum_k \sum_i \left[t_{ik}^r c_{kr}^\dagger a_i + t_{ik}^{r*} a_i^\dagger c_{kr} \right]$$

$$H_{vib} = \sum_\alpha \Omega_\alpha B_\alpha^\dagger B_\alpha$$

$$H_{el-vib} = \sum_i \sum_\alpha \lambda_0\, a_i^\dagger a_i (B_\alpha + B_\alpha^\dagger)$$

$$+ \sum_i \sum_{j \neq i} \sum_\alpha \lambda_{ij}\, a_i^\dagger a_j (B_\alpha + B_\alpha^\dagger) \tag{11.1}$$

with the index $r = L, R$ describing the left and the right electrode. Here, $H_{L/R}$
models the non-interacting electrodes with a flat density of states ρ_e in the left
and right electrode, respectively, and $(c_{k\,L/R}^\dagger, c_{k\,L/R}$ are the Fermi operators
for the states in the electrodes). The term H_{el} describes the DNA chain in
a single orbital tight-binding representation with on-site energies ϵ_i of the
base pairs and hopping t_{ij} between the neighboring base pairs ($a_i^\dagger, a_i$ are the
Fermi operators creating/destroying electrons on the base pairs). Both the on-
site energies and the hopping are sequence dependent, e.g. the on-site energy
of a GC base pair is different from the on-site energy of a AT base pair. For
the hopping matrix elements t_{ij}, we used the values calculated by Siebbeles
et al. [13] who studied intra- and interstrand hopping between the bases in
DNA-dimers by the density functional theory (DFT). They computed the
direction-dependent values for all possible hopping matrix elements in such
dimers. Adapting these results to our simplified model of the base pairs,
we obtain the hopping elements listed in Table 11.1[3]. The number in the
G row and the A column denotes the hopping matrix element from a GC
base pair to an AT base pair to its "right" (to the 3′ direction), for example.
The above parameters are overall consistent with the values obtained by
other methods, in particular the work by us [4]. For the homogeneous GC

---

[3] We assume that the holes can only reside on the purine bases, G or A. The
hopping integrals between two purines depend on the specific bases involved and
to which strands these two bases belong to. The values we use are the hopping
integral J of the first, second and fifth row in the Table 3 of Ref. 10. They are
exactly reproduced in our Table 11.1.

Table 11.1. Hopping integrals t_{ij} taken from [13] and adapted to our model. The notation 5′-XY-3′ indicates the direction along the DNA strand

	5′-XY-3′ (all in eV)			
X\ Y	G	C	A	T
G	0.119	0.046	−0.186	−0.048
C	−0.075	0.119	−0.037	−0.013
A	−0.013	−0.048	−0.038	0.122
T	−0.037	−0.186	0.148	−0.038

sequences considered below, we only use the top left element in this table ($t_{\mathrm{GG}} = 0.119\,\mathrm{eV}$).

The vibronic degrees of freedom are described by H_{vib}, where B_α and $B_\alpha^\dagger$ are the boson destruction and creation operators for the vibron mode α with frequency Ω_α. $H_{\mathrm{el-vib}}$ couples the electrons (or holes) on the DNA to the vibrational mode, where λ_0 and λ_{ij} are the strengths for the local and non-local electron-vibron coupling, respectively. We further restrict the non-local coupling terms to the nearest neighbors, $\lambda_{ij} = \lambda_1 \delta_{i,j=i\pm1}$. Note that the frequency of the vibron mode and the magnitude of the couplings to the electrons in the model is independent of the considered base pair.

The vibrons are coupled to the environment, the microscopic details of which do not matter. We model it by a harmonic oscillator bath H_{bath}, whose relevant properties are summarized by its linear ('Ohmic') power spectrum (or spectral function) up to a high-frequency cut-off ω_c [19]. The coupling of the vibrons to the bath changes the vibrons spectra from discrete (Einstein) modes to continuous spectra with a peak around the vibron frequency. Physically, the coupling to a bath allows for dissipation of electronic and vibronic energy. This dissipation is crucial for the stability of the DNA molecule in a situation where inelastic contributions to the current dissipate a substantial amount of power on the DNA itself.

As mentioned before, we only consider a single vibrational mode when performing the numerical calculations. This vibrational mode with resonance frequency ω_0 coupled to the bath is then described by a spectral density

$$D(\omega) = \frac{1}{\pi} \left(\frac{\eta(\omega)}{(\omega - \omega_0)^2 + \eta(\omega)^2} - \frac{\eta(\omega)}{(\omega + \omega_0)^2 + \eta(\omega)^2} \right) , \tag{11.2}$$

with a frequency dependent broadening $\eta(\omega)$ which arises from the vibron-bath coupling. For the 'Ohmic' bath with weak vibron-bath coupling and cut-off ω_c we consider $\eta(\omega) = 0.05\,\omega\,\theta(\omega_c - \omega)$. Mathematically the crossover from the discrete vibrational modes to a continuous spectrum of a single mode is done by substituting $\sum_\alpha \delta(\omega - \Omega_\alpha) \rightarrow \int d\omega D(\omega)$.

11.2.2 Derivation of the Vibronic Parameters

Below we describe the quantum-chemical method to estimate the electron-vibron coefficients and to learn which of the DNA conformational variables are most coupled with the hole motion in DNA (the details about the method, as well as the results obtained using it, are published in [4]).

(1) We take the stacked trimers of AT or GC base pairs to investigate the consequences of the motions within a single base pair (changes in shear, stretch, stagger, buckle, propeller and opening; for detailed definitions of these DNA structural parameters see [20]). We systematically change one of the above six variables for the central base pair, whereas the two flanking base pairs always retain their averaged (equilibrium) geometry corresponding to the B-form of DNA.

(2) We carry out analogous manipulations to explore the changes in shift, slide, rise, tilt, roll and the twist [20], but in this case we consider the stacked tetramers of the AT or the GC base pairs, and the changes in question take place in the central BPs. Here, we take into account the flanking base pairs in order to simulate the influence of the DNA duplex environment.

(3) The parameters for the equilibrium B-DNA geometry, as well as the accessible intervals for changing the twelve conformational variables, have been taken from the work of Olson et al. [20].

(4) For each of the twelve variables, five conformations have been generated, namely, the equilibrium and the two maximum 'strained' conformations, as well as the two conformations on the half-way between the equilibrium and the corresponding maximum 'strained' ones, we use the 3DNA software package [21] to generate atomistic models corresponding to the above five conformations.

(5) For each of the five conformations, the energies of the HOMO and the HOMO-1 for the corresponding AT and GC trimers/tetramers have been calculated using the semiempirical quantum-chemical PM3 method implemented in the MOPAC routine package [22].

(6) According to the Koopmans theorem for a homogeneous dimer (see, for example [23]) the absolute value of the HOMO energy is approximately equal to the ionization potential or the on-site energy (ϵ) in the tight-binding approximation, whereas one half of the difference between the energies of the HOMO and the HOMO-1 is a plausible estimate for the hopping integral (t) in the tight-binding approximation. Since we are considering homogeneous tri- and tetramers here, the Koopmans estimates for them must be properly corrected (see the Appendix in [4] for details).

(7) We have tried to estimate the linear regression coefficients for ϵ and t as a function of the changes in each of the 12 conformational variables both for the AT and the GC systems. If the linear regression ($y = ax + b$) describes well the functional interdependence between the electronic (y) and conformational (x) variables, its slope corresponds to the electron-vibron coupling coefficient, $a = g$, where g is the coupling in the first quantization representation of the

electron-vibron coupling [4], whereas its intercept stands for the equilibrium value of the tight-binding parameter involved ($b = t_0$ or ϵ_0). It should be pointed out that despite the completely different physical origin of the 'on-site energy' and the 'hopping-integral' electron-vibron coupling, we choose here the same notation g for both, in order to stress the generic character of the electron-vibron coupling [24].

To decide whether the coupling under study is significant or not, we compare the maximum possible change in the tight-binding parameters due to the conformational alterations with the energy of a thermal quantum, $k_B T$, where k_B is the Boltzmann constant and T is at the room temperature. If this maximum possible change is less than $k_B T$, the linear electron-vibrational correlation corresponding to the g value involved is considered insignificant. Moreover, we compare the absolute values of these coupling constants among themselves by taking the whole set of g values into two groups, namely, those defined for the displacements (measured in Angstroms) and those defined for the angular changes (measured in degrees), and only performing our comparisons within each of these two groups.

The correlations between some of the 12 conformational variables are known to exist from the X-ray experiments [20]. We have employed the approach described above in the points (1)–(7) to estimate the coupling between these four concerted modes and the DNA tight-binding parameters.

Using the above technique, we were able to reveal the main vibronic degree of freedom in the DNA duplexes – the 'stretch' variable i.e., the periodical stretching and squeezing of the H-bonds within the canonical Watson-Crick base pairs (AT and GC). Our choice is in full accordance with the results obtained by other authors using a different quantum-chemical method [18]. Whereas we can immediately adopt the natural frequency of the 'stretch' mode as the frequency ω_0 of the single vibration considered in H_{vib}, the matching of the couplings g with the parameters λ_0 and λ_1 is, strictly speaking, not possible without knowing more details about the actual potential that leads to the considered vibrations. However, considering the simplicity of our DNA model, here we can get away with a rough estimate. Assuming that the 'average' extension of the 'stretch' vibration is of the order of a few picometers, the couplings λ_0 and λ_1 ought to be of the order of $1 - 10$ meV.

For the specific case of the 'stretch' mode in a homogeneous GC sequence, λ_0 turns out to be positive and λ_1 negative, where $|\lambda_0| > |\lambda_1|$. Whereas the sign of λ_0 is irrelevant for the polaron shift (see below) which is $\sim \lambda_0^2$, the signs of λ_0 and λ_1 do matter for the terms obtained within the equation-of-motion method due to the non-local coupling λ_1 as some of these terms are $\sim \lambda_0 \lambda_1$ (for details, see [25]).

11.2.3 Computation of Transport

For the strong electron-vibron coupling predicted for DNA [4], one expects the polaron formation, with a polaron size of a few base pairs. To describe these polarons (a combined electron-vibron "particle") theoretically we apply the Lang-Firsov unitary transformation with the generator function S to our Hamiltonian (see e.g. [26])

$$\bar{H} = e^S H e^{-S} \; ; \; S = -\sum_i \sum_\alpha \frac{\lambda_0}{\Omega_\alpha} a_i^\dagger a_i \left[B_\alpha - B_\alpha^\dagger \right] \; . \tag{11.3}$$

By introducing the transformed electron and the vibron operators according to

$$\bar{B}_\alpha = B_\alpha - \sum_i \frac{\lambda_0}{\Omega_\alpha} a_i^\dagger a_i \tag{11.4}$$

$$\bar{a}_i = a_i \chi \, , \tag{11.5}$$

$$\chi = \exp\left[\sum_\alpha \frac{\lambda_0}{\Omega_\alpha} \left(B_\alpha - B_\alpha^\dagger \right) \right] \, , \tag{11.6}$$

the new Hamiltonian now reads (with $\chi\chi^\dagger = \chi^\dagger\chi = 1$)

$$\begin{aligned}
\bar{H} = &\sum_i (\epsilon_i - \Delta) a_i^\dagger a_i - \sum_i \sum_{j \neq i} t_{ij} a_i^\dagger a_j \\
&+ \sum_r \sum_k \sum_i \left[t_{ik}^r c_{kr}^\dagger a_i \chi + t_{ik}^{r*} a_i^\dagger \chi^\dagger c_{kr} \right] \\
&+ \sum_\alpha \Omega_\alpha B_\alpha^\dagger B_\alpha + H_{\mathrm{L/R}} \\
&+ \sum_i \sum_{j \neq i} \sum_\alpha \lambda_1 \, a_i^\dagger a_j (B_\alpha + B_\alpha^\dagger)
\end{aligned} \tag{11.7}$$

$$\Delta = \int d\omega D(\omega) \frac{\lambda_0^2}{\omega} \, , \tag{11.8}$$

where we neglected terms with the vibron-mediated electron-electron interaction. This is a reasonable approximation for the low electron (hole) density in DNA. The purpose of the Lang-Firsov transformation is to remove the local electron-vibron coupling term from the transformed Hamiltonian in exchange for the transformed operators and the so-called polaron shift Δ, describing the lower on-site energy of the polaron as compared to the bare electron (hole). However, the non-local coupling term remains unchanged and has to be dealt with in a different way than the local term (see below). There is an additional electron-vibron coupling due to the vibron shift generator χ in the transformed tunnel Hamiltonian from the leads. In this study, we neglect the effects arising from this additional coupling (a valid approximation for the wide band limit, i.e. $\Gamma^{\mathrm{L,R}} \gg \lambda_0$. (This is an usual approximation adapted in the literature [9, 11]).

We now define the retarded electron Green function in the usual way as

$$G_{kl}^{\text{ret}}(t) = -i\theta(t) \left\langle \left\{ a_k(t)\chi(t), a_l^\dagger \chi^\dagger \right\} \right\rangle$$

$$= \underbrace{-i\theta(t) \left\langle a_k(t)\chi(t)a_l^\dagger \chi^\dagger \right\rangle}_{G_{kl}^{(1)}(t)}$$

$$\underbrace{-i\theta(t) \left\langle a_l^\dagger \chi^\dagger a_k(t)\chi(t) \right\rangle}_{G_{kl}^{(2)}(t)}, \tag{11.9}$$

where the thermal average is taken with respect to the transformed Hamiltonian, which does not explicitly include the local electron-vibron interaction. The separation of the retarded Green function into $G_{kl}^{(1)}(t)$ and $G_{kl}^{(2)}(t)$ is necessary because only for these objects self-consistency equations can be derived via the equation-of-motion technique (EOM) (the EOM for $G_{kl}^{\text{ret}}(t)$ leads to an equation not only including $G_{kl}^{\text{ret}}(t)$ itself, but also other Green functions). The EOM technique for an interacting system generates the correlation functions of higher order than initially considered, resulting in a hierarchy of equations that does not close in itself. Therefore, an appropriate truncation scheme needs to be applied. In our case, we close the hierarchy on the first possible level, i.e. we neglect all higher order Green functions beyond the one defined above.

In applying the EOM technique we encounter expressions of the type $\left\langle a_j(t)B_\alpha(t)\chi(t)a_l^\dagger \chi^\dagger \right\rangle$. As the thermal average involves the transformed Hamiltonian with a non-local electron-vibron coupling, these averages can not be evaluated exactly. Instead, we use the approximation

$$\left\langle a_j(t)B_\alpha(t)\chi(t)a_l^\dagger \chi^\dagger \right\rangle_{\bar{H}} \approx F(t) \left\langle a_j(t)\chi(t)a_l^\dagger \chi^\dagger \right\rangle_{\bar{H}} \tag{11.10}$$

where the function $F(t)$ is taken from the equality

$$\left\langle a_j(t)B_\alpha(t)\chi(t)a_l^\dagger \chi^\dagger \right\rangle_{H_0} = F(t) \left\langle a_j(t)\chi(t)a_l^\dagger \chi^\dagger \right\rangle_{H_0}. \tag{11.11}$$

In contrast to the (11.10), here the thermal average is taken with respect to the non-interacting Hamiltonian H_0 without any electron-vibron coupling. In this case the function $F(t)$ can be calculated in a straightforward manner. We take the same function $F(t)$ to approximate the effect of the $B_\alpha(t)$ for our situation where there is a non-local coupling of order λ_1.

Terms involving four electron operators such as $\left\langle a_l^\dagger \chi^\dagger a_k(t)a_i^\dagger(t)a_j(t)\chi(t) \right\rangle$ are treated in a mean-field approximation (for details see [25]). In matrix notation we get

$$G^1(E) = (\hat{I} - \hat{f})G_0^{\text{ret}}(E) + G_0^{\text{ret}}(E) \cdot \mathcal{F}_1\left(G^1(E)\right), \tag{11.12}$$

$$G^2(E) = \hat{f}\, G_0^{\text{ret}}(E) + G_0^{\text{ret}}(E) \cdot \mathcal{F}_2\left(G^2(E)\right), \tag{11.13}$$

where $\hat{I}$ is the identity matrix and the elements of the matrix $\hat{f}$ are $\left\langle a_k a_l^\dagger \right\rangle$. $G_0^{\mathrm{ret}}(E)$ is the retarded Green function for the isolated DNA without electron-vibron interaction. The expressions $\mathcal{F}_1\left(G^1(E)\right)$ and $\mathcal{F}_2\left(G^2(E)\right)$ stand for certain sets of matrix multiplications and integrations acting upon $G^1(E)$ and $G^2(E)$. They are too lengthy to be displayed here, and we refer the reader to [25] for full details.

For a DNA chain with N bases the density of states is

$$A(E) = -\frac{1}{\pi N} \sum_{i=1}^{N} \Im\left(G_{ii}^{\mathrm{ret}}(E)\right) . \tag{11.14}$$

In the wide band limit, the electrode self-energies are constant and purely imaginary: $\Sigma_{ij}^{\mathrm{L}} = i\Gamma^{\mathrm{L}}\delta_{i1}\delta_{j1}$ and $\Sigma_{ij}^{\mathrm{R}} = i\Gamma^{\mathrm{R}}\delta_{iN}\delta_{jN}$.

To calculate the current we use the general relation derived by Meir and Wingreen [28],

$$
\begin{aligned}
I = \frac{ie}{h} \int \mathrm{d}\epsilon \Big(&\mathrm{tr}\left\{\left[f_{\mathrm{L}}(\epsilon)\Gamma^{\mathrm{L}} - f_{\mathrm{R}}(\epsilon)\Gamma^{\mathrm{R}}\right]\left(G^{\mathrm{ret}}(\epsilon) - G^{\mathrm{adv}}(\epsilon)\right)\right\} \\
&+ \mathrm{tr}\left\{\left[\Gamma^{\mathrm{L}} - \Gamma^{\mathrm{R}}\right] G^{<}(\epsilon)\right\}\Big),
\end{aligned}
\tag{11.15}
$$

where $f_{\mathrm{L}}(\epsilon)$ and $f_{\mathrm{R}}(\epsilon)$ are the Fermi distributions in the left and right lead, respectively.

To compute the 'lesser' Green function $G^{<}(\epsilon)$, we use the well known relation [26]

$$G^{<}(\epsilon) = G^{\mathrm{ret}}(\epsilon)\left[\Sigma^{\mathrm{L}<} + \Sigma^{\mathrm{R}<} + \Sigma_{\mathrm{vib}}^{<}(\epsilon)\right] G^{\mathrm{adv}} \tag{11.16}$$

While the lesser electrode self-energies can be a determined easily within the above approximation for any applied bias, we have to approximate the behavior of the lesser self-energy due to the vibrons $\Sigma_{\mathrm{vib}}^{<}$. Extending the known relation for the equilibrium situation, we write

$$\Sigma_{\mathrm{vib}}^{<}(\epsilon) = -f_{\mathrm{eff}}(\epsilon)\left(\Sigma_{\mathrm{vib}}^{\mathrm{ret}}(\epsilon) - \Sigma_{\mathrm{vib}}^{\mathrm{adv}}(\epsilon)\right) \tag{11.17}$$

$$\text{with } f_{\mathrm{eff}} = \frac{1}{2}\left(f_{\mathrm{L}}(\epsilon) + f_{\mathrm{R}}(\epsilon)\right) \tag{11.18}$$

i.e., we assume an effective Fermi distribution f_{eff} that multiplies the (computed) equilibrium expressions for $\Sigma_{\mathrm{vib}}^{\mathrm{ret}}, \Sigma_{\mathrm{vib}}^{\mathrm{adv}}$. Combining all the terms we obtain a concise expression for the current which can be separated into the "elastic" and the "inelastic" part as

$$I = \frac{2e}{h} \int \mathrm{d}\epsilon \left[f_{\mathrm{L}}(\epsilon) - f_{\mathrm{R}}(\epsilon)\right]\left(T_{\mathrm{el}}(\epsilon) + T_{\mathrm{inel}}(\epsilon)\right) , \tag{11.19}$$

where we identify the "elastic" and the "inelastic" transmission functions [29, 30]

$$T_{\text{el}}(\epsilon) = 2\text{tr}\left\{\Gamma^{\text{R}}G^{\text{ret}}(\epsilon)\Gamma^{\text{L}}G^{\text{adv}}(\epsilon)\right\} \tag{11.20}$$

$$T_{\text{inel}}(\epsilon) = \frac{\text{i}}{4}\text{tr}\left\{\left(\Gamma^{\text{R}} + \Gamma^{\text{L}}\right)G^{\text{ret}}(\epsilon)\right.$$
$$\left.\times \left[\Sigma_{\text{vib}}^{\text{ret}}(\epsilon) - \Sigma_{\text{vib}}^{\text{adv}}(\epsilon)\right]G^{\text{adv}}(\epsilon)\right\} . \tag{11.21}$$

Note that the term "elastic" transmission does *not* imply that no effects of vibrons are included there. As the Green function is self consistently computed in the presence of the vibrons and the environment this term is very much affected by the vibrons.

11.3 Results: Poly(dG)-Poly(dC) DNA

We analyze now the effect of vibrations on the electronic properties of DNA, i.e. the density of states, the transmission and the current. We consider a single vibrational mode for all base pairs as described in Sect. 11.2.1. We used the DNA sequences of 26 base pairs. For simplicity, we couple the left and right electrodes symmetrically to the DNA, so that $\Gamma^{\text{L}} = \Gamma^{\text{R}} \equiv \Gamma$, where we choose $\Gamma = 0.1$ eV. Also note that the bias V_{b} is dropped symmetrically at the electrode-DNA interfaces.

For a homogeneous DNA consisting only of the repeated GC base pairs we obtain a band-like density of states (Fig. 11.1). With the fairly small hopping element of 0.119 eV (see Table 11.1) for this small system one can still resolve the peaks due to the single electronic resonances, especially near the van-Hove-like pile up of states near the band edges. All states are delocalized over the entire system. The inset displays the elastic transmission, showing that the states have a high transmission of ~ 0.5, with the states at the upper band edge showing the highest values. Both the density of states and the elastic transmission show a strong asymmetry which is a direct consequence of the non-local electron-vibron coupling in this model.

To further elucidate this connection we take a closer look at the upper and lower band edge of the density of states (see Fig. 11.2). Without the electron-vibron coupling (solid curve) we see the electronic resonances of equal height, positioned at the energies corresponding to the "Bloch"-like states of this finite size tight-binding chain. If we include only the local electron-vibron coupling (dashed line), vibron satellites states are visible as well as a decrease of the spectral weight of the original electronic resonances (the spectral sum rule is fulfilled). Note that the displayed vibron satellite is not a satellite from the displayed electronic state but from another at higher (lower) energy. This can be deduced by the difference in peak positions that is not equal to ω_0.

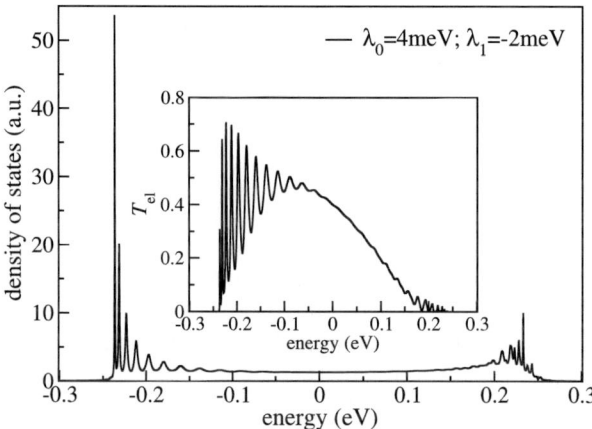

Fig. 11.1. The density of states and the transmission of poly(dG)-poly(dC) with 26 base pairs for the parameters: base pair on-site energy $\epsilon_G = 0\,\text{eV}$, the Fermi energy $E_F = 0.35\,\text{eV}$, the vibrational energy $\hbar\omega_0 = 0.01\,\text{eV}$, the cutoff $\hbar\omega_c = 0.03\,\text{eV}$, the linewidth $\Gamma = 0.1\,\text{eV}$ and the temperature $k_B T = 0.025\,\text{eV}$ (room temperature). The strong asymmetry of the curves with respect to the band center is a consequence of the non-local electron-vibron coupling λ_1

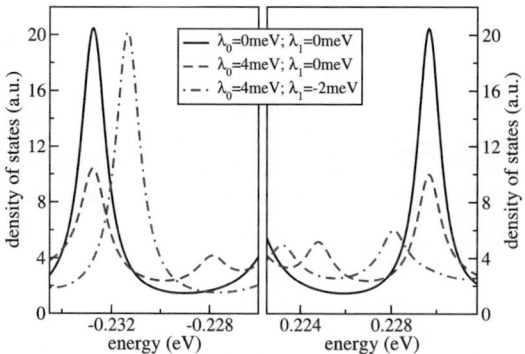

Fig. 11.2. The density of states of poly(dG)-poly(dC) with 26 base pairs for the parameters as in Fig. 11.1. The *solid line* shows the purely electronic resonances. Inclusion of only a local electron-vibron coupling λ_0 reduces the weight at the original electronic resonance in favor of 'vibron satellites' (*dashed line*). The addition of a non-local electron-vibron coupling λ_1 (*dash-dotted line*) introduces differential shifts of the resonance peaks (leading to an effective band narrowing) as well as a strong asymmetry in the height of the resonances

With a finite non-local coupling λ_1 the resonances shift positions, which leads to an effective change in bandwidth, depending on the sign of $\lambda_0\lambda_1$ (in the present case, the bandwidth is decreased). Furthermore, a distinct asymmetry of the resonances is observed, i.e. the lower band edge states have a larger peak height than the upper band edge states (again, this would be reversed for the opposite sign of $\lambda_0 \lambda_1$). This asymmetry in the density of states induces a corresponding asymmetry in the elastic transmission, see Fig. 11.1 for the overall view.

In Fig. 11.3 we see that the local coupling to vibrations (dashed line) leads to a strong increase in the zero-bias conductance at low temperatures, whereas the conductance is decreased at high temperatures. This effect has been observed before, e.g. in [10]. For low temperatures, the conductance is larger due to an effectively enhanced density of states at the Fermi energy due to the broadened vibronic resonances. The transport remains "elastic", i.e. electrons enter and leave the DNA at the same energy (first contribution to the current (11.19)). At sufficiently high temperatures, however, the back scattering of electrons due to vibrons reduces the conductance in comparison to the purely electronic system (no electron-vibron coupling, solid line), where the conductance depends on the temperature exclusively via the Fermi distribution function.

Including the non-local electron-vibron coupling (dash-dotted line) further increases the low temperature conductance. Although the bandwidth is effectively narrowed in the present case (electronic resonances move away from the Fermi energy, see Fig. 11.2, right panel), the density of states induced by the vibronic resonances right at the Fermi energy is enhanced by the non-local coupling. However, at room temperature the zero bias conductance is reduced. The corresponds to further enhanced back scattering of the states around the Fermi energy when compared to the case of only local coupling.

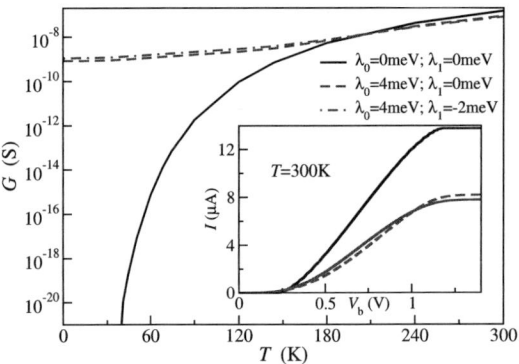

Fig. 11.3. The zero-bias conductance and I–V characteristics for poly(dG)-poly(dC) with 26 base pairs. Parameters are the same as in Fig. 11.1. The inclusion of the vibrons increases the zero-bias conductance at low temperatures (roughly below ω_0 by several orders of magnitude). At room temperature, the zero bias conductance is however reduced if a electron-vibron coupling is included. *Inset:* the I–V characteristics shows a "semiconducting" behavior at room temperature. The non-local electron-vibron coupling λ_1 leads to a decrease in the non-linear conductance in the gap and around the threshold, leading to a smaller current in the bias range below ~ 1 V

The inset shows a typical I–V-characteristic for the system (only the positive bias is shown as the I–Vs are antisymmetric). A quasi-semiconducting behavior is observed, where the size of the conductance gap is determined by the energetic distance of the Fermi energy to the (closest) band edge. After crossing this threshold, the current increases basically linearly with the voltage, until at larger bias it saturates when the right chemical potential drops below the lower transmission band edge. Small step-like wiggles due to the 'discrete' electronic states are visible at low temperature (not shown), but are smeared out at room temperature. The current is dominated by the elastic transmission, as expected for a homogeneous system.

Just above the threshold the differential conductance (slope of the I–V) is essentially proportional to the density of states at the upper band edge (see Fig. 11.2, right panel). Correspondingly, the differential conductance at the threshold is highest for the case without any vibrons (solid line, highest peak in the density of states) and lowest for the case of both local and non-local coupling (dash-dotted line). At a bias of $\sim 1\,V$ the lower band edge comes within the bias window. The differential conductance of the case with non-local coupling is correspondingly enhanced. At the plateau (above $\sim 1.2\,V$) the currents of the curves with electron-vibron coupling are nearly identical, as the asymmetry induced by the non-local coupling has been 'averaged out'. Note that the size of the current for the chosen value of $\Gamma = 0.1\,eV$ is quite large, also when compared to experiments on short poly(dG)-poly(dC) chains [3].

11.4 Conclusions

To summarize, we have presented a technique that allows the computation of the non-equilibrium electron transport through short sequences of DNA including a dissipative environment and the non-local coupling to single base pair vibrations. The parameters of the DNA model as well as the vibrational modes and their coupling to the electrons have been adapted from the ab-initio quantum chemistry methods. Using an equation-of-motion approach we identify the elastic and the inelastic contributions to the current. For the homogeneous DNA sequences, the transport is dominated by the elastic quasi-ballistic contributions through a band-like density of states (Figs. 11.1 and 11.2), which display an asymmetry due to the non-local electron-vibron coupling. The coupling to the vibrations strongly enhances the zero bias conductance at low temperatures. The current at a finite bias (above the 'semiconducting' gap), however, is only quantitatively modified by the non-local electron-vibron coupling (Fig. 11.3).

Acknowledgement. We wish to acknowledge discussions and correspondence with Janne Viljas, Anti-Pekka Jauho and Elke Scheer. We also thank the Landesstiftung Baden-Württemberg for financial support via the Kompetenznetz "Funktionelle Nanostrukturen".

References

1. E. Braun, Y. Eichen, U. Sivan and G. Ben-Yoseph, Nature **391**, 775 (1998).
2. D. Porath, A. Bezryadin, S. de Vries and C. Dekker, Nature **403**, 635 (2000).
3. B. Q. Xu, P. M. Zhang, X. L. Li and N. J. Tao, Nano Lett. **4**, 1105 (2004).
4. E.B. Starikov, Phil. Mag. **85**, 3435 (2005).
5. E. M. Conwell and S. V. Rakhmanova, Proc. Natl. Acad. Sci. USA **97**, 4556 (2000).
6. P. T. Henderson, D. Jones. G. Hampikian, Y. Kan and G. B. Schuster, Proc. Natl. Acad. Sci. USA **96**, 8353 (1999).
7. K. H. Yoo, D. H. Ha, J. O. Lee, J. W. Park, J. Kim, J. J. Kim, H. Y. Lee, T. Kawai and H. Y. Choi, Phys. Rev. Lett. **87**, 198102 (2001).
8. Y. Asai, Phys. Rev. Lett. **93**, 246102 (2004).
9. M. Galperin, A. Nitzan and M. A. Ratner, Phys. Rev. B. **73**, 045314 (2006).
10. R. Gutierrez, S. Mandal and G. Cuniberti, Phys. Rev. B **71**, 235116 (2005).
11. R. Gutierrez, S. Mohapatra, H. Cohen, D. Porath and G. Cuniberti, Phys. Rev. B **74**, 235105 (2006).
12. A. A. Voityuk, J. Jortner, M. Bixon and N. Rösch, Chem. Phys. Lett. **324**, 430 (2000).
13. K. Senthilkumar, F. C. Grozema, C. F. Guerra, F. M. Bickelhaupt, F. D. Lewis, Y. A. Berlin, M. A. Ratner and L. D. A. Siebbeles, J. Am. Chem. Soc. **127**, 14894 (2005).
14. D. Hennig, E. B. Starikov, J. F. R. Archilla and F. Palmero, J. Biol. Phys. **30**, 227 (2004).
15. H. Yamada, E. B. Starikov, D. Hennig and J. F. R. Archilla, Eur. Phys. J. E **17**, 149 (2005).
16. E. Artacho, M. Machado, D. Sanchez-Portal, P. Ordejon and J. M. Soler, Mol. Phys. **101**, 1587 (2003).
17. J. P. Lewis, T. E. Cheatham, E. B. Starikov, H. Wang and O. F. Sankey, J. Phys. Chem. B **107**, 2581 (2003).
18. S. S. Alexandre, E. Artacho, J. M. Soler and H. Chacham, Phys. Rev. Lett. **91**, 108105 (2003).
19. U. Weiss, *Quantum Dissipative Systems* (World Scientific, 1999).
20. W. K. Olson, M. Bansal and S. K. Burley, J. Mol. Biol. **313**, 229 (2001).
21. X.-J. Lu and W. K. Olson, Nucl. Acids Res. **31**, 5108 (2003).
22. J. J. P. Stewart, J. Comput. Chem. **10**, 209 (1989).
23. A. Fortunelli and A. Painelli, Phys. Rev. B **55**, 16088 (1997).
24. M. Capone, W. Stephan and M. Grilli, Phys. Rev. B **56**, 4484 (1997).
25. B.B. Schmidt, M.H. Hettler and G. Schön, Phys. Rev. B **75**, 115125 (2007).
26. G.D. Mahan, *Many-Particle Physics* (Kluwer Academic/Plenum Publishers, 2000).
27. H. Boettger and V. V. Bryksin, *Hopping Conduction in Solids* (Akademie Verlag Berlin, 1985).
28. Y. Meir and N. S. Wingreen, Phys. Rev. Lett. **68**, 2512 (1992).
29. M. Galperin, M. A. Ratner and A. Nitzan, J. Chem. Phys. **121**, (2004).
30. J. K. Viljas, J. C. Cuevas, F. Pauly and M. Häfner, Phys. Rev. B **72**, 245415 (2005).

12 DNA-Based Assembly of Metal Nanoparticles: Structure and Functionality

Monika Fischler and Ulrich Simon

RWTH Aachen, Institute of Inorganic Chemistry, Landoltweg 1, 52074 Aachen, Germany
ulrich.simon@ac.rwth-aachen.de

12.1 Introduction

Nanoparticles of various sizes and shapes have been synthesized and widely studied within the last decades. The exploration of their physical and chemical properties and the elaboration of theoretical methods suitable for mesoscopic systems inspired scientists to develop strategies how to utilize the fascinating effects that occur in the nanometer scale for the creation of new materials for nanoelectronic, diagnostic or sensing devices [1–7]. Materials in the nanoscale exhibit properties that are situated between those of solid state materials and those of single atoms or molecules, described by the rules of quantum physics. If small enough in diameter these particles show quantization of the electronic states. The resulting size-dependent change of physical properties is called the *quantum size effect* (QSE) or the *size quantization effect* [8]. Nanoparticle assemblies are of particular interest as the design of "artificial molecules" or "artificial solids" built up from nanoscale subunits may lead to a new state of matter, where optical and electronic coupling effects between the nanoscale subunits influence the material properties and offer new, and promising application fields in a variety of future technologies [6, 9].

Therefore, controlled assembly mechanisms in one, two and three dimensions are highly desirable. Here DNA plays an important role as template and interconnector between the particles. Figure 12.1 represents how DNA can serve as an assembly tool for one-, two-, and three-dimensional nanoparticle arrays.

DNA has been chosen as an assembly tool as it offers a variety of useful prerequisites: The molecule bears a high degree of information which arises with the self recognition process, the Watson-Crick base pairing scheme, unachieved by any other natural or artificial polymer, concerning its selectivity. Noteworthy are also the mechanical properties of the DNA molecule: the relative stiffness of short DNA sequences enables their use as spacers which function like rigid rods between the attached nanoobjects. Modern synthetic methods open up the access to virtually any sequence up to a length of approximately 120 base pairs, if desired equipped with sticky ends to allow predictable association of the strands to complicated geometries [10, 11]. Furthermore, the well known chemical structure of the sequences allows further

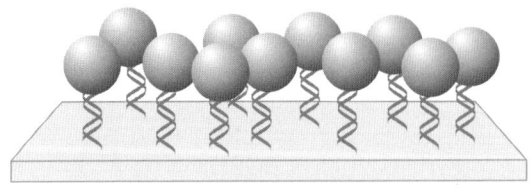

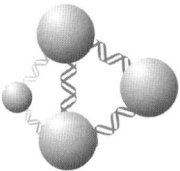

Fig. 12.1. Scheme of one-, two-, and three-dimensional nanoparticle assemblies constructed by means of DNA molecules

modification with functional groups, either at the 5′- or 3′-terminus of the strands or at the functionalities of the DNA bases within the strands. Additionally, the negative charge of the sugar phosphate backbone of the DNA can be exploited in order to decorate strands with positively charged objects like cations, multiply charged molecules or nanoparticles.

In the following, we will introduce the principles of DNA mediated assembly of nanoparticles. A brief overview about the synthesis of gold nanoparticles and their surface modification with DNA strands will be given. Furthermore, detailed information on DNA mediated assembly processes of nanoparticles will be explained. The structural features and the electrical properties will be exemplarily described together with emerging applications.

12.2 Materials Synthesis

12.2.1 Liquid Phase Synthesis of Metal Nanoparticles

Many different synthetic approaches have been pursued in order to obtain metal nanoparticles of different sizes and shapes. Thereby it has to be distinguished between the top-down and the bottom-up fabrication processes. Top-down approaches follow the strategy to disperse a macroscopic material until the particle size in the nanometer range is reached (e.g. milling or lithographic processes) while in "bottom-up" approaches nanoparticles are assembled from smaller subunits, like atoms or molecules, mostly via wet-chemical routes. As the most common processes in the discussed field follow the bottom-up approach, we will focus only on this aspect in the following. The general way for their preparation is the reduction of a soluble metal precursor, i.e. $HAuCl_4$, in the presence of stabilizing ligand molecules, either by the reducing agents (like hydrogen, boron hydride, citric acid, alcohols, and others), through an electrochemical setup [12,13], or by the physical assisted methods like sonochemistry, thermolysis, or photochemistry [14–16]. The general procedure for the preparation of metal nanoparticles is depicted in Fig. 12.2.

The ligand molecules that are added in order to stabilize the nanoparticles serve as a protecting shield for the particles and prevent them from aggregation and growth. This effect is either based on the electrostatic repulsion if the ligand molecules are charged, or through sterical shielding which occurs when bulky molecules, like polymers, are added. In most cases the ligand shell fulfills both of the afore-mentioned properties and the shielding effect is a combination of the electrostatic and the steric repulsion. Thereby the group which binds to the gold surface has to be an electron donor, like $-NR_2$, $-PR_3$ or R–SH groups. In various cases a post-synthetic ligand exchange is possible, depending on the binding strength of the ligand which was used in the initial nanoparticle synthesis. Through this procedure the nanoparticle properties, e.g. solubility, or the chemical functionalization can flexibly be adjusted to the requirements of the respective system [17].

The choice of metal precursors, the great variety of different reducing agents and the huge number of possible ligand molecules have lead to a broad diversity of metal nanoparticles with different sizes, shapes and chemical functionality. For detailed information the reader might refer to review articles in [4,18–20]. In the following, some illustrative examples of nanoparticle synthesis will be presented.

The most popular route for the synthesis of gold nanoparticles is the reduction of $HAuCl_4$ with sodium citrate in aqueous solution, a route that was developed by Turkevich et al. in 1951 [21]. Depending on the concentration of the citrate which serves as a reducing agent as well as stabilizing ligand particles with sizes ranging from 14.5 ± 1.4 nm to 24 ± 2.9 nm can be obtained. Due to the weak binding of the citrate to the gold the particles undergo ligand exchange after synthesis very easily. The red color of the particle solutions resulting from the plasmon resonance serves as a good example for the difference

Fig. 12.2. General scheme for fabrication of the nanoparticles via the wet chemical route: A soluble metal salt is reduced in presence of stabilizing ligand molecules

in the physical properties between the nano-sized and the bulk gold. Many other syntheses have been applied, involving the use of the block-copolymer micelles as the microreactors for the nanoparticles [22–24]. The various synthetic methods, the different metal precursors, the great variety of different reducing agents, and the huge number of the possible ligand molecules have lead to a broad diversity of metal nanoparticles with different sizes, shapes, and chemical functionality that cannot be described in detail here.

However, all of these preparation methods have one thing in common that is the size and the size distribution of the nanoparticle products are of great importance. Thus, the most prominent example for the synthesis of the gold clusters, meaning aggregates with a defined number of gold atoms, given by Schmid et al. in 1981 should be mentioned in this context [25]. As a precursor $Au(PPh_3)Cl$ was utilized, which was reduced by diborane, generated in situ from $NaBH_4$ and $BF_3 \cdot Et_2O$. The resulting $Au_{55}(PPh_3)_{12}Cl_6$ Cluster could be isolated as a black microcrystalline solid which was soluble in dichloromethane. Characterization by means of the small angle X-ray diffraction and the TEM revealed an extremely narrow size distribution $(1.4\,nm \pm 0.4\,nm)$ and a cubic packing of the gold atoms, following the model of a so called full-shell cluster. The number of atoms in each shell thereby can be determined by the rule $10n^2 + 2$ atoms (n = number of shells) [4,26,27]. As further examples for full-shell clusters $Pt_{309}phen_{36}^*O_{30}$ and $Pd_{561}phen_{36}O_{200}$ (phen* = bathophenantroline and phen = 1,10-phenantroline) can be mentioned [28–30]. The full-shell-clusters with few atoms such as the Au_{55}-Cluster exhibit an extremely narrow size distribution and therefore a highly defined electronic structure, making them especially interesting for the design of materials with novel electronic properties [31].

Furthermore, the nanoparticles that are functionalized with DNA as a ligand molecule are particularly important in the construction of one, two and three-dimensional assemblies, therefore the following section will describe their synthesis and characterization.

12.2.2 Preparation of the DNA-Functionalized Metal Nanoparticles

In the last decades, synthesis of the artificial DNA strands has become a routine technology so that nowadays single and double stranded DNA oligomers up to a length of 120 base pairs and with designable sequences and variable functionalities at the 3'- or 5'- terminus are commercially available. This opens up the possibility to decorate the nanoparticles with DNA sequences and exploit the high specifity of the Watson-Crick base pairing which allows for the self assembly and construction through self recognition processes between the DNA single strands, and hybridization of the complementary sequences to the double helix.

Besides electrostatic binding of the DNA backbone to the nanoparticles surface [32], mainly 3'- and 5'-thiol or amino-modified DNA strands have been used to modify gold nanoparticles with DNA as illustrated in Fig. 12.3. The initial protocol was described by Mirkin and coworkers who performed ligand exchange of citrate stabilized nanoparticles against thiol- and amino-modified oligomers [33]. Briefly, the oligomers were added to the nanoparticle solution, and stirred for prolonged times up to several days. After purification through repeated centrifugation the DNA functionalized nanoparticles were obtained as stable solutions.

In the following years the above described procedure was slightly varied by Mirkin and coworkers and by Niemeyer and coworkers [34–36]. As an important step towards a selective assembly of the nanoparticles one may thereby judge the synthesis of the gold nanoparticles that carry more than one single stranded oligomer-sequence, as it was presented by Niemeyer et al. The latter describes the preparation of two-up to heptafunctional particles as depicted in Fig. 12.4.

Compared to the conventional monofunctional DNA-nanoparticles these conjugates show almost unaltered hybridization capabilities and can therefore be individually and selectively addressed with the complementary DNA single strands. The application of such multifunctionalized particles will be described in Sect. 12.3.2 in detail.

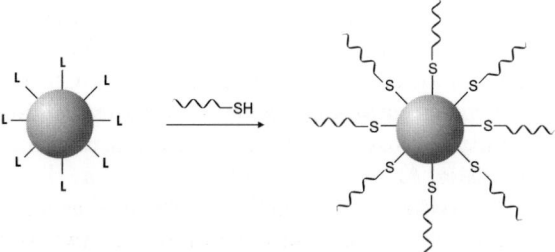

Fig. 12.3. Scheme for the modification of gold nanoparticles with thio-modified DNA-oligomers via the ligand exchange

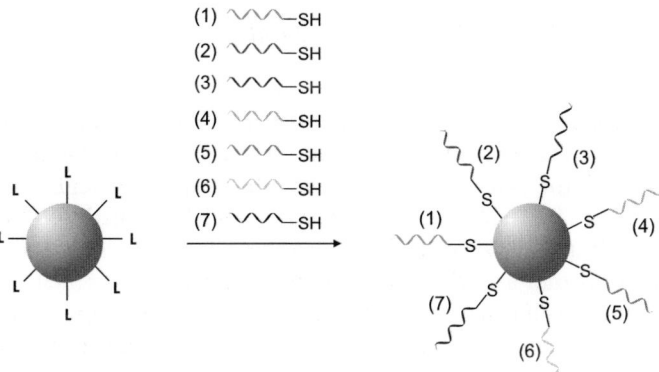

Fig. 12.4. Fabrication of multi-functionalized gold nanoparticles via ligand exchange reaction with different oligomer sequences

In several works, the linker molecules (for example di- or trithiols [37,38]) between the nanoparticle surface and the DNA moiety have been used. Other methods take advantage of the highly specific molecular recognition between the biotin and the streptavidin for the coupling of DNA oligomers and a large variety of other biomolecules to the nanoparticle surfaces [6,39,40]. The number of different protocols demonstrates that modifying the nanoparticles with the DNA oligomers and through that creating a biomolecule-nanoparticle interface has now a days become a routine prodecure. Even particles with a defined number of oligomers present at the particle surface can be prepared, as demonstrated by Chen et al. through the use of a polymer-linker [41]. To quantify the density of the coverage and purify the DNA-modified gold nanoparticles, gel electrophoresis can be utilized as described by Alivisatos and coworkers [42,43]. Another possibility to quantify the oligomer coverage density is the fluorescence-based assay of Demers et al. [44].

12.3 Nanoparticle Assemblies and Properties

12.3.1 Three-dimensional Assemblies

The three-dimensional assembly of the metal nanoparticles can be regarded as the simplest way for construction of the nanoparticle architectures. In principle, any crystallization process of the nanoparticles describes a three-dimensional assembly. However, the spontaneous formation of a crystalline superstructure built from nanoparticles only occurs if the nanoparticles exhibit a monodisperse size distribution. As an example of a cluster that shows this crystallization, the $Au_{55}(PPh_3)_{12}Cl_6$-cluster which was already described in Sect. 12.2.1 can be mentioned here [27]. Schmid and coworkers have also reported that the use of a dendrimer supports the crystallization of the product [45]. We will not describe the phenomenon of the nanoparticle crystallization in any more detail, but rather focus on those examples where the nanoparticle assemblies are built up with the help of the interconnecting molecules.

Various methods for the linkage of the nanoparticles have been developed, most of them utilize the bifunctional linker molecules, i.e. dithiols, to form the nanoparticle aggregates. This approach has been pursued by Brust et al. who proved the formation of the aggregate structures as a result of the particle linkage by means of the TEM [46]. Other methods describe the use of the biotin/streptavidin system [47]. The drawback of these methods is that the processes are irreversible and are often difficult to control.

The use of DNA as a linker between the nanoparticle building blocks circumvents these disadvantages. Mirkin and coworkers presented the DNA-mediated aggregation of the Au nanoparticles in solution for the first time [33, 48]. In their approach they prepared two sets of 13 nm nanoparticles that were modified with two non-complementary sets of the thiol-terminated, single stranded oligonucleotides, according to the method described in Sect. 12.2.2. Upon addition of linking a DNA duplex which contains the respective sequences, complementary to those that were bound to the particles before, as "sticky ends", aggregation of the nanoparticle due to hybridization of the strands occurred. The procedure is depicted in Fig. 12.5a. The aggregation became visible in the change of the spectral properties of the colloid solutions: a significant red-shift of the plasmon resonance of the DNA-modified particles could be detected as a consequence of the addition of the linking strand (Fig. 12.5b). Furthermore, slow precipitation of a pinkish-grey solid after seven hours, resulting from the formation of a macroscopic solid comprised of the networked colloids, was observed. The effect was shown to be reversible upon heating: the precipitate dissolves again yielding the original red particle solutions. The aggregation and melting effects were monitored by means of the UV/vis measurements of the characteristic absorption of

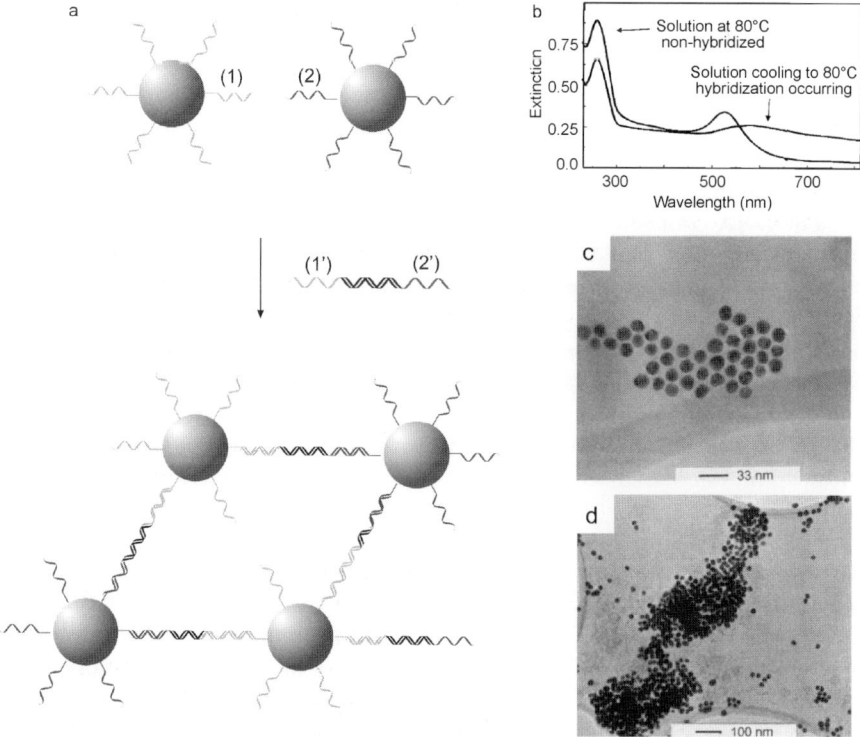

Fig. 12.5. a Three-dimensional linkage of the DNA-modified gold nanoparticles induced by hybridization. **b** The UV/vis spectra of the gold nanoparticles in the hybridized state and after heating to $80°C$ (unhybridized state) (reprinted with permission from [48], Copyright 1999 American Chemical Society). **c,d** TEM micrographs of the two- and three-dimensional aggregates precipitation from the solution (reprinted from [33] with permission of the Nature Publishing Group)

the DNA hydribization at 260 nm as well as of the plasmon resonance of the colloids between 500 nm and 700 nm. Furthermore, the two and three-dimensionally aggregated colloids were investigated by TEM, where a clear coherence of the interparticle distance and the length of the DNA linker was observed (Fig. 12.5c,d).

With the same method Mirkin and coworkers demonstrated the assembly of the binary structures linked by DNA meaning different kinds of particles connected by DNA linkers, i.e. the gold clusters with a diameter of 8 and 30 nm or the CdSe quantum dots coupled with 30 nm gold particles [49,50]. Furthermore, the protocol was extended to the anisotropic shaped nanoparticles [51, 52]. Mirkin and coworkers also pointed out recently that the deoxyguanine-rich DNA sequences should be avoided in the nanoparticle assembly design as the formation of G-quartet structures may initiate aggregation of the nanoparticles as a consequence of the increasing salt concentrations, especially when the potassium salts are used [53].

The optical properties of the gold nanoparticle-DNA networks with varying linker lengths were investigated by Storhoff et al. [54]. It was shown that the optical differences observed for the DNA aggregates with different linker-lengths are not only due to the varying interparticle distances but also due to the aggregate size. This was explained with the different rates of the hybridization of the respective linkers which provide the kinetic control over the aggregates size.

The DNA connected nanoparticle-aggregates exhibit extraordinary sharp melting transitions due to the cooperative binding of the oligonucleotides between the particles. This effect is of great advantage in the differentiation of the perfect complementary strands and those with base mismatches in the colorimetric DNA detection. In a systematic study, Jin et al. investigated the influencing parameters on the melting point of the DNA network. They observed that the melting point increased with the length of the DNA-linker, the salt concentration and the density of oligomer-coverage on the particle surface and decreased with the size of the nanoparticles used in the experiment [55].

Studies concerning the electrical properties of the dried nanoparticle networks were conducted by Park et al. [56]. Thereby the linker length in the nanoparticle networks was varied. In the electrical characterization, the particle networks showed semiconductor behavior in comparison to the aggregates of the citrate stabilized nanoparticles that behave like metals. However, as for the conductivity of the dried nanoparticle-DNA networks, virtually no difference in the systems with varying linker lengths could be observed. This effect was attributed to a collapse of the network after dehydration. Nevertheless, it has to be mentioned that this collapse was found out to be reversible and once solved in buffer solution again the particles exhibited the same melting and optical properties as before.

The described methods for the assembly of nanoparticles lead to one major field of application, the diagnostic of the nucleic acids as for example, described by Storhoff et al. [34]. The basic principle is the following: The nanoparticles are functionalized with the DNA sequences that are complementary to the sequences that are to be detected. Upon presence of the target, a rapid colorimetric change can be utilized as a detection signal. For further information on these methods the reader can refer to several review articles about this topic [6, 7, 40, 57–59].

12.3.2 Two-dimensional Assemblies

The two dimensional assembly of the metal nanoparticles requires their binding to a surface. For this purpose, again the DNA molecule is suited well as a variety of surfaces like glass, silicon and metals can be modified with DNA molecules in a similar way as it was already shown for the modification of nanoparticles in Sect. 12.2.2. The challenge in the assembly of the nanoparticles in two dimensions is the formation of well ordered, densely covered layers which allow the study of the electrical coupling and the transport between the particles. In this section some recent approaches for the generation of such layers, their specific electrical properties, and their application will be described as examples.

Park et al. demonstrated a two-dimensionally structured device for the electrical detection of the specific DNA strands [60]. The latter is depicted in Fig. 12.6. In a gap between two microelectrodes on silicon, DNA oligomers were bound to the surface. Gold nanoparticles were modified with a set of single-stranded DNA oligomers as well. In the presence of a target strand, which carried contiguous recognition elements to the single strands on the surface and on the particles, hybridization and through this the surface coverage with the gold nanoparticles in-between the gap occurred. Particle densities of ≥ 420 particles per μm^2 were achieved which is still too low to give a measurable electrical signal. In order to close the gap, the nanoparticles were plated with silver. Using this method, Mirkin and co-workers were able to detect specific DNA-strands down to concentrations of $0.5 \times 10^{-12}\,mol \cdot L^{-1}$

Surface coverage with higher particle densities which showed sufficient conductivity without the silver plating step could be achieved with a protocol that was developed recently by Koplin et al. [61]. Herein, the citrate stabilized gold nanoparticles of a diameter of 15 nm were immobilized on the IDC structures on silicon which was previously treated with a dendrite linker

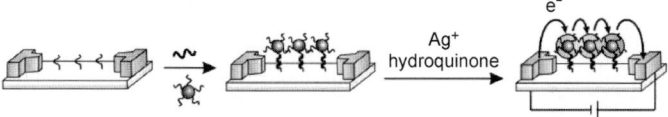

Fig. 12.6. Scheme of the DNA-nanoparticle array for the electrical detection of specific nucleic acid targets (reprinted with permission from [60] Copyright 2002 AAAS)

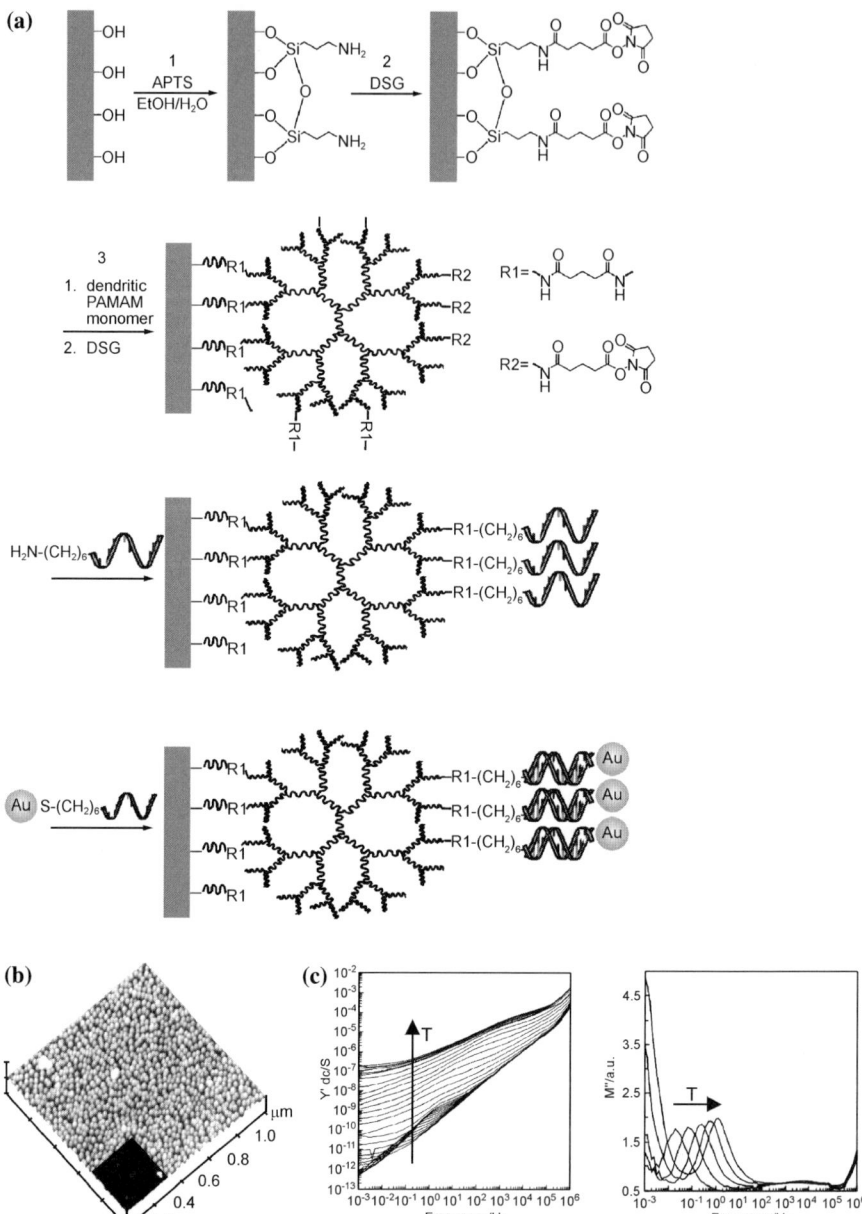

Fig. 12.7. a Stepwise surface functionalization for the immobilization of a densely packed nanoparticle monolayer on glass. **b** AFM micrograph showing the dense covering of the surface. **c** Electrical characterization showing the admittance spectra (plot of Y' vs. v) (*left*) and modulus spectra (plot of M' vs. v) (*right*) for temperatures between 75 K and 300 K (reproduced from [61] by permission of The Royal Society of Chemistry)

system. The surface functionalization scheme is depicted in Fig. 12.7. In a first step the surface was modified with an aminonsilane. After subsequent treatment with the homobifunctinal linker reagent disuccinimdyl glutarate (DSG), a dendritic PAMAM-starburst monomer was attached to the surface. This polymeric dendrimer layer was again modified with the DSG linker, which enables the binding of 5′-amynofunctionalized oligomers to the modified surface. Gold nanoparticles that were functionalized with a complementary DNA single-stranded oligomer could now be bound by hybridization to the activated support, resulting in a coverage of more than 850 particles μm^{-2}, determined by the AFM measurements (Fig. 12.7a). The arrays were characterized by the I(V)-measurements and the temperature-dependent impedance spectroscopy (IS) (Fig. 12.7b). Thereby the particle layers showed pronounced

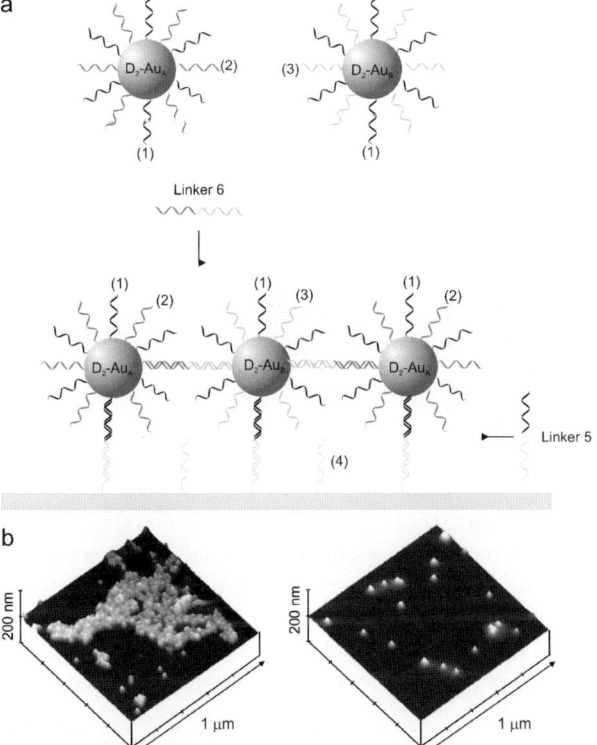

Fig. 12.8. Schematic representation of a two-dimensional assembly process for the oligofunctionalized nanoparticles. *Above*: two sets of nanoparticles modified with the oligomer 1 and 2 and 1 and 3, respectively. *Below*: Immobilization by hybridization of the surface bound oligomer 4 with oligomer 1 upon addition of the linker 5 and the interparticle cross linking upon addition of the linker 6. **b** AFM analysis of the surface with a cross-linker **c** Control experiment in the absence of the cross-linker 6 (**b** and **c** reprinted from [62] with permission from Elsevier)

field dependence as well as a thermal activation of the conductivity, reflecting the classical hopping transport.

Niemeyer et al. reported recently on the use of the oligofunctional DNA-nanoparticle conjugates in two-dimensional assemblies. The synthesis of the particles which carry different nucleic acid sequences was already described in Sect. 12.2.2. Due to the specifity of the Watson-Crick base pairing, such a system allows the selective addressing of these oligomers and therefore opens the possibility to generate densely packed and well ordered systems with tunable interparticle distances. The approach is depicted schematically in Fig. 12.8.

Two sets of gold nanoparticles were synthesized, called D_2Au_a and D_2Au_b in the following. Each of them contained two different, independently addressable oligomers: D_2Au_a is stabilized by the sequence 1 and 2, and D_2Au_b by the sequence 1 and 3. Oligomer 1 was used for the immobilization purposes by connecting the particles to the surface bound captive oligomers 4 with the complementary linking strand 5. The particle bound oligomers 2 and 3 were used for the cross linking of the adjacent particles by means of the oligomer 6, complementary to 2 and 3, respectively. In-situ AFM studies and statistical evaluation of the surface coverage and morphology revealed that the self-assembly of the gold nanoparticles on the solid substrates is influenced by the linker length and that particle layers with programmable inter-particle spacing can be achieved [62, 63].

Another method to assemble the nanoparticles on surfaces by means of DNA is the construction of the complex DNA geometries by hybridization of the oligo sequences. Subsequent binding of the gold nanoparticles to these DNA scaffold structures yields well ordered two-dimensional nanoparticle assemblies on the surface [64–66].

12.3.3 One-dimensional Assemblies

DNA has also become an important tool for the one-dimensional assembly of the metal nanoparticles as it may serve as a programmable and selective template for the nanowires. Pioneering work on the DNA templated nanowire fabrication was reported by Braun et al. in 1998, who exchanged the native charge compensation cations of the DNA backbone against silver and by a reduction and further electroless plating process generated nanowires which are depicted in Fig. 12.9 [67].

The DNA template strands were assembled on the electrode structures prior to the described metallization process, and thus the conductivity measurements on the generated wires could be accomplished. The densely metallized DNA wires showed ohmic behavior and a resistance between 7 and $30\,M\Omega$, depending on the duration of the silver deposition. These works were later improved by Keren et al. who recently published their work on the first DNA templated, self-assembled transistor [68, 69]. Various other syntheses of the metallized DNA strands have been reported, i.e. the photoreduction of

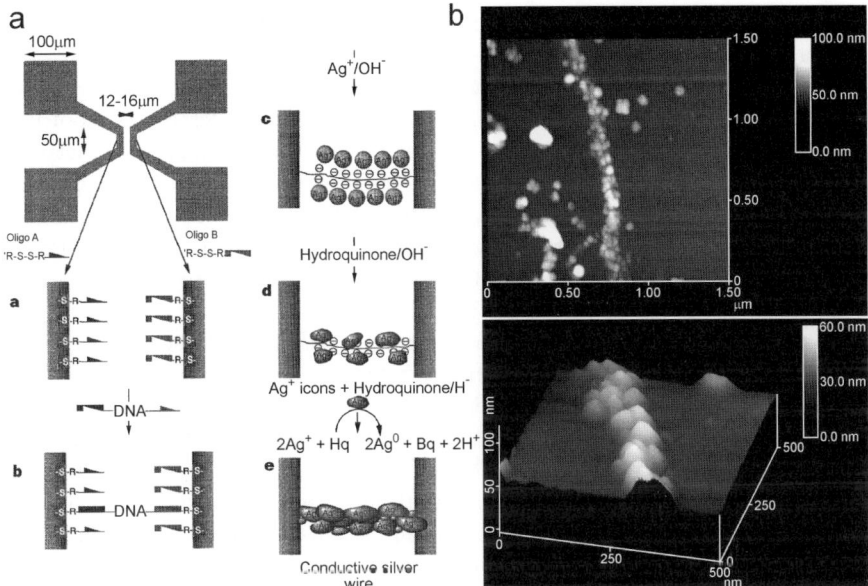

Fig. 12.9. Scheme for generation and placement between an electrode, the structure of the first DNA templated nanowire. **b** AFM micrographs of the nanowires (Reprinted from [67] with permission of the Nature Publishing Group)

the electrostatically bound Ag-ions and chemical reduction of the electrostatically bound Pt-, Pd- or Cu-ions [70–73]. Another strategy uses the affinity of the Pt^{2+}-complexes towards the DNA strands and subsequent electroless plating to generate Pt-nanowires [74–76]. A review about this topic was presented by Richter [77]. A selective concept for continuous DNA metallization was presented by Burley et al. who utilized the DNA duplexes which incorporate the modified DNA bases equipped with the alkyne reporter groups [78]. After conversion with sugarazides and subsequent *Tollens* reaction and a silver enhancement step, silver wires could be obtained.

In spite of all these developments, the so-called full metallization of DNA bears problems as the reduction on the DNA strand in most cases forms particles with inhomogeneous size distributions and distances. Therefore, the size specific electronic properties of the nanoparticles can not be exploited in the structures whose electronic properties are mainly dominated by the defects. A solution for this problem might be the binding of the preformed metal clusters to the DNA backbone. Different binding mechanisms between the DNA and the metal nanoparticles can be utilized for this metallization route, involving hybridization of the DNA-functionalized nanoparticles to the template strands, the use of specific binding sites of the DNA or the electrostatic interaction between the positively charged nanoparticles and the negative charges at the backbone. Further, *cis*-diamminedichloroplatinum (*cis* platin) was used as an inter-connector between the DNA bases and the nanoparticles.

As described before for two-and three-dimensional nanoparticle assemblies, the hybridization of the DNA-modified nanoparticles to the complementary template strands has been applied as a method to align the nanoparticles in one dimension. Alivisatos et al. demonstrated the spatially defined alignment of the nanoparticles on a DNA template with the hybridization method [79, 80]. In this approach, they assembled the nanoparticles on DNA to form the head-to-tail and head-to-head dimers as well as parallel trimers on the template strands, proven by the TEM analysis. In an extension of their work they applied the method for the formation of hetero-dimers and trimers, meaning two kinds of different particles were selectively bound to a strand carrying different sequences complementary to those at the nanoparticle surface. Deng et al. developed this method further, yielding long chains of metal clusters along a DNA template [81].

Noyong et al. utilized the affinity of *cis* platin to the specific binding sites at the DNA to decorate the DNA densely with the nanoparticles [82, 83]. Thereby the planar platinum-complex is incubated with the natural DNA. It preferably occupies the neighboring GG basepairs and has two free ammonia ligands that were utilized as the anchoring groups for the amine-terminated nanoparticles. Through exchange of the ammonia groups at the *cis* platin against the amino functions in the nanoparticle ligand shell, the strands could be covered densely with the nanoparticles. By means of the AFM, string-of-pearl-like arrangement of the particles on the DNA could be proven. In the TEM images, the structures appear as continuous gold wires, which results from melting of the particles under the high energy impact of the TEM conditions.

Several examples for the electrostatic binding of the nanoparticles that are protected by the positively charged ligands have been presented in the last years. Hutchison and coworkers utilized particles that were modified with three different ligands, (2-mercaptoethyl)trimethylammonium iodide (TMAT, 1), [2-(2-mercaptoethoxy)ethyl]trimethyl-ammonium toluene-4-sulfonate (MEMA, 6), and {2-[2-(2-mercaptoethoxy)ethoxy]ethyl}-trimethyl-ammonium-toluene-4-sulfonate (PEGNME,11), respectively. The ammonia groups in the nanoparticle ligand shell were bound electrostatically to λ-DNA in solution. TEM studies revealed that with this method DNA could be densely covered with the nanoparticles over long ranges, and that the interparticle distance strongly depends on the chain length of the utilized ligands, resulting in the interparticle distances of 1.5, 2.1, and 2.8 nm, respectively. Thereby the interparticle distance resembles the double length of the ligand, assuming that they are in a fully extended configuration [84, 85].

A similar approach was pursued by Reich and coworkers, who used thiocholine as the stabilizing ligand for the nanoparticles [86]. The modification of the DNA with the particles was done on Si/SiO_2 surfaces that were modified with different aminosilanes before. The density of the particle coverage on the DNA here was found to be strongly dependent on the incubation time as well as on the surface modification.

Schmid and coworkers studied the interactions of λ-DNA with the water soluble $[Au_{55}(Ph_2C_6H_4SO_3H)_{12}Cl_6]$ cluster by TEM an AFM methods and observed a dense covering of the DNA strands with the nanoparticles [87,88]. Surprisingly, TEM studies of the DNA-nanoparticle assembly revealed a dramatic size degradation of the clusters in the linear arrangement. By modelling experiments this phenomenon could be explained with a conformational change of the DNA structure from the B-DNA to the A-DNA which is formed under the ultra high vacuum conditions and initiated by dehydration. The calculated compression of the length of the groove from 1.43 nm in the B-DNA to 0.73 nm in the A-DNA corresponds with the change of the cluster diameter from 1.4 nm to 0.6 nm that could be observed during the microscopy studies. Thus, it is assumed that the change of DNA morphology induces the degradation of Au_{55} clusters to Au_{13} clusters which fit into the major groove of the new DNA structure.

Kretschmer and Fritzsche presented an interesting approach towards device fabrication based on nanowires that were assembled between two metallic electrodes by means of dielectrophoresis. This approach is expected also to be applicable to DNA-functionalized nanoparticles [89]. The resulting nanowires were characterized by means of the SEM. The electrical characterization of these structures together with the SEM images of the nanowires before and after characterization is presented in Fig. 12.10. Figure 12.10c shows the preservation of the individual particles. The melting of the particles to continuous wire structures during the high voltage measurements can be seen in Fig. 12.10e–g.

Another approach towards a DNA templated electronic device was introduced by Fitzmaurice and coworkers [90]. First, double-stranded DNA was deposited between the conventionally patterned gold electrodes on a silicon wafer. Then the wafer was exposed to an aqueous dispersion of the dimethyaminopyridine (DMAP)-stabilized gold nanoparticles which due to their positively charged ligand shell selectively assembled on the negatively charged DNA backbone. By electroless deposition a continuous nanowire was formed in the next step. The resulting wires between the 1 µm gaps of the electrode structure were electrically characterized and found to exhibit ohmic behavior with a single wire resistance of $2 \times 10^4\,\Omega m$. The wires after metallization and the corresponding I(V) characteristic for the geometry is presented in Fig. 12.11:

In an extension of their work, Fitzmaurice and coworkers presented a nano-gap structure assembled of a thiolated DNA strand with a biotin moiety, placed centrally on the DNA strand [91]. Upon incubation with the DMAP-stabilized nanoparticle dispersion, the strand was densely decorated with the particles. In a further step, streptavidin was added which replaced the nanoparticle adsorbed weakly at the biotin moiety, forming a gap in the

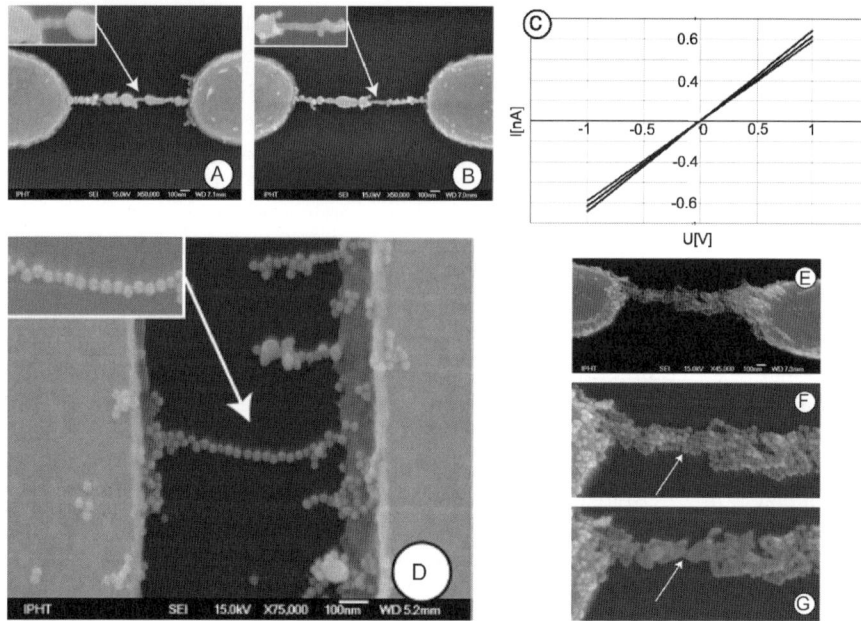

Fig. 12.10. Nanowires constructed of individual gold nanoparticles. **a,b** SEM micrographs of the nanowires between the gold electrodes. **c** depicts the I–V curves of these wires, showing ohmic behavior. **d** shows a magnification of a single wire, clearly representing the preservation of individual particles. **e–g** illustrate the melting of the particles which occurs after application of the high voltages during the electrical characterization (reprinted with permission from [89], Copyright 2004 American Chemical Society)

nanoparticle chain. After electroless plating, a nano-gap in-between a continuous gold wire could be formed in this way.

12.4 Conclusion

In this chapter we have elucidated the important steps of DNA-based assembly of metal nanoparticles, starting with the synthesis and modification of the nanoparticles material leading over to the one-, two- and three-dimensional DNA-based assembly techniques. Thereby we describe how the combination of the enormous binding specifity of the nucleic acids together with the virtue of the organic and the inorganic chemistry has created a fascinating new field in materials science.

Although there are some barriers yet to overcome until sophisticated nanoarchitectures find their way into the technological application, promising examples for the first nanoelectronic devices have already been introduced. Here one may mention the works of Williams et al. and of Keren et al. who

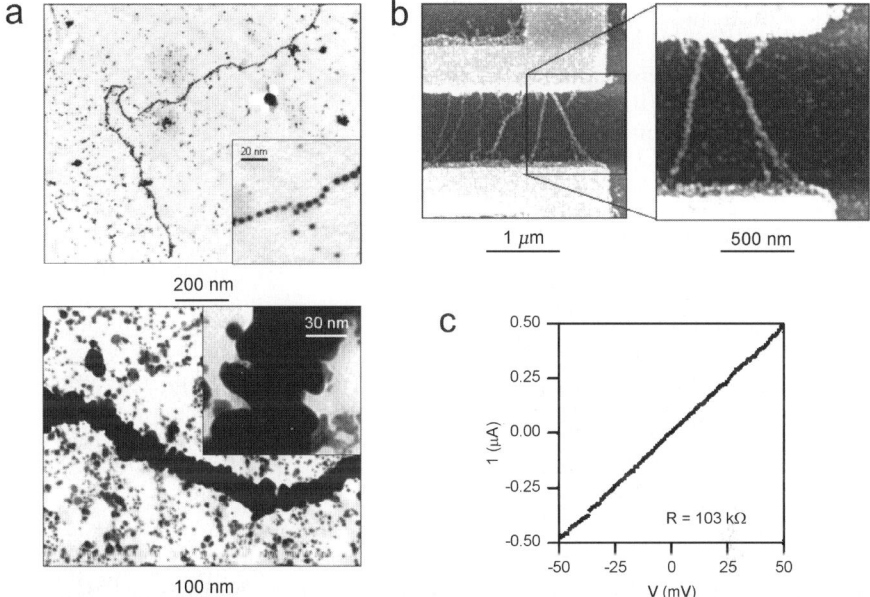

Fig. 12.11. Gold nanowires fabricated by immobilization of the DMAP-modified nanoparticles on the DNA strands followed by electroless plating. **a**: TEM of the wire before and after the electroless plating. **b**: SEM images of the wires between two litographically formed gold electrodes **c** I(V) characteristics of the setup shown in **b** (reprinted with permission from [90], Copyright 2005 American Chemical Society)

recently presented promising examples for the devices of higher complexity [69, 92]. While Williams et al. utilized the recognition of DNA to bind the carbon nanotubes to metallic contacts by self-assembly, Keren et al. described the self-assembly of the segmented nanowires in a multi-step process to build up a molecular-based transistor element.

However, the challenge remains to develop a reliable technology based on the selective assembly processes which enable the use of such nanoelectronic building blocks routinely in the nanoelectronic circuitry.

References

1. G. Schön and U. Simon, Coll. Polym. Sci. **273**, 101 (1995).
2. D.L. Feldheim and C.A. Foss, *Metal nanoparticles: Synthesis, characterization, and applications* (Mercel Dekker, New York, 2002).
3. U. Simon, Adv. Mater. **10**, 1487 (1998).
4. G. Schmid, *Nanoparticles: From theory to applications* (Wiley-VCH, Weinheim, 2004).
5. L.J. de Jongh, in *Physics and Chemistry of materials with low-dimensional structures* (Springer, New York, 1994).

6. I. Willner and E. Katz, Angew. Chem. Int. Ed. **43**, 6042 (2004).
7. N.L. Rosi and C.A. Mirkin, Chem. Rev. **105**, 1547 (2005).
8. W.P. Halperin, Rev. Mod. Phys. **58**, 533 (1986).
9. F. Remacle and R.D. Levine, Chem. Phys. Chem. **2**,20 (2001).
10. N.C. Seeman, Angew. Chem. Int. Ed. **37**, 3220 (1998).
11. N.C. Seeman, Nature **421**, 427 (2003).
12. M.T. Reetz and W. Helbig, J. Am. Chem. Soc. **116**, 7401 (1994).
13. H. Ma, B. Yin, S. Wang, Y. Jiao, W. Pan, S. Huang, S. Chen and F. Meng, ChemPhysChem. **5**, 68 (2004).
14. K. Okitsu, Y. Mizukoshi, H. Bandow, Y. Maeda, T. Yamamote and Y. Nagata, Ultrason. Sonochem. **3**, S249 (1996).
15. M. Nakamoto, Y. Kahiwagi and M. Yamamoto, Inorg. Chim. Acta **358**, 4229 (2005).
16. K. Mallick, M.J. Witcomb and M.S. Scurrel, Appl. Phys. A **80**, 395 (2005).
17. G. Schmid and A. Lehnert, Angew. Chem. Int. Ed. **28**,780 (1989).
18. M.C. Daniel and D. Astruc, Chem. Rev. **104**, 293 (2004).
19. R.M. Richards and H. Boennemann, edited by C.S.S.R. Kumar, J. Hormes and C. Leuschner Wiley-VCH, Weinheim, (2005).
20. C. Burda, X. Chen, R. Narayanan and M.A. El-Sayed, Chem. Rev. **105**, 1025 (2005).
21. J. Turkevich, P.C. Stevenson and J. Hillier, Faraday Soc. **11**, 55 (1951).
22. J.P. Spatz, A. Roescher, S. Sheiko, G. Krausch and M. Moeller, Adv. Mater. **7**, 731 (1995).
23. J.P. Spatx, A. Roescher and M. Moeller, Adv. Mater. **8**, 337 (1996).
24. M.P. Pileni, Nat. Mat. **2**, 145 (2003).
25. G. Schmid, R. Boese, R. Pfeil, F. Bandermann, S. Meyer, G.H.M. Calis and J.W.A. van der Velden, Chem. Ber. **114**, 3634 (1981).
26. G. Schmid, A. Lehnert, U. Kreibig, Z. Damczyk and P. Belouschek, Z. Naturforsch. **45**b, 989 (1990).
27. G. Schmid, R. Pugin, T. Sawitowski, U. Simon and B. Marler, Chem. Comm. **14**. 1303 (1999).
28. G. Schmid, B. Morun and J.O. Malm, Angew. Chem. Int. Ed. **28**, 778 (1989).
29. M.N. Vargaftik, V.P. Zagorodnikov, I.P. Stolyarov, I.I. Moiseev, V.A. Likholobov, D.I. Kochubey, A.L. Chuvilin, V.I. Zaikovsky, K.I. Zamaraev and G.I. Timofeeva, J. Chem. Soc. Chem. Comm. **14**, 937 (1985).
30. I.I. Moiseev, M.N.Vargaftik, T.V. Cernysheva, T.A. Stromnova, A.E. Gekhman, G.A. Tsirkov and A.M. Makhlina, J. Mol. Cat. A **108**, 77 (1996).
31. G. Schmid, Adv. Eng. Mater. **3**, 737 (2001).
32. L.A. Gearheart, H.J. Ploehn and C.J. Murphy, J. Phys. Chem. B **105**, 12609 (2001).
33. C.A. Mirkin, R.L. Letsinger, R.C. Mucic and J.J. Storhoff, Nature **382** 607 (1996).
34. J.J. Storhoff, R. Elghanian, R.C. Mucic, C.A. Mirkin and R.L. Letsinger, J. Am. Chem. Soc. **120**, 1959 (1998).
35. P. Hazarika, T. Giorgi, M. Beibner, B. Ceyhan and C.M. Niemeyer, in *Bioconjugation protocols: Strategies and Methods* (Humana Press, Totowa New York, 2004).
36. C.M. Niemeyer, B. Ceyhan and P. Hazarika, Angew. Chem. **115**, 5944 (2003).
37. R.L. Letsinger, R. Elghanian, G. Viswanadham and C.A. Mirkin, Bioconjug. Chem. **11**, 289 (2000).

38. Z. Li, R. Jin, C.A. Mirkin and R.L. Letsinger, Nucleic Acids Res. **30**, 1558 (2002).
39. S. Cobbe, S. Connoly, D. Ryan, L. Nagle, R. Eritja and D. Fitzmaurice, J. Phys. Chem. B **107**, 470 (2003).
40. C.M. Niemeyer, Angew. Chem. Int. Ed. **40**, 4128 (2001).
41. Y. Chen, J. Aveyard and R. Wilson, Chem. Comm. **24**, 2804 (2004).
42. D. Zanchet, C.M. Micheel, W.J. Parak, D. Gerion and A.P. Alivisatos, Nano Lett. **1**, 32 (2001).
43. W.J. Parak, T. Pellegrino, C.M. Micheel, D. Gerion, S.C. Williams and A.P. Alivisatos, Nano Lett. **3**, 33 (2003).
44. L.M. Demers, C.A. Mirkin, R.C. Mucic, R.A. Reynolds, R.L. Letsinger, R. Elghanian and G. Viswanadham, Anal. Chem. **72**, 5535 (2000).
45. G. Schmid, W. Meyer-Zaika, R. Pugin, T. Sawitowski, J.P. Majoral, A.M. Caminade and C.O. Turrin, Chemistry - European Journal, **6**, 1693 (2000).
46. M. Brust, D. Schiffrin, D.J. Bethell and C.J. Kiely, Adv. Mater. **7**, 795 (1995).
47. D. Fitzmaurice and S. Connolly, Adv. Mater. **11**, 1202 (1999).
48. J.J. Storhoff and C.A. Mirkin, Chem. Rev. **99**, 1849 (1999).
49. G.P. Mitchell, C.A. Mirkin and R.L. Letsinger, J. Am. Chem. Soc. **121**, 8122 (1999).
50. R.C. Mucic, J.J. Storhoff, C.A. Mirkin and R.L. Letsinger, J. Am. Chem. Soc. **120**,12674 (1998).
51. J.K.N. Mbindyo, B.R. Reiss, B.R. Martin, C.D. Keating, M.J. Natan and T.E. Mallouk, Adv. Mater, **13**, 249 (2001).
52. E. Dujardin, L.B. Hsin, C.R.C. Wang, S. Mann, Chem. Comm. **14**, 1264 (2001).
53. U. Li and C.A. Mirkin, J. Am. Chem. Soc. **127**, 11568 (2005).
54. J.J. Storhoff, A.A. Lazarides, R.C. Mucic, C.A. Mirkin, R.L. Letsinger and G.C. Schatz, J. Am. Chem. Soc. **122**, 4640 (2000).
55. R. Jin, G. Wu, Z. Li, C.A. Mirkin and G.C. Schatz, J. Am. Chem. Soc. **125**, 1643 (2003).
56. S.J. Park, A.A. Lazarides, C.A. Mirkin, P.W. Brazis, C.R. Kannewurf and R.L. Letsinger, Angew. Chem. Int. Ed. **39**, 3845 (2000).
57. C.A. Mirkin, Inorg. Chem. **39**, 2258 (2000).
58. A.N. Shipway, E. Katz and I. Willner, Chem. Phys. Chem. **1**, 19 (2000).
59. C.M. Niemeyer and C.A. Mirkin, *NanoBiotechnology: Concepts, Methods and Applications* (Wiley-VCH, Weinheim, 2004).
60. S.J. Park, T.A. Taton and C.A. Mirkin, Science **295**, 1503 (2002).
61. E. Koplin, C.M. Niemeyer and U. Simon, J. Mater. Chem. **16**, 1338 (2006).
62. C.M. Niemeyer, B. Ceyhan, M. Noyong and U. Simon, Biochem. Biophys Res. Comm. **311**, 995 (2003).
63. B. Zou, B. Ceyhan, U. Simon and C.M. Niemeyer, Adv. Mater. **17**, 1643 (2005).
64. J.D. Le, Y. Pinto, N.C. Seeman, K. Musier-Forsyth, T.A. Taton and R.A. Kiehl, Nano Lett. **4**, 2343 (2004).
65. J. Sharma, R. Chhabra, Y. Liu, Y. Ke and H. Yan, Angew. Chem. **118**, 744 (2006).
66. S. Xiao, F. Liu, A.E. Rosen, J.F. Hainfeld, N.C. Seeman, K. Musier-Forsyth and R.A. Kiehl, J. Nanoparticle Res. **4**, 313 (2002).
67. E. Braun, Y. Eichen, U. Sivan and G. Ben-Yoseph, Nature **39**, 775 (1998).
68. K. Keren, M. Krueger, R. Gilad, G. Ben-Yoseph, U. Sivan and E. Braun, Science **297**, 72 (2002).

69. K. Keren, R.S. Berman, E. Buchstab, U. Sivan and E. Braun, Science **302**, 1380 (2003).
70. L. Berti, A. Alessandrini and P. Facci, J. Am. Chem. Soc. **127**, 11216 (2005).
71. J. Richter, M. Mertig, W. Pompe, I. Mönch and H.K. Schackert, Appl. Phys. Lett. **78**, 536 (2001).
72. R.M. Stoltenberg and A.T. Woolley, Biomed. Microdevices **6**, 105 (2004).
73. J. Richter, R. Seidel, R. Kirsch, M. Mertig, W. Pompe, J. Plaschke and H.K. Schackert, Adv. Mater. **12**, 507 (2000).
74. W.E. Ford, O. Harnack, A. Yasuda and J.M. Wessels, Adv. Mater. **13**, 1793 (2001).
75. R. Seidel, L.C. Ciacchi, M. Weigel, W. Pompe and M. Mertig, J. Phys. Chem. B **108**, 10801 (2004).
76. M. Mertig, L.C. Ciacchi, R. Seidel, W. Pompe and A. De Vita, Nano Lett. **2**, 841 (2002).
77. J. Richter, Physica E **16**, 157 (2003).
78. G.A. Burley, J. Gierlich, M.R. Mofid, H. Nir, S. Tal, Y. Eichen and T. Carell, J. Am. Chem. Soc. **128**, 1398 (2006).
79. A.P. Alivisatos, K.P. Johnsson, X. Peng, T.E. Wilson, C.J. Loweth, M.P. Bruchez Jr. and P.G. Schultz, Nature **382**, 609 (1996).
80. C.J. Loweth, W.B. Candwell, X. Peng, A.P. Alivisatos and P.G. Schultz, Angew. Chem. **111**, 1925 (1999).
81. Z. Deng, Y. Tian, S.H. Lee, A.E. Ribbe and C. Mao, Angew. Chem. **72**, 5535 (2005).
82. M. Noyong, K. Gloddek and U. Simon, in *Bioinspired Nanoscale Hybrid Systems*, Symposium Proceedings Vol. 735, Materials Research Society, Warrendale, Pennsylvania (2003), p. 153.
83. M. Noyong, K. Gloddek, J. Mayer, T. Weirich and U. Simon, J. Cluster Sci. **18**, 193 (2006).
84. M.G. Warner and J.E. Hutchison, Nat. Mater. **2**, 272 (2003).
85. G.H. Woehrle, M.G. Warner and J.E. Hutchison, Langmuir **20**, 5982 (2004).
86. G. Braun, K. Inagaki, R.A. Estabrook, D.K. Wood, E. Levy, A.N. Cleland, G.F. Strouse and N.O. Reich, Langmuir **21**, 10699 (2005).
87. M. Tsoli, H. Kuhn, W. Brandau, H. Esche and G. Schmid, Small **1**, 841 (2005).
88. Y. Liu, W. Meyer-Zaika, S. Franzka, G. Schmid, M. Tsoli and H. Kuhn, Angew. Chem. **115**, 2959 (2003).
89. R. Kretschmer and W. Fritzsche, Langmuir **20**, 11797 (2004).
90. A. Ongaro, F. Griffin, P. Beecher, L. Nagle, D. Iacopino, A. Quinn, G. Redmond and D. Fitzmaurice, Chem. Mater. **17**, 1959 (2005).
91. A. Ongaro, F. Griffin, L. Nagle, D. Iacopino, R. Eritja and D. Fitzmaurice, Adv. Mater. **16**, 1799 (2004).
92. K.A. Williams, P.T.M. Veenhuizen, B.G. de la Torre, R. Eritja and C. Dekker, Nature **420**, 761 (2002).

Index

Printing: Krips bv, Meppel
Binding: Stürtz, Würzburg